Adaptive Structural Systems with Piezoelectric Transducer Circuitry

K.W. Wang • J. Tang

Adaptive Structural Systems with Piezoelectric Transducer Circuitry

 Springer

K. W. Wang
Department of Mechanical Engineering
University of Michigan
2350 Hayward Street, G. G. Brown Building
Ann Arbor, MI 48109

J. Tang
Department of Mechanical Engineering
University of Connecticut
191 Auditorium Road, Unit 3139
Storrs, CT 06269

ISBN: 978-0-387-78750-3   e-ISBN: 978-0-387-78751-0

Library of Congress Control Number: 2008928849

Printed on acid-free paper

9 8 7 6 5 4 3 2 1

springer.com

To our families

Preface

The objective of this book is to present a general description of an emerging technology, the *piezoelectric transducer circuitry*, for adaptive structure development. The book provides basic understanding of the subject and discusses some of the research ideas investigated in recent years for structural control and identification enhancement. It is intended for researchers and engineers that are interested in adaptive structural systems. As this emerging technology has been explored for a wide variety of applications, individual technical papers on this topic are scattered in journals and conference proceedings with different foci. This research monograph is an outgrowth and organized compilation of the selected papers published by the authors during the recent years; it aims to providing a coherent platform for a systematic illustration of the subject, both quantitatively and qualitatively. It encompasses multidisciplinary research in the areas of dynamics, control, materials, and electronics, and consists of comprehensive descriptions of algorithm development, theoretical analysis, hardware implementation, and experimental investigation.

An introduction of the concept and related technical background is presented in Chapter 1. In a piezoelectric transducer circuitry configuration, the transducer serves as the interface between the mechanical and electrical regimes, and the electronic elements in the circuitry allow one to favorably alter the dynamics of the integrated system for control and identification enhancement. Essentially, these elements offer additional design freedom and lead to energy relocation within the electro-mechanically integrated system. Through a series of studies, this book discusses the piezoelectric transducer circuitry's local actuation/damping capability as well as its global control and identification features. At the local level, Chapters 2 and 3 demonstrate that properly designed active-passive hybrid circuitry can amplify the actuation authority and simultaneously increase the passive damping. This approach will thus outperform a purely active method and at the same time require less control effort for vibration control. Based on the basic active-passive piezoelectric circuitry architecture, adaptive action is then developed for variable frequency disturbance rejection (Chapter 4). Combined with nonlinear robust control method, the piezoelectric transducer circuitry can facilitate the cancellation of material hysteresis, which will lead to high precision actuation, as described in Chapter 5. At the global level, with an eigenstructure assignment methodology, the piezoelectric transducer circuitry can confine the system energy (to circuitry elements or unimportant region of the structure) to realize vibration control and isolation (Chapter 6). On the other hand, a network with interconnected electronic elements between individual piezoelectric transducer circuits can be developed to propagate the otherwise localized vibratory energy in a mistuned periodic structure. Concurrently, such a circuitry network can be synthesized to become a global vibration absorber and ensure multi-spatial-harmonics vibration suppression. Such a system is discussed in Chapter 7. Piezoelectric transducer circuitry has also shown promising features in vibration-based structural damage identification enhancement. It is presented in Chapter 8 that by integrating a tunable piezoelectric circuitry to the host structure to be monitored, more resonant frequencies can be induced within the sensitive frequency range, and multiple families of frequency response functions can be generated. This will provide much more information of the dynamics of the damaged structure, and leads to improved damage detection and identification results.

ACKNOWLEDGMENTS

The former graduate students and co-workers (Lijun Jiang, Ronald Morgan, Meng-Shiun Tsai, Tian-Yau Wu, Xingjian Xue, Hongbiao Yu, and Jianhua Zhang) of the authors at the Pennsylvania State University and the University of Connecticut deserve enormous credit for their contributions. This research monograph would not have been possible without the innovation and hard work of these co-authors of the various technical papers that form the basis of the individual chapters in the book. We also want to thank our peers and colleagues in the adaptive structure community for years of technical interactions; they inspired us to explore new ideas and be persistent in our research. Finally, our deepest appreciations go to our families and loved ones, for their tremendous support throughout the years.

Kon-Well Wang   Jiong Tang
The Pennsylvania State University   University of Connecticut
University of Michigan

Contents

1 Background and Introduction

The recent significant advances in transducer materials, micro-electronics, signal processing, and system integration technology have led to rapid progress in the field of *adaptive structural systems*. The vision is to create advanced structures with various built-in functions, such as sensing, actuation, control, self-power, and self-diagnosis, aiming at satisfying the ever increasing demands on modern engineering systems.

Because of their electro-mechanical coupling characteristics, piezoelectric materials have been widely used in structural actuation and sensing applications. Some of the advantages of piezoelectric transducers include high bandwidth, high precision, compactness, and easy integration with existing host structures. Utilizing the piezoelectric transducer's electromechanical feature with circuitry and network design is a relatively recent concept for adaptive structural systems. Particularly, this concept leads to a system-level integral design of a class of adaptive structures that consist of the host mechanical structure, the piezoelectric transducers, and the electrical circuitry network that can favorably and adaptively modify the overall system dynamics. In such cases, the piezoelectric transducers serve as the interface between the mechanical and electrical regimes. The electronic elements in the circuitry, which can be passive and/or active components, allow one to alter the dynamics of the integrated system for control or identification enhancement. Essentially, these elements lead to energy relocation within the electro-mechanically integrated system. Compared to conventional control and sensing configurations using piezoelectric actuators/sensors, this concept employs circuitry elements as additional design variables, which adds a dimension to the development of adaptive structural systems and offers great amount of new opportunities for performance improvement.

K.W. Wang and J. Tang, *Adaptive Structural Systems with Piezoelectric Transducer Circuitry*, doi: 10.1007/978-0-387-78751-0_1,
© Springer Science + Business Media, LLC 2008

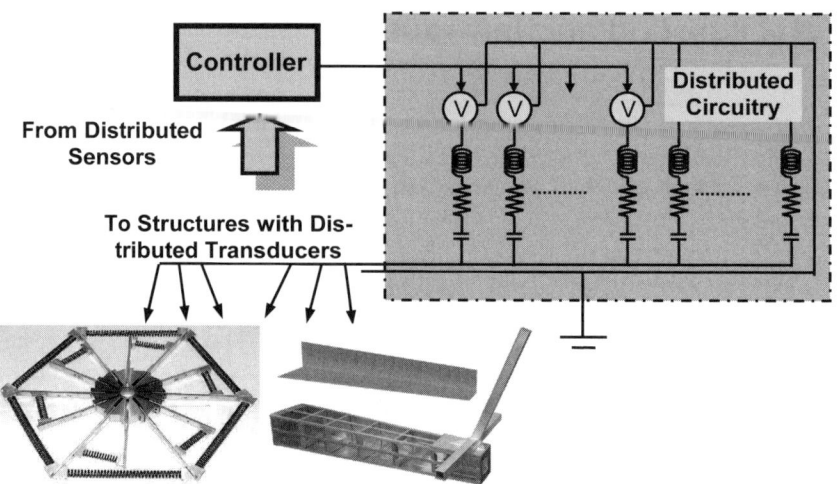

Fig. 1.1. Schematic of piezoelectric transducer circuitry integrated with mechanical structures. Illustrative examples shown are a periodic structure testbed (left) and a scaled helicopter semi-monocoque tailboom testbed (right) developed at Penn State University (Yu and Wang, 2007b; Heverly *et al.*, 2002; Belasco and Wang, 2003).

Fig. 1.1 illustrates mechanical structures integrated with distributed piezoelectric transducers and a circuitry network that consists of both passive (inductor/resistor/capacitor) and active (voltage or current sources) elements. In this book, the combination of the embedded piezoelectric transducers and the electrical circuitry network is referred to as the piezoelectric transducer circuitry. The equation of motion for a structure integrated with such a piezoelectric transducer circuitry can be expressed as (Tang and Wang, 2003)

$$\begin{bmatrix} \mathbf{M} & \\ & \mathbf{L} \end{bmatrix} \begin{Bmatrix} \ddot{\mathbf{q}} \\ \ddot{\mathbf{Q}} \end{Bmatrix} + \begin{bmatrix} \mathbf{C_d} & \\ & \mathbf{R} \end{bmatrix} \begin{Bmatrix} \dot{\mathbf{q}} \\ \dot{\mathbf{Q}} \end{Bmatrix} + \begin{bmatrix} \mathbf{K} & \mathbf{K_1} \\ \mathbf{K_1^T} & \mathbf{K_2} \end{bmatrix} \begin{Bmatrix} \mathbf{q} \\ \mathbf{Q} \end{Bmatrix} = \begin{Bmatrix} \mathbf{F_d} \\ \mathbf{0} \end{Bmatrix} + \begin{Bmatrix} \mathbf{0} \\ \mathbf{B_0 V} \end{Bmatrix} \qquad (1.1)$$

The detail of the derivation will be illustrated in the subsequent chapters on a variety of applications. In the above equation, $\mathbf{M}$, $\mathbf{C_d}$, and $\mathbf{K}$ are the mechanical mass, damping, and stiffness matrices, $\mathbf{q}$ is the generalized displacement vector, $\mathbf{Q}$ is the electrical charge flow vector in the multi-branch circuitry, $\mathbf{L}$, $\mathbf{R}$, and $\mathbf{K_2}$ are the inductance, resistance, and inverse capacitance matrices of the circuitry, $\mathbf{K_1}$ is related to the electro-mechanical coupling due to the piezoelectric effect, $\mathbf{F_d}$ is the external disturbance, $\mathbf{B}_0$ is the control input coefficient matrix, and $\mathbf{V}$ is the active control input (voltage source) vector of the circuitry.

Observing Fig. 1.1 and Eq. (1.1), we can more specifically recognize the potential advantages of utilizing piezoelectric transducer circuitry for structural control and identification purposes. Essentially, additional degrees-of-freedom (*i.e.*, the charge flow $\mathbf{Q}$ with second order dynamic effect) are introduced into the system, and these electrical degrees-of-freedom are coupled to those of the original mechanical structure through the electro-mechanical coupling term $\mathbf{K}_1$ due to the piezoelectric transducers. Indeed, the underlying principle of the concept of piezoelectric transducer circuitry is that the additional electrical degrees-of-freedom and the associated dynamic effect can be tailored/controlled to favorably alter the overall dynamics of the integrated system and enhance the system performance. To a certain extent, this approach is analogous to mechanical tailoring (changing structural inertial, stiffness, and damping properties) through adding/altering mechanical components or substructures. However, electrical tailoring is much easier to implement than mechanical tailoring, in terms of design space limitations, weight constraints, and system integration. The electronic elements are discrete and compact, and can be routinely measured/ examined to ensure high accuracy. Besides traditional passive electronic components (such as capacitance, inductance, or resistance), synthetic elements (e.g., synthetic inductance or capacitance) using operational amplifier (op-amp) circuits or power-electronic designs can be utilized that will greatly reduce the size and weight of the treatments. In addition, these electronic elements can be easily adjusted by varying the synthetic circuits, which leads to another significant advantage: online tunability. That is, a piezoelectric transducer circuitry can be made real-time adaptable to specific conditions.

Building upon the above foundational idea of piezoelectric transducer circuitry, in this book we present a series of generic adaptive structural system concepts with the dual-field electro-mechanical tailoring. These concepts include simultaneous active-passive damping augmentation, adaptive disturbance rejection, high precision robust control, vibration delocalization, energy confinement, and damage identification enhancement. The background and research highlights in these subjects are briefly reviewed and summarized in the following sections.

1.1 ACTIVE-PASSIVE MODAL DAMPING AUGMENTATION

Considerable amount of work has been performed to utilize piezoelectric transducers for active-passive hybrid modal damping augmentation. The idea here is to incorporate active vibration control into a classical piezo-electric shunt circuit (Hagood and von Flotow, 1991) that also provides passive modal damping. Such method could have the advantages of both the passive (stable, fail-safe, low power requirement) and active (high performance, feedback or feedforward actions) approaches. Agnes (1994, 1995) and Tsai and Wang (1999) examined the open-loop passive damping and active control authority of a resonant *RL* (resistance-inductance) shunt circuit connected in series with a piezoelectric transducer. It was observed that with this serial arrangement, not only the passive damping effect can be significantly enhanced; the response of the structure actuated by voltage or current inputs can also be greatly amplified. In the case of using voltage as the driving source, also referred to as the Active-Passive Piezoelectric Actuator Network (APPN) in relevant literature (Tsai and Wang, 1999; Tang *et al.*, 2000), the shunted system frequency response was shown to be similar to the non-shunted response below the tuned (shunted mode) frequency, but exhibit greater roll-off above the tuned frequency (Agnes, 1995). For broadband control, this would help prevent spillover (Meirovitch, 1990) since the magnitude of the response is in general lower for higher modes. When a current source was used, the shunted system active action was shown to be less effective below the tuned frequency when compared to the non-shunted case, but no roll-off was observed in the high frequency region. Tsai (1998) also performed experimental investigations to illustrate the serial shunt circuit's passive damping ability as well as the active authority amplification ability of the APPN.

It is easy to envision that various topologies of active-passive hybrid piezoelectric circuitry configurations exist for modal damping augmentation. For example, the circuitry elements can be in series or in parallel to the piezoelectric transducer, and the power source can be voltage source or current source. Tang and Wang (2001) compared the open-loop passive damping and active authority amplification effects of different circuitry configurations. In their study, the circuitry parameters were selected as those providing the optimal passive damping. This selection gives the best fail-safe situation, and more importantly, the passive damping comparison results are independent of the control law. The comparison was performed in a nondimensionalized manner, and the role of the generalized electro-mechanical coupling coefficient between the mechanical and the electrical sub-systems was highlighted. Moreover, a method of improving this electro-mechanical coupling coefficient by using an op-amp based negative capacitance circuit was developed, which was shown to further improve the damping and active authority amplification effects of the circuit.

With those open-loop studies as basis, the next question in line is on how to determine the active control and passive parameters to achieve effective hybrid vibration control for modal damping augmentation. While Hagood and von Flotow's tuning method (1991) can minimize the maximum of frequency response for a passive system, their circuitry elements are not necessarily the best choices for an active-passive hybrid system. From the driving voltage (control input) standpoint, the circuit inductance value will determine the electrical resonant frequency around which the active control authority will be amplified, and, although appropriate resistance is needed to achieve broadband passive damping, resistance in general reduces the voltage amplification effect (Niezrecki and Cudney, 1994). To address such issues, Kahn and Wang (1994, 1995) and Tsai and Wang (1996, 1999) developed methods to optimally synthesize the circuitry and control parameters and performed analysis to examine the system performance under closed-loop vibration control.

1.2 ADAPTIVE NARROW BAND DISTURBANCE REJECTION

As demonstrated by Hagood and von Flotow (1991), an electro-mechanical vibration absorber effect can be created by connecting a piezoelectric transducer with a circuit containing resistance and inductance elements. Such device is tuned to a specific structural natural frequency with proper inductance, and absorber damping (resistance) is introduced in order to increase their robustness. The idea can be extended to multi-mode vibration suppression (Wu, 1998; Fleming *et al.*, 2002; Moheimani and Behrens, 2004). The absorbers can also be tuned to suppress a harmonic excitation at a given excitation frequency (*i.e.*, tonally tuned). The major drawback of tonally tuned absorbers is that they require the absorber damping to be very low to achieve good performance. The low damping yields a very small effective bandwidth of the absorber (von Flotow *et al.*, 1994). Consequently, the system performance is extremely sensitive to tuning accuracy and the device can in fact increase the structural vibration if mistuned. For this reason, passive absorbers are not generally useful in off-resonance situations, especially when the excitation frequency is unsteady or varying.

The piezoelectric transducer circuitry absorber can be adapted to harmonic excitations with time-varying frequencies. An example of this type of excitation is rotating unbalance in machinery. The frequency variation could be a slow drift due to changes in operating conditions or a rapid spin-up/spin-down when the machine is turned on/off. One class of such adaptations is referred to as the semi-active absorber which requires either a variable inductance or a variable capacitance element (Hollkamp and Starchville, 1994; Davis and Lesieutre, 2000). Although effective in many cases, these methods have some inherent limitations. It is generally difficult to tune the variable elements rapidly with high accuracy. The variable capacitance method limits the tuning of the piezoelectric absorber to a relatively small frequency range. The variable inductance approach is normally built upon a synthetic inductance circuit which has certain amount of parasitic resistance that is undesirable for narrow-band off-resonance applications.

Morgan and Wang (2002a, 2002b) developed an alternative approach for achieving an adaptive piezoelectric absorber: an active-passive hybrid configuration which consists of a piezoelectric inductive shunt with an active voltage source. The inductance is tuned to the value that corresponds to the nominal or the steady-state excitation frequency. The active control consists of three actions. The first part of the control action is designed to imitate a variable inductance so that the absorber can be adaptively tuned to the correct frequency. The advantages of this active inductance tuning include fast and accurate adjustment, no parasitic resistance, and easier implementation compared to a semi-active approach. The second control action is the negative resistance, which is used to reduce the absorber damping (inherent resistance in the circuit) and increase the performance. Finally, an active coupling enhancement action is used to enhance the robustness and to further increase the performance of the absorber. To ensure that the active inductance is properly tuned, an expression for the optimal tuning on a general multiple-degrees-of-freedom (MDOF) structure is derived. The closed-loop inductance is achieved using this optimal tuning law in conjunction with an algorithm that estimates the fundamental frequency of the measured excitation. In a later study, Morgan and Wang (2002c) made further extension to the adaptive active-passive piezoelectric absorber configuration so that it can track and suppress multiple harmonic excitations.

1.3 HIGH-PRECISION CONTROL WITH HYSTERESIS COMPENSATION

While high control precision is usually claimed as one of the advantages of piezoelectric actuators, the actual performance in this regard is clearly dependent upon the modeling accuracy and the control algorithms. Most of the studies related to piezoelectric actuators have been based upon a linear strain-field constitutive relation assumption. The presence of nonlinearities in the response of piezoelectric transducers, however, has been well documented since the early description of ferroelectrics (Devonshire, 1954). The physics involved in piezoelectric theory may be regarded as a coupling between the Maxwell's equations of electromagnetism and the elastic stress equations of motion. Normally, in practical applications the electrical field (against the poling direction) applied to the piezoelectric actuator should be kept below the coercive field to avoid depoling. On the other hand, experiments have revealed that even in cases where the applied fields are not sufficient to completely re-orient the remnant polarization in the entire actuator, a small number of domains can still be switched. Thus both the material states and the electro-mechanical coupling are changed, giving rise to the nonlinear hysteretic strain-field behavior even at low field level.

The hysteresis phenomenon obviously affects the piezoelectric control performance especially for applications that require high precision. There has been a significant number of recent activities on modeling this behavior using such as Preisach model (Ge and Jouaneh, 1997) or its variant, the Maxwell resistive capacitor (MRC) model (Goldfarb and Celanovic, 1997), polynomial approximation (Chonan et al., 1996), and time delay process (Tsai and Chen, 2003). Several control strategies have also been proposed, which include hysteresis cancellation using inverse model (Ge and Jouaneh, 1996), feedback linearization (Choi et al., 2002), and Smith predictor (Tsai and Chen, 2003). Ge and Jouaneh (1996) developed an inverse linearized Preisach model to offset the hysteresis of piezoelectric actuator in the feedforward loop. Similarly, based on a polynomial approximation of hysteresis effect, an inverse polynomial approach was proposed and studied (Croft and Devasia, 1998). In these approaches, however, some factors may deteriorate the control performance. For example, very complex coupling effects exist among the stress, strain, electrical field, and electrical displacement of a piezoelectric actuator. Thus the hysteretic strain-field relation of the actuator actually also depends on the electrical displacement/charge and stress which are typically treated as internal variables in the aforementioned control designs. The estimation of these internal variables and hence the characterization of hysteresis become extremely complicated and unreliable in practical applications when the actuator is bonded to a host structure and thus also undergoes deformation. Moreover, differential operation is involved in the hysteresis inverse calculation, which may introduce error when the hysteresis measurement data contains noise.

Tang and Wang (2000) explored a different approach to deal with the piezoelectric nonlinearity and hysteresis. Their idea was to incorporate an *RL* (resistance-inductance) circuit with the piezoelectric actuator to introduce dynamics to the electrical part of the controlled system. This essentially leads to two coupled dynamic sub-systems, the mechanical subsystem to be controlled and the electrical sub-system with second-order effect. The system equation can then be directly cast into a standard statespace format for control design. One advantage of introducing circuitry dynamics to the piezoelectric actuator is that the charge and/or current in the transducer circuitry now become independent state variables that can be directly measured and fed back. In their robust sliding mode control formulation, Tang and Wang used the linear constitutive relation as the baseline and treated all nonlinearities and hysteresis as uncertainties. Such approach, however, has not utilized the progress in hysteresis modeling, is conservative in nature, and may compromise the system control performance. To further advance, Xue and Tang (2006) enhanced the approach by explicitly incorporating the hysteresis modeling in the control design. With this piezoelectric circuitry configuration, not only the real-time hysteresis prediction accuracy can be improved, but also the nonlinear control design can be greatly simplified as compared to the inverse cancellation methods. Using a continuous sliding mode technique, Xue and Tang (2006) developed a robust and high precision control methodology for piezoelectric actuator with direct hysteresis compensation.

1.4 VIBRATION CONFINEMENT AND ISOLATION

Vibration control is often realized through either relocation or dissipation of the vibration energy. The underlying principle of the concept of vibration confinement is to alter the structural vibration modes in such a manner that the corresponding modal components have much smaller amplitude in concerned region than in other regions of the structure. As a result, the vibration energy will be relocated and confined to regions that are 'less important' or where damping can be applied more easily. Such confinement techniques could suppress vibration in the area of concern more effectively than many traditional methods, and could reduce control power requirement when active means is involved (Shelley and Clark, 2000a). Earlier vibration confinement methods were mostly passive ones. It was found that the presence of irregularities in a nominally periodic structure could inhibit the propagations of wave/vibration energy within the structure, causing normal mode localization. (The vibration localization will be further discussed in the next section.) Vakakis (1994) showed that the impulse response could be localized in weakly coupled geometrically nonlinear beams. Using a similar idea, Vakakis *et al.* (1999) further developed a nonlinear spring-mass device to isolate structures from earthquake-induced motions. Keane (1995) proposed a passive scheme that was based on redesigning a nominally periodic structure so that it had intrinsic characteristics of vibration/noise isolation within a frequency band. For more general structures (*i.e.*, non-periodic), additional stiffening elements were proposed to alter the structural modal distribution (Allaei and Tarnowski, 1997).

While the ideas of the various passive vibration confinement techniques are interesting, their performance however has been limited. This is mainly because passive alteration of vibration modes requires significant modification of the structure in most cases and thus oftentimes might not be practical. On the other hand, researchers have been resorting to active feedback controls to realize closed-loop vibration confinement by using the inverse eigenvalue problem approach (Choura and Yigit, 1995). Shelley and Clark (1996) demonstrated that, for a special class of discrete systems where coupling only occurs between neighboring elements, only two actuators are needed to shape all the eigenvectors. The flexibility offered by state feedback in multi-input systems beyond closed-loop eigenvalue assignment was first recognized by Moore (1976). It was shown that in such systems, not only the eigenvalues could be assigned; one also has the freedom to adjust the eigenvectors, which is now generally referred to as eigenstructure assignment. Andry *et al.* (1983) discussed some key issues in eigenstructure assignment in linear control systems. In general, the assignment of eigenvector is possible only with multiple inputs, and the assigned eigenvectors must fall into an admissible space. They paid special attention to feedback gain solutions yielding the achievable eigenvectors that are the closest to the desired ones. Song and Jayasuriya (1993) practiced using an eigenstructure assignment algorithm on multi-input-multi-output systems to realize active vibration confinement. Their method requires that the number of actuators equals to the number of the structural degrees of freedom. Shelley and Clark (2000b) used a singular value decomposition based method for vibration confinement, where the closed-loop eigenvectors are pre-selected and the feedback control will lead to the achievable approximation that are closest to these desired eigenvectors. Corr and Clark (1999) further suggested an active-passive hybrid scheme to realize vibration confinement, where passive mechanical absorbers were used to reduce active power requirements and cost.

Tang and Wang (2004) advanced the vibration confinement technique in two correlated ways. First, they proposed to apply the active control input through a piezoelectric transducer circuitry. One of the main limitations of the many active vibration confinement methods is due to the restricted choice of the closed-loop eigenvectors. With the introduction of circuitry elements which is much easier to implement than changing or adding mechanical components, the state matrices of the system can be augmented (see Eq. (1.1)) and the design space for eigenstructure assignment can be greatly enlarged. Second, a new eigenstructure assignment algorithm was developed to suppress vibration more directly in regions of interest. This algorithm features the optimal selection of achievable eigenvectors that minimizes the eigenvector components at concerned regions by using the Rayleigh Principle. To maximize the system performance, a simultaneous optimization/optimal eigenvector assignment approach to concurrently determine the circuitry elements and active control parameters was formulated. Using this energy confinement idea and the related algorithm, Wu and Wang (2007) practiced the design of vibration isolation systems by confining vibration energy in the circuitry and in areas away from the attenuated end (*i.e.*, the end that requires small vibration) of the isolator.

1.5 VIBRATION DELOCALIZATION AND CONTROL IN PERIODIC STRUCTURES

Periodic structures are commonly employed in industrial applications, such as turbo-machinery bladed-disks, space antennae and other space structures. These structures are designed to be comprised of identical substructures that form a spatial periodicity. In an ideal situation where the substructures are perfectly identical, the vibration energy is uniformly distributed among the substructures, and the mode shapes are extended throughout the entire structure. In realistic situations, however, there are always small differences among the substructures, hereafter referred to as mistuning, such as the blade-to-blade differences in geometry and material properties due to manufacturing tolerances or in-service degradations.

Such mistuning could change the dynamic behavior of the periodic structures (which now become nearly periodic or so-called mistuned structures) drastically. A phenomenon known as *vibration localization* could occur under certain circumstances (Pierre *et al.*, 1996; Slater *et al.*, 1999), especially when the mechanical couplings among substructures are weak. In a localized vibration situation, the vibration energy is confined to a small region of the structure, resulting in increased stresses and amplitudes locally. The occurrence of vibration localization could cause severe damage to the structure.

Most of the relevant research has concentrated on exploring the cause of vibration localization and predicting the localized response. Fewer studies, however, have been performed in developing methods to reduce or control vibration localization in mistuned structures. Castanier and Pierre (2002) explored the feasibility of using intentional mistuning to decrease the localization effects in turbo-machinery rotors. Following the principle that stronger coupling among substructures could reduce localization, Cox and Agnes (1999) and Gordon and Hollkamp (2000) attempted using embedded piezoelectric transducers to increase the mechanical coupling ratio between blades in a laboratory bladed-disk. It was found that directly shorting the piezoelectric transducers had little effect on reducing localization. An inductance element was later added to the shorted piezoelectric transducers, where it was observed that this addition could enhance the system coupling (Agnes, 1999); nevertheless, the illustrative example only showed improvement in some of the localized modes. Recently, Tang and Wang (2003) proposed a new approach for vibration delocalization via piezoelectric circuitry. They used local piezoelectric resonant circuit on each substructure to absorb the vibration energy from the substructure and store that portion of energy in electrical form. These local circuits are then connected together through electrical coupling. Such a networked piezoelectric transducer circuitry preserves the nominal periodicity, and essentially establishes an additional electrical coupling in the system. While in most cases the direct increase of the mechanical coupling between the substructures is difficult to achieve due to design limitations, one can easily introduce strong electrical coupling. This creates a new energy channel which can propagate the otherwise localized energy in a nearly periodic structure throughout the entire structure. Yu *et al.* (2006) investigated the using of a negative capacitance to increase the electro-mechanical coupling as well as the energy transfer capability of the piezoelectric transducer circuitry for vibration delocalization.

While reducing vibration localization in nearly periodic structures has practical significance, an equally important issue in such structures is the suppression of overall vibration under external disturbances such as the engine-order excitation in bladed disks. Tang and Wang (1999) explored the feasibility of utilizing piezoelectric absorber circuits applied on individual substructures for the vibration control of periodic structures, where they derived an active compensation law to deal with the inter-blade coupling effect and achieved multi-harmonic excitation suppression. Yu and Wang (2007a) further advanced such a vibration control idea by utilizing only simple passive coupling elements to create a networked architecture and achieved the optimal design goal. In addition, an active coupling enhancement concept via negative capacitance is studied for performance and robustness improvement. The network's vibration suppression ability for the mistuned bladed-disk system was systematically investigated through Monte Carlo simulation. Experimental investigations have also been carried out to verify this concept (Yu and Wang, 2007a).

1.6 STRUCTRUAL DAMAGE IDENTIFICATION ENHANCEMENT

The timely detection of structural damage is of vital importance to structural reliability and durability and everybody's daily life/safety. Among the various damage detection approaches, the vibration-based methods have been quite popular. Some structural damages will change the properties of the structure (mass, stiffness or damping) and these changes will result in changes in the dynamic characteristics of the global system response, such as natural frequencies, damping ratio, and mode shapes. Since the measurement of natural frequencies and frequency response functions is quite straightforward, the damage detection schemes that require only the measurement of natural frequencies (hereafter referred to as the frequency-shift-based methods) are considered to be the easiest to implement, which is a significant advantage especially for complex structures (Doebling et al., 1996, 1998).

The current practice of frequency-shift-based damage identification methods, however, has severe limitations. One problem is that the natural frequencies can be relatively insensitive to damage occurrence. To address this issue, Ray and Tian (1999) and Jiang et al. (2007) introduced the concept of sensitivity enhancing control to increase modal frequency sensitivity to damage. Another serious limitation of the traditional frequency-shift-based damage identification method is that the number of natural frequencies that can be accurately measured is normally much smaller than the number of system parameters required to completely and accurately characterize the damage. Several methods have been proposed in the literature to address this issue. Cha and Gu (2000) and Nalitolela *et al.* (1992) introduced a mass/stiffness addition technique to enrich the modal information measurement. However, the direct addition of mass/stiffness to a structure might be difficult to implement for many practical applications. To overcome this difficulty, Lew and Juang (2002) proposed an active control approach using a virtual passive controller to enrich the modal frequency measurement, where a series of feedback controllers are incorporated to generate additional closed-loop natural frequencies. Koh and Ray (2004) also addressed this issue by utilizing multiple closed-loop systems that can lead to a much enlarged dataset of frequency measurement.

An alternative approach that utilizes the piezoelectric transducer circuitry with tunable inductance to enrich frequency measurement data has been proposed by Jiang *et al* (2006). The key idea is to use a tunable piezoelectric transducer circuitry coupled to the mechanical structure to favorably alter the dynamics of the electro-mechanically integrated system. First, the circuitry can be tailored to change the system frequency/modal distribution by introducing additional resonant frequencies and vibration modes. Second, taking advantage of the property that the circuitry elements (e.g., inductors) can be easily tuned/adjusted, one can obtain a much enlarged dataset consisting of a family of frequency response functions (under different circuitry tunings) as compared to the original single frequency response of the mechanical structure without the circuitry. It was shown that this method could have the potential of more completely and accurately characterizing the variation of system dynamic response due to damage. Jiang *et al.* (2006, 2008) discussed the circuitry tuning criterion that can fully utilize the frequency measurement enrichment feature of such idea. Analysis showed that in the region of eigenvalue curve veering, high sensitivity of damaged-induced eigenvalue changes with respect to circuitry tuning can be expected. Thus, by tuning the circuitry around the curve veering region, one may obtain a family of frequency response functions that could reflect the damage occurrence most effectively.

1.7 SUMMARY

The preceding sections outline the versatile means of synthesizing piezo-electric transducer circuitry for a variety of adaptive structural systems. As can be seen, the foundational idea is to take full advantage of the two-way electro-mechanical energy transfer capability of the piezoelectric trans-ducer by coupling electronic circuitry elements to the integrated system. This facilitates the possibility of favorably altering the system dynamics to improve the control and monitoring performances. At the local transducer level, this idea can provide simultaneous passive damping and active au-thority amplification effects, lead to enhancement and adaptiveness in vi-bration absorbing, and yield high-precision robust control. At the global level, this idea can result in vibration confinement for more effective vibra-tion control as well as robust vibration delocalization and suppression in nearly periodic structures, both based on the energy relocation feature of the piezoelectric transducer circuitry. Moreover, the tunable additional degrees-of-freedom of the circuitry can lead to more complete reflection of structural damage in the system frequency response, which can enhance our capability to achieve more accurate damage identification. There have been a large number of papers published by the authors and their graduate students and by fellow researchers in the adaptive structure community that explore piezoelectric transducer circuitry.

This research monograph is an outgrowth and organized compilation of the selected papers published by the authors and their co-workers during the recent years. The authors thank the ASME, Elsevier, Sage Publica-tions, and Institute of Physics Publishing for allowing them to use the ma-terials in the relevant papers (please see Copyright Permission Acknowl-edgment at the end of this book).

The following chapters will provide detailed discussions of the various piezoelectric transducer circuitry concepts in the order that is outlined in this chapter. The individual chapters will be presented in a self-contained manner in terms of model development, analysis, and design/synthesis, so interested readers may follow the derivations and contents selectively.

It is worth noting that this book is by no means a complete coverage of this subject. Indeed, as in all other promising technologies being devel-oped, rapid advancements are occurring in integrating piezoelectric trans-ducer with electrical circuits for various other applications; one interesting example being energy harvesting. For such investigation, interested readers may refer to the review papers by Sodano *et al.* (2004) and Anton and Sodano (2007).

2 Open-Loop Comparison and Enhancement of Active-Passive Piezoelectric Circuitry

Hagood and von Flotow (1991) were among the first to explore passive vibration control utilizing piezoelectric materials. By shunting the piezoelectric transducer with serially-connected inductor and resistor, they demonstrated the damping and vibration absorbing effects of such passive shunt. Moreover, the integration of the passive shunt and active source for modal damping/control has shown promising results. This configuration not only preserves the passive damping effect of the circuitry, but also, has been found to amplify the active control authority around the circuit tuned frequency (Agnes, 1995; Kahn and Wang, 1994; Tsai and Wang, 1999).

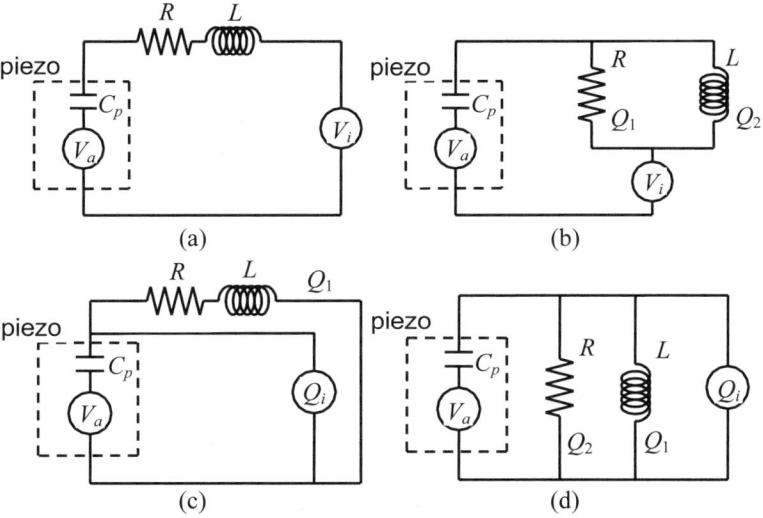

Fig. 2.1. Schematics of active-passive hybrid piezoelectric circuitry networks. V_a: equivalent voltage generator attributed to the piezoelectric effect; V_i: voltage source; Q_i: charge or current source; Q_1 and Q_2: charge flow in branches; C_p: piezoelectric capacitance; R: resistance; L: inductance.

K.W. Wang and J. Tang, *Adaptive Structural Systems with Piezoelectric Transducer Circuitry*, doi: 10.1007/978-0-387-78751-0_2,
© Springer Science + Business Media, LLC 2008

Earlier investigations regarding the active-passive hybrid piezoelectric circuitry mainly focus on the serial configuration. That is, the resistor, the inductor, and the power source (e.g., voltage source) are all connected in series with the piezoelectric transducer as shown in Fig. 2.1a. Wu (1996) found that by connecting the resistor and inductor elements in parallel with the piezoelectric transducer, one may achieve the passive vibration absorbing/damping effect similar to that of the serial configuration. Combining the parallel and serial passive configurations with parallel and serial active driving sources, one can envision a few possible configurations of piezoelectric circuitry, some of which are shown in Figs. 2.1b-2.1d (Tang and Wang, 2001). It is worth noting that for the passive effect to function normally in the absence of active input, one should use charge or current control when the power source is in parallel with the shunting elements, such as those shown in Figs. 2.1c and 2.1d. On the other hand, voltage source should be used in series with the shunts (Figs. 2.1a and 2.1b).

2.1 PROBLEM STATEMENT AND OBJECTIVE

The objective of this chapter is to present a comprehensive study of the open-loop characteristics of piezoelectric transducer circuitries with both active and passive functions. More specifically, the questions to be addressed are:

(a) What is the fundamental difference between the different configurations of piezoelectric transducer circuitry, in terms of open-loop performance? What roles do the circuitry elements play to affect the control authority and passive damping performance respectively?

(b) It is well known that high electro-mechanical coupling would be beneficial to transducer performance in general. Is it possible to increase the system electro-mechanical coupling effect through circuitry enhancement, to yield better open-loop performance?

In what follows, we compare the open-loop passive damping and active authority amplification effects of different circuitry configurations, which provides basic understandings to the piezoelectric transducer circuitry concept and also leads to some guidelines for various applications in the succeeding chapters. As explained later in this chapter, in all cases the circuitry parameters are selected as those offering the optimal passive damping. This selection yields the best fail-safe situation, and more importantly, the passive damping comparison results are independent of the control law/algorithm. The comparison is performed in a nondimensional-ized manner, and the role of the generalized electro-mechanical coupling coefficient is highlighted in the analysis. Based on such analysis, a method of improving this electro-mechanical coupling coefficient by using an operational amplifier (op-amp) circuit based negative capacitance (Tang and Wang, 2001) is presented. As will be shown later in this book, this improvement plays an important role in many applications utilizing the piezoelectric transducer circuitry.

2.2 OPEN-LOOP COMPARISON OF ACTIVE-PASSIVE PIEZOELECTRIC CIRCUITRY

2.2.1 Transfer Functions

The linear constitutive relation for a piezoelectric transducer has various equivalent forms. Following is one of these forms, which is frequently used in one dimensional applications (IEEE, 1988),

$$\tau = E_p \varepsilon - h_{31} D \qquad (2.1)$$
$$E = -h_{31} \varepsilon + \beta_{33} D$$

where τ, ε, D and E represent the stress, strain, electrical displacement (charge/area) and electrical field (voltage/length along the transverse direction) in the piezoelectric transducer, respectively, and E_p, h_{31} and β_{33} are its Young's modulus, piezoelectric constant and dielectric constant. In this chapter, a generic cantilevered beam is used as the host structure for illustration. The beam is surface-bonded with a single piezoelectric transducer that is connected to a shunt circuit and a power source. All relevant notations are listed in Nomenclature.

One of our goals here is to evaluate the performance of the basic active-passive hybrid piezoelectric circuitry. Previous studies have provided two kinds of resonant shunt circuits for passive damping, *i.e.*, the serial configuration where the resistance (R) and inductance (L) elements are connected in series with the piezoelectric transducer (Hagood and von Flotow, 1991), and the parallel configuration where the R and L elements are connected in parallel with the piezoelectric transducer (Wu, 1996). On the other hand, Niezrecki and Cudney (1994) suggested that a resonant shunt could correct the power factor of the piezoelectric transducer, and the apparent power could be significantly reduced around the shunt circuit frequency by implementing the inductor either in series with the power source (amplifying voltage input) or in parallel (reducing current consumption). The combination of the serial or parallel shunt circuits with the serial or parallel driving power source leads to a number of different network configurations (Tang and Wang, 2001). Here we examine the performance of four representative circuitry configurations shown in Figs. 2.1a-2.1d.

Under the linear piezoelectric constitutive relation assumption, the systems under consideration are all linear, and the structural response is the summation of that caused by the external disturbance and that caused by the control input (voltage or current/charge input). Hence, the system damping ability can be evaluated from the transfer function between the structural response and the external disturbance. The actuation authority evaluation, however, is more complicated. The structural response caused by the control input is affected by other factors, such as structural damping as well as the additional damping introduced by the shunt circuit. As suggested by Niezrecki and Cudney (1994), the circuitry could have the function of power factor correction, thereby amplifying the voltage or charge/current applied to the piezoelectric material around the circuit frequency. Therefore, the effect of the circuitry to the actuation authority should be judged and compared using the ratio between the voltage or charge/current that the piezoelectric transducer is subject to and the voltage or charge/current applied from the driving source.

The system equations of motion are derived using the Hamilton's principle, and discretized by applying the assumed mode method (see Appendix 2.1 for details). We only focus on a single-mode model in this study. Thus, the transverse displacement of the beam is assumed as

$$w(x,t) = \phi(x)q(t) \tag{2.2}$$

where ϕ is the first vibration mode of the cantilevered beam without the piezoelectric transducer and the shunt circuit, and q is the generalized mechanical displacement.

Configuration (a): the piezoelectric transducer is connected in series with the RL elements and the voltage source. The equations of motion are given as (A2.14), and the transfer functions between the structural response and the external disturbance and that between the structural response and the input voltage are, respectively,

$$\frac{\overline{q}}{\overline{F}_m} = \frac{-\omega^2 L + Ri\omega + k_2}{(-\omega^2 L + Ri\omega + k_2)(-\omega^2 m + gi\omega + k) - k_1^2} \qquad (2.3a)$$

$$\frac{\overline{q}}{\overline{V}_i} = \frac{k_1}{k_1^2 - (-\omega^2 L + Ri\omega + k_2)(-\omega^2 m + gi\omega + k)} \qquad (2.3b)$$

The over-bar is used to indicate the magnitude of the corresponding variable. The electrical property of a piezoelectric transducer is similar to that of a capacitor, which leads to a reactive current that provides only an electromagnetic field and does not perform work or result in useful power being delivered to the load. The power factor of the piezoelectric actuator is thus approximately zero. The addition of an inductor forms a resonant circuit, and around the resonant frequency the reactive elements will cancel and the phase between the current and the voltage becomes zero, resulting in a unity power factor (Niezrecki and Cudney, 1994). Therefore, the effect of the circuitry on the driving voltage can be evaluated from the transfer function between the voltage across the piezoelectric transducer with the piezoelectric transducer being blocked (before it is coupled to the host structure) and the input voltage from the source, which is given as

$$\frac{\overline{V}_a}{\overline{V}_i} = \frac{1}{-LC_p\omega^2 + RC_p i\omega + 1} \qquad (2.3c)$$

Denote

$$\omega_m = \sqrt{\frac{k}{m}}, \qquad \omega_e = \sqrt{\frac{k_2}{L}}, \qquad \varsigma = \frac{g}{2m\omega_m}, \qquad \xi = \frac{k_1}{\sqrt{kk_2}} \qquad (2.4a\text{-}g)$$

$$\Omega = \frac{\omega}{\omega_m}, \qquad r = \frac{R}{k_2}\omega_m, \qquad \delta = \frac{\omega_e}{\omega_m}$$

Here r and δ are often referred to as the resistance and inductance tuning ratios, and ξ is called the generalized electro-mechanical coupling coefficient which is further discussed later in this chapter. The nondimensionalized transfer functions for configuration (a) are,

$$\frac{\overline{q}}{\overline{q}_{f0}} = \frac{(\delta^2 - \Omega^2) + r\delta^2 i\Omega}{(1 + 2\varsigma i\Omega - \Omega^2)(\delta^2 + r\delta^2 i\Omega - \Omega^2) - \delta^2 \xi^2} \qquad (2.5a)$$

$$\frac{\overline{q}}{\overline{q}_{V0}} = \frac{\delta^2(\xi^2 - 1)}{\delta^2 \xi^2 - (1 + 2\varsigma i\Omega - \Omega^2)(\delta^2 + r\delta^2 i\Omega - \Omega^2)} \qquad (2.5b)$$

$$\frac{\overline{V}_a}{\overline{V}_i} = \frac{\delta^2}{\delta^2 + r\delta^2 i\Omega - \Omega^2} \qquad (2.5c)$$

where $\overline{q}_{f0}$ and $\overline{q}_{V0}$ are the static displacements of the beam (without the circuit) under the external disturbance force and under the voltage input, respectively,

$$\overline{q}_{f0} = \frac{\overline{F}_m}{k} \qquad (2.6a)$$

$$\overline{q}_{V0} = \frac{\overline{V}_i}{k_1 - \dfrac{kk_2}{k_1}} = \frac{\overline{V}_i}{k_1(1 - \dfrac{1}{\xi^2})} \qquad (2.6b)$$

Note here $\overline{q}_{V0}$ is a function of k_1 and the coupling coefficient ξ.

Configuration (b): the RL elements are in parallel, and the voltage source is in series with the shunting elements. The equations of motion are given as (A2.12) and (A2.15). The nondimensionalized transfer functions are respectively, after derivations,

$$\frac{\overline{q}}{\overline{q}_{f0}} = \frac{(\delta^2 - \Omega^2)r + i\Omega}{(1 + 2\varsigma i\Omega - \Omega^2)(r\delta^2 + i\Omega - r\Omega^2) - \xi^2(r\delta^2 + i\Omega)} \qquad (2.7a)$$

$$\frac{\overline{q}}{\overline{q}_{V0}} = \frac{(r\delta^2 + i\Omega)(\xi^2 - 1)}{\xi^2(r\delta^2 + i\Omega) - (1 + 2\varsigma i\Omega - \Omega^2)(r\delta^2 + i\Omega - r\Omega^2)} \tag{2.7b}$$

$$\frac{\overline{V}_a}{\overline{V}_i} = \frac{r\delta^2 + i\Omega}{r\delta^2 + i\Omega - r\Omega^2} \tag{2.7c}$$

Configuration (c): the piezoelectric transducer is connected in series with the RL elements, and the charge/current source is in parallel with the shunting elements. The equations of motion are given as (A2.12) and (A2.16). This configuration obviously has the same passive damping ability as configuration (a), as both of them degenerate to the serial passive shunt with the absence of active driving source. The nondimensionalized transfer function between the structural response and the charge input is such that

$$\frac{\overline{q}}{\overline{q}_{Q0}} = \frac{r\delta^2 i\Omega - \Omega^2}{\xi^2(r\delta^2 i\Omega - \Omega^2) + (1 + 2\varsigma i\Omega - \Omega^2 - \xi^2)(\delta^2 + r\delta^2 i\Omega - \Omega^2)} \tag{2.8a}$$

where $\overline{q}_{Q0}$ is the static displacement of the beam (without the shunt circuit) under charge input,

$$\overline{q}_{Q0} = \frac{k_1}{k} Q_i \tag{2.8b}$$

The shunt circuit effect on the driving charge can be evaluated from the transfer function between the applied charge and the charge that the piezoelectric transducer is subject to with the piezoelectric transducer being blocked (before it is coupled to the host structure), which is given as

$$\frac{\overline{Q}}{\overline{Q}_i} = \frac{r\delta^2 i\Omega - \Omega^2}{\delta^2 + r\delta^2 i\Omega - \Omega^2} \tag{2.8c}$$

Configuration (d): the RL elements and the charge source are all in parallel. The equations of motion are given as (A2.12) and (A2.17). This configuration has the same passive damping ability as configuration (b), as both of them degenerate to the parallel passive shunt (Wu, 1996) with the absence of active driving source. The transfer function between the structural response and the charge input and that between the charge that the piezoelectric transducer is subject to and the applied charge are, respectively,

$$\frac{\overline{q}}{\overline{q}_{Q0}} = \frac{r\Omega^2}{\xi^2 r\Omega^2 - (1 + 2\varsigma i\Omega - \Omega^2 - \xi^2)(r\delta^2 + i\Omega - r\Omega^2)} \quad (2.9a)$$

$$\frac{\overline{Q}}{\overline{Q}_i} = \frac{r\Omega^2}{r\Omega^2 - i\Omega - r\delta^2} \quad (2.9b)$$

2.2.2 Optimal Passive Damping Comparison

In both the serial and the parallel circuitry configurations, a second order dynamics is augmented to the original system, and the optimal values for the inductance and resistance can be found based upon the analogy with the damped vibration absorber concept. The details of the derivations are given in Appendix 2.2. In general, the frequency response of the system with the shunt circuit intersects that of the baseline system (without the shunt circuit) at two invariant points. The locations of the intersection points depend on the inductance tuning. If the amplitudes of these two points are equated, the optimal inductance tuning can be calculated to be, respectively,

$$\delta^* = 1 \qquad \text{serial } RL \text{ configuration} \qquad (2.10a)$$

$$\delta^* = \sqrt{1 - \frac{3}{2}\xi^2} \qquad \text{parallel } RL \text{ configuration} \qquad (2.10b)$$

Under the optimal inductance tuning, the aforementioned two invariant points are, respectively,

$$\Omega_{1,2}^2 = 1 \pm \frac{\sqrt{2}}{2}\xi \qquad \text{serial } RL \text{ configuration} \qquad (2.11a)$$

$$\Omega_{1,2}^2 = 1 - \xi^2 \pm \frac{1}{2}\sqrt{2\xi^2(1-\xi^2)} \quad \text{parallel } RL \text{ configuration} \qquad (2.11b)$$

The frequency response for any given value of the resistance tuning, r, passes through these two intersection points. An optimal value of r can be found by equating the amplitude of the frequency response at $\Omega = \delta^*$ and the amplitude of the two invariant points

$$r^* = \sqrt{2}\xi \qquad \text{serial } RL \text{ configuration} \qquad (2.12a)$$

$$r^* = \frac{1}{\sqrt{2}\xi} \qquad \text{parallel } RL \text{ configuration} \qquad (2.12b)$$

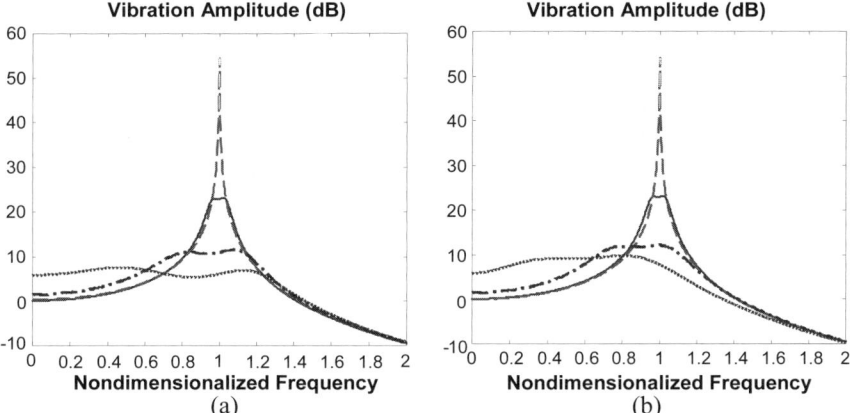

Fig. 2.2. Passive damping performance comparison: (a) serial RL configuration, (b) parallel RL configuration.
— — — — : open circuit; ————: shunt, ξ=0.1;
— . — . — : shunt, ξ=0.4; ⋯⋯⋯⋯ : shunt, ξ=0.7.

The frequency responses (between the structural response and the external disturbance) of the integrated system with the serial and the parallel shunts are shown in Fig. 2.2a and Fig. 2.2b, respectively, where results corresponding to different electro-mechanical coupling coefficients are plotted. The frequency responses of the integrated system under the optimal tunings all exhibit the plateau shape around the structural resonant frequency. The response amplitudes at the two invariant points (the two corners of the plateau) for the two configurations, which can be used to assess the vibration damping/absorbing performance, are respectively,

$$\left.\frac{\overline{q}}{\overline{q}_{f0}}\right|_{\Omega_{1,2}} = \frac{\sqrt{2}}{\xi} \qquad \text{serial } RL \text{ configuration} \qquad (2.13a)$$

$$\left.\frac{\overline{q}}{\overline{q}_{f0}}\right|_{\Omega_{1,2}} = \frac{\sqrt{2}}{\xi\sqrt{(1-\xi^2)}} \qquad \text{parallel } RL \text{ configuration} \qquad (2.13b)$$

From Figs. 2.2a and 2.2b, one may conclude that these two configurations have almost the same passive damping ability.

At this point, the role of ξ, the electro-mechanical coupling coefficient defined in Eq. (2.4d), deserves further discussion. At the material level, the piezoelectric electro-mechanical coupling constant is defined as the square root of the ratio between the open circuit (constant electrical displacement) and short circuit (constant electrical field) stiffness difference and the open circuit stiffness, or equivalently, the square root of the ratio between the constant stress and constant strain permittivity difference and the constant stress permittivity (IEEE, 1988). This concept can be extended to the structure or device level (Lesieutre and Davis, 1997). In dynamic analyses, the electro-mechanical coupling coefficient of the piezoelectric transducer bonded to or embedded in a structure is often defined as the square root of the ratio between the open circuit and short circuit structural frequency difference and the open circuit structural frequency. An equivalent definition of the coupling coefficient is the square root of the ratio between the difference between the square of circuitry frequencies corresponding to free and blocked piezoelectric and the square of blocked circuitry frequency. Recalling Eq. (2.4d), one can readily find that

$$\xi^2 = \frac{k_1^2}{kk_2} = \frac{\left(\dfrac{k}{m}\right) - \dfrac{1}{m}\left(k - \dfrac{k_1^2}{k}\right)}{\dfrac{k}{m}} = \frac{\left(\dfrac{k_2}{L}\right) - \dfrac{1}{L}\left(k_2 - \dfrac{k_1^2}{k}\right)}{\dfrac{k_2}{L}} \qquad (2.14)$$

This means that the ξ defined is consistent with the usual definition of electro-mechanical coupling coefficient. In fact, ξ^2 represents the ratio of the energy converted to that imposed to the piezoelectric material. The largest coupling constants for piezoelectric ceramic materials are at the order of 0.7 for 3-3 applications (Lesieutre, 1998), corresponding to energy conversion factors of about 50%. The configurations in this chapter all use the 3-1 direction piezoelectric effect, which has a smaller coupling constant. Clearly, the structure or device level electro-mechanical coupling coefficient ξ is smaller than the coupling constant of the piezoelectric material, because the integrated structure contains non-active elements.

The increase of the electro-mechanical coupling coefficient in general increases the passive damping ability, as shown in Figs. 2.2a and 2.2b. For the case where coupling coefficient is smaller than 0.7 (which is the usual case due to the material property limitation), larger coupling coefficient means further separation of the two invariant points, as indicated by the following expressions:

$$\Omega_1^2 - \Omega_2^2 = \sqrt{2}\xi \qquad \text{serial } RL \text{ configuration} \qquad (2.15a)$$

$$\Omega_1^2 - \Omega_2^2 = \sqrt{2(1-\xi^2)}\xi \quad \text{parallel } RL \text{ configuration} \qquad (2.15b)$$

The peak amplitudes of the frequency response plateau also decrease for both the parallel and series configuration as ξ increases, as shown in Eqs. (2.13a) and (2.13b). Therefore, more significant vibration suppression and broader bandwidth can be achieved with larger electro-mechanical coupling. The reason behind this is clear, as larger electro-mechanical coupling means more vibration energy can be converted to electrical energy and stored/dissipated in the circuit.

2.2.3 Active Authority Amplification Comparison

Voltage Driving
For configurations (a) and (b) shown in Fig. 2.1, the active driving source is a voltage source. The inductor in the shunt circuit has a leading power factor, which cancels the lagging power factor of the piezoelectric capacitance around the circuit resonant frequency. From Eqs. (2.5c) and (2.7c), one can readily find that, in each case, around the circuit resonant fre-

quency the real part of the denominator of the transfer function goes to zero. This indicates that the voltage across the piezoelectric transducer can become much higher than the input voltage. This effect is shown in Fig. 2.3. Clearly, the two configurations have virtually the same voltage amplification effect.

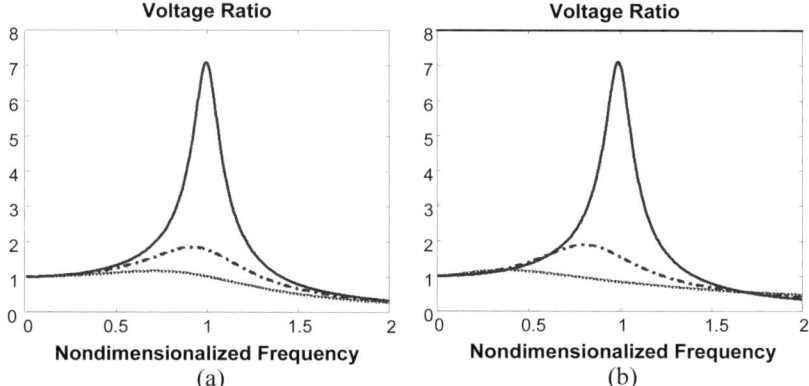

Fig. 2.3. Voltage amplification comparison. (a) serial RL configuration (b) parallel RL configuration.

———— : $\xi=0.1$; – · – · – · – : $\xi=0.4$; ················ : $\xi=0.7$.

At zero frequency, the voltage ratio is unity, because the circuitry elements have no effect for static input and the voltage across the piezoelectric transducer is the same as the input voltage. The amplification effect is most significant around the resonant frequency, which, for a typical electro-mechanical coupling coefficient of $\xi = 0.1$, exhibits a more than 600% increase. Above the circuit resonant frequency, the voltage ratio has a significant roll-off. The reason is that the inertia effect of the inductor tends to block the voltage input from the power source at high frequencies. For broadband control this would help prevent spillover (Meirovitch, 1990) since the high frequency components of the control voltage will be cut off in a manner similar to adding a low pass filter.

While the resistance in the shunt circuit tends to reduce the voltage amplification effect by dissipating a portion of the control power, it is still needed to provide passive damping to the integrated system. The increase of the electro-mechanical coupling coefficient necessitates the increase of the optimal nondimensionalized resistance in Eqs. (2.5c) and (2.7c) and, as a result, the voltage amplification effect reduces. Nevertheless, one should note that a larger coupling coefficient increases the static driving response amplitude, and thus the active authority of a piezoelectric transducer circuitry with larger coupling coefficients could outperform that with smaller coupling coefficient, which will be demonstrated in the next section. On the other hand, in real applications the electro-mechanical coupling at the structure level could hardly reach a value that is higher than 0.3, as the 3-1 direction coupling constant for a typical piezoelectric material is lower than 0.35. These observations provide the rationale for developing a device to increase the generalized electro-mechanical coefficient which is shown in the next section.

Charge/Current Driving

For configurations (c) and (d) in Fig. 2.1, the active driving source is a charge/current source. Similar to the voltage source case, the power factor is corrected to unity around the circuit resonant frequency due to the circuitry dynamics. The charge/current amplification effects (Eq. (2.8c) and Eq. (2.9b)) are shown in Fig. 2.4. Again, the two configurations have virtually the same charge amplification effect.

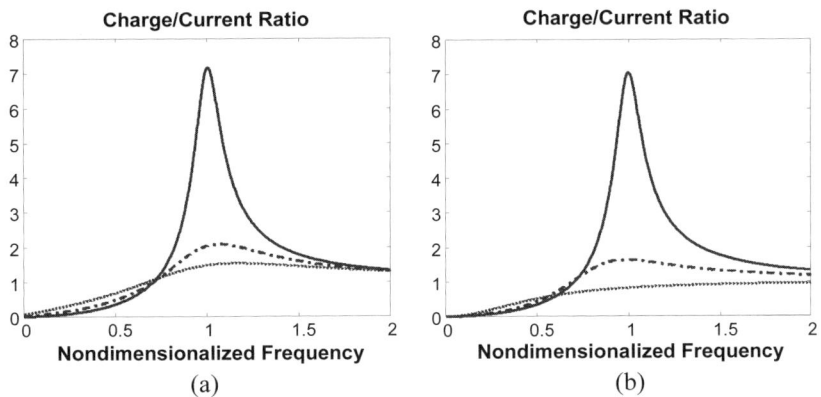

Fig. 2.4. Charge amplification comparison. (a) serial *RL* configuration (b) parallel *RL* configuration.

———— : $\xi=0.1$; – · – · – · – : $\xi=0.4$; ················· : $\xi=0.7$.

At zero frequency, the charge ratio is zero, as the circuitry elements have no effect at static and the charge input directly goes to ground instead of going to the piezoelectric transducer. The amplification effect is most significant around the resonant frequency, which, for a typical electro-mechanical coefficient of $\xi = 0.1$, exhibits more than 600% increase. The charge ratio approaches unity when frequency goes to infinity. This is because at high frequency the inertial effect of the inductor tends to block charge flow in that branch. The increase of electro-mechanical coupling coefficients requires larger resistance in the circuit for optimal passive tuning, and similar to the voltage driving case the charge amplification effect is reduced. In virtue of Eq. (2.8b), one can see that, unlike the voltage driving case where the static driving response is a function of the electromechanical coupling coefficient ξ, the static response for the charge/current driving case is not directly related to ξ. Therefore, if one selects the inductance and resistance based on the optimal passive damping criterion (i.e., the best fail-safe situation), the increase of the electromechanical coupling coefficient does not necessarily increase the overall active control authority.

2.3 ENHANCING OPEN-LOOP PERFORMANCE USING NEGATIVE CAPACITANCE

From the comparison in the previous section, one can see that the four active-passive hybrid piezoelectric circuitry configurations shown in Fig. 2.1 have almost the same passive damping ability and driving source amplification characteristics. However, as discussed at the end of the last section, unlike the voltage driving configuration, increasing the coupling coefficient does not necessarily increase the active authority when charge/current sources are used (under optimal passive damping or fail-safe design). In other words, under the best fail-safe consideration, it is in the voltage driving cases that one could potentially simultaneously increase the passive and active system performance by solely increasing the generalized electro-mechanical coupling coefficient ξ. Without loss of generality, the study in this section is illustrated through the serial circuitry configuration with voltage driving (configuration (a)).

The increase of the generalized electro-mechanical coupling coefficient, as shown in the previous section, increases the circuitry passive damping performance. It also increases the overall active control authority for the voltage driving case, as will be further demonstrated in this section. The coupling coefficient, as defined in Eq. (2.4d), is a function of the host structure stiffness k, the cross coupling term k_1, and the inverse of the capacitance of the piezoelectric patch k_2. Lesieutre and Davis (1997) proposed to apply mechanical pre-loads to the host structure, thus reducing its stiffness and increasing the coupling. Although the result is promising, it might not be easy to implement, because such approach usually requires pre-conditioning or design modification of the mechanical structure.

The generalized electro-mechanical coupling coefficient, actually, can be increased by electrical tailoring, that is, by introducing a negative capacitance. By connecting a negative capacitance in series with the piezoelectric material, the overall effective capacitance of the circuit is increased, resulting in a larger electro-mechanical coupling coefficient. Such negative capacitance cannot be realized passively, and one needs to use an op-amp to form a negative impedance converter circuit (Horowitz and Hill, 1989), which requires a power source. However, the power consumption is normally very low and can be easily compensated as explained in the following. In many applications the piezoelectric capacitance is quite small, which leads to large k_2. Therefore, oftentimes the optimal passive tuning criterion requires large inductance. In that case one has to use a synthetic inductor that also requires a power source (Hagood and Crawley, 1991; Edberg et al., 1992). The negative capacitance increases the overall capacitance value of the piezoelectric circuitry, and directly reduces the required optimal inductance value. Either it can make the synthetic inductor unnecessary when the inductance value is reduced to a sufficiently low level, or it can share the same power source with a synthetic inductor and the power it consumes can be partly compensated for by the power consumption reduction in the synthetic inductor with lower inductance requirement.

As stated previously, the effect of inductance in the resonant shunt is to cancel the inherent capacitive reactance of the piezoelectric transducer. Bondoux (1996) explored the possibility of canceling the reactance using a negative capacitance. He compared the resonant shunt with the negative capacitance shunting. Although the negative capacitance provides a broadband efficiency possibly allowing multiple-mode damping, the passive damping performance is worse than the resonant shunt because it is difficult to cancel the piezoelectric capacitive reactance by using a negative capacitance alone, due to stability problem. Bruneau et al. (1999) also discussed increasing the piezoelectric capacitance with a negative capacitance circuit. The following section will illustrate how the negative capacitance helps improve both the passive damping and active authority amplification in a typical active-passive hybrid network (Tang and Wang, 2001).

2.3.1 Effect of Negative Capacitance: Analysis

2.3.1.1 Passive damping performance improvement

Note that the equations of motion for the serial active-passive hybrid pie-zoelectric circuitry are given in (A2.14a,b). When the negative capacitance is added in series to the piezoelectric transducer in the circuitry, the system equations become,

$$m\ddot{q} + g\dot{q} + kq + k_1 Q = F_m \qquad (2.16)$$
$$L\ddot{Q} + R\dot{Q} + \hat{k}_2 Q + k_1 q = V_i$$

where $\hat{k}_2 = k_2 - \tilde{k}_2$ is the inverse of the overall capacitance in the circuit. The apparent electro-mechanical coupling coefficient is now defined as

$$\hat{\xi} = \frac{k_1}{\sqrt{k\hat{k}_2}} \qquad (2.17)$$

and the optimal resistance tuning is

$$\hat{r} = \frac{R}{\hat{k}_2}\omega_m \qquad (2.18)$$

The performance increase due to the negative capacitance is shown in Fig. 2.5a for various negative capacitance values, where the original coupling coefficient without the negative capacitance is $\xi = 0.1$.

From the system stability point of view, there is a limit on adding negative capacitance to the system. That is, the system generalized stiffness matrix has to maintain its positive definiteness,

$$\begin{bmatrix} k & k_1 \\ k_1 & \hat{k}_2 \end{bmatrix} > 0$$

which means $k\hat{k}_2 > k_1^2$, or, $\hat{\xi} < 1$. Within this limit, adding negative capacitance in general benefits the system passive performance.

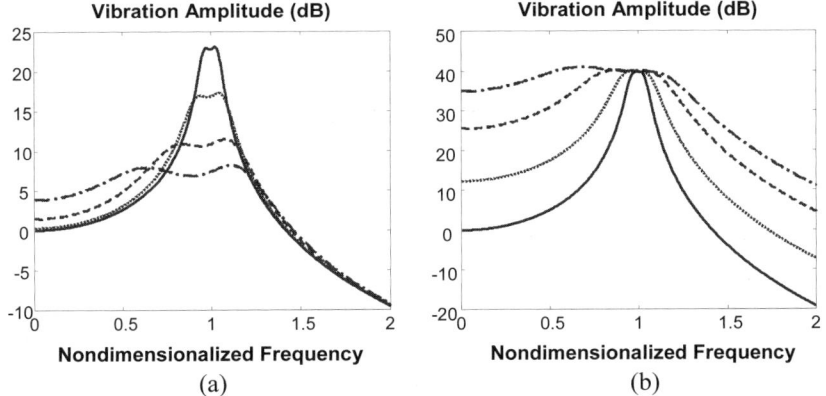

Fig. 2.5. Performance improvement due to negative capacitance (a) passive damping (b) voltage driving response.
————— : $\xi = 0.1$, original piezoelectric shunt without negative capacitance;
·············· : $\tilde{k}_2 / k_2 = 3/4$, $\hat{\xi} = 0.2$;   ------ : $\tilde{k}_2 / k_2 = 15/16$, $\hat{\xi} = 0.4$;
— · — · — : $\tilde{k}_2 / k_2 = 35/36$; $\hat{\xi} = 0.6$.

2.3.1.2 Open-loop active performance improvement

With the negative capacitance, the transfer function between the structural response and the input voltage now becomes

$$\frac{\overline{q}}{\overline{V_i}} = \frac{k_1}{k_1^2 - (-\omega^2 L + Ri\omega + \hat{k}_2)(-\omega^2 m + gi\omega + k)} \tag{2.19}$$

As discussed in the previous section, the increase of coupling coefficient requires larger resistance in optimal passive tuning, thus reducing the voltage amplification effect of the shunt circuit. However, as one can see from Eq. (2.6b), this also increases the static driving voltage response. The overall effect of increasing the coupling coefficient by negative capacitance can be evaluated from the following nondimensionalized transfer function,

$$\frac{\overline{q}}{\overline{q}_{V0}} = \frac{\delta^2(\hat{\xi}^2 - \hat{\xi}^2 / \xi^2)}{\delta^2\hat{\xi}^2 - (1 + 2\varsigma i\Omega - \Omega^2)(\delta^2 + \hat{r}\delta^2 i\Omega - \Omega^2)} \qquad (2.20)$$

where the baseline response is the static voltage response of the original system without the negative capacitance. The simulation results for the above transfer function corresponding to various negative capacitance values are shown in Fig. 2.5b.

In the simulations, the structure is assumed to be lightly damped, with structural damping ratio $\varsigma = 0.001$. Therefore, no matter what the apparent coupling ratio $\hat{\xi}$ is, at resonant frequency the active authority index is approximately

$$\frac{\overline{q}}{\overline{q}_{V0}} = \frac{\hat{\xi}^2 - \hat{\xi}^2 / \xi^2}{\hat{\xi}^2} = 1 - \frac{1}{\xi^2} \qquad (2.21)$$

as shown in Fig. 2.5b. Therefore, the introduction of the negative capacitance does not change the voltage driving response amplification at the resonant frequency. Its effect on active authority is to increase the bandwidth of the voltage driving response.

2.3.2 Experimental Verification

Experimental investigations are performed on a cantilevered beam to demonstrate the performance enhancement. The beam is made of aluminum attached with two piezoelectric transducers (PZT 5A, supplied by Piezo Systems, Inc.) attached close to the root. The top piezoelectric transducer is used as the transducer in the serial piezoelectric circuitry configuration (configuration (a)) for active-passive control, and the bottom one is used to provide disturbance excitation. The dimensions of the beam and the piezoelectric transducers are listed in Table 2.1. A fiber optical sensor (D125, Philtech, Inc.) is used to collect the beam tip displacement signal, and an

HP35665A dynamic signal analyzer is used to extract the frequency responses. The experimental setup is shown in Fig. 2.6a.

Table 2.1. Parameters used in experimental verification of negative capacitance

Length of beam $l_b = 0.30$ m	Left end of PZT $x_l = 0.02$ m
Right end of PZT $x_r = 0.0724$ m	Beam width $w_b = 0.0381$ m
Width of PZT $w_p = 0.0343$ m	Beam thickness $t_b = 0.003175$ m
Thickness of PZT $t_p = 0.000267$ m	

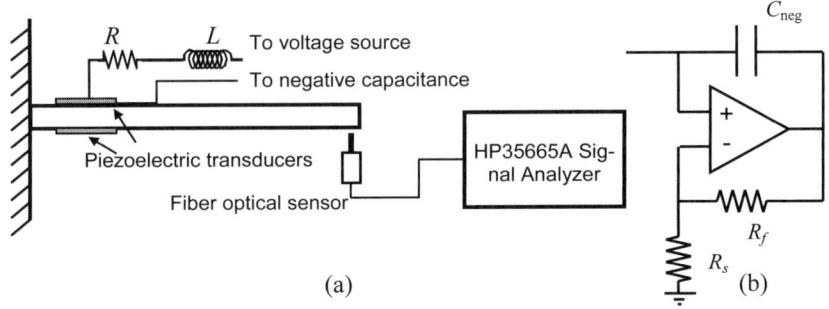

(a)   (b)

Fig. 2.6. Experimental setup. (a) Beam with piezoelectric circuitry and negative capacitance; (b) Circuit diagram of negative capacitance.

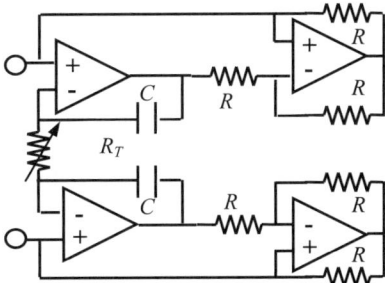

Fig. 2.7. Circuit diagram of synthetic inductor.

The circuit diagram of the negative capacitance (Horowitz and Hill, 1989) is shown in Fig. 2.6b. Since the negative capacitance circuit is non-floating, the top piezoelectric transducer needs to be electrically insulated from the beam, and if a synthetic inductor is used it needs to be a floating one (Agnes, 1995). In piezoelectric circuitry, often the inductance required is large or needs to be adaptable. In many implementations a synthetic inductor (Fig. 2.7) is used. As the purpose of this experiment is to illustrate the effect of negative capacitance, only passive inductors (UTC W-3881 and UTC HVC-9, ELTECH) are used throughout the experiment.

The open circuit and short circuit frequencies of the test beam are shown in Fig. 2.8. The resonant frequency of the open circuit case is 32Hz, and that of the short circuit case is 31.7813Hz. Therefore, according to Eq. (2.14), the electro-mechanical coupling coefficient is $\xi = 0.1167$. The optimal inductance is obtained experimentally (*i.e.*, by adjusting the inductance to yield the plateau shape frequency response around the structural resonant frequency) as $L^* = 249.5$ H.

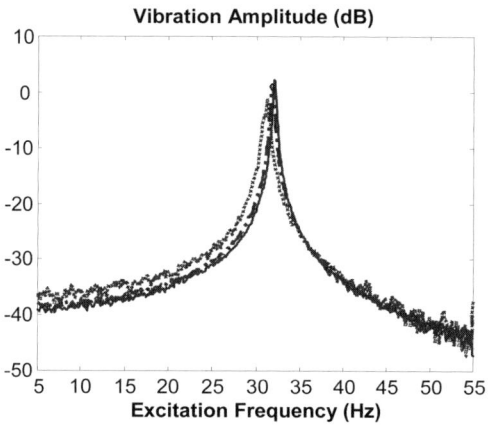

Fig. 2.8. Beam response under external excitation.
——————— : open circuit response; — · — · —: short circuit response (without negative capacitance); ················ : short circuit response with negative capacitance.

The piezoelectric capacitance is measured as 99.14nF. An HA 2645 operational amplifier (op-amp, Harris Co.) is used to form the negative capacitance. In this experiment, the maximum negative capacitance that the circuit can provide is 113.8nF. The primary limitation on the negative capacitance in the experiment is the current that the op-amp can sustain, which, for this specific operational amplifier, should be kept below 5mA. One may achieve larger negative capacitance by using higher current op-amps. The short circuit frequency response with the negative capacitance is also shown in Fig. 2.8, and the resonant frequency now becomes 30.31Hz, which clearly indicates a larger apparent electro-mechanical coupling, $\hat{\xi} = 0.3207$. It should be noted that the short circuit status with negative capacitance refers to the situation that the positive electrode of the piezoelectric transducer is grounded. In other words, the capacitance of this short circuit is the combined effect of the piezoelectric transducer capacitance and the negative capacitance. The optimal inductance value for the shunt circuit with negative capacitance is found to be 34.3H.

The frequency responses (structural response versus disturbance excitation) of the optimal passive shunt circuit with or without the negative capacitance are compared in Fig. 2.9a. Clearly, the introduction of the negative capacitance results in an additional 7dB vibration reduction near the resonant peak. The frequency responses between the structural response and the input voltage from the source (with or without the negative capacitance) are compared in Fig. 2.9b. As predicted in the analysis, the peak amplitude around the resonant frequency remains the same, but the bandwidth of the amplification effect is greatly increased due to the negative capacitance, thus resulting in much larger overall control authority.

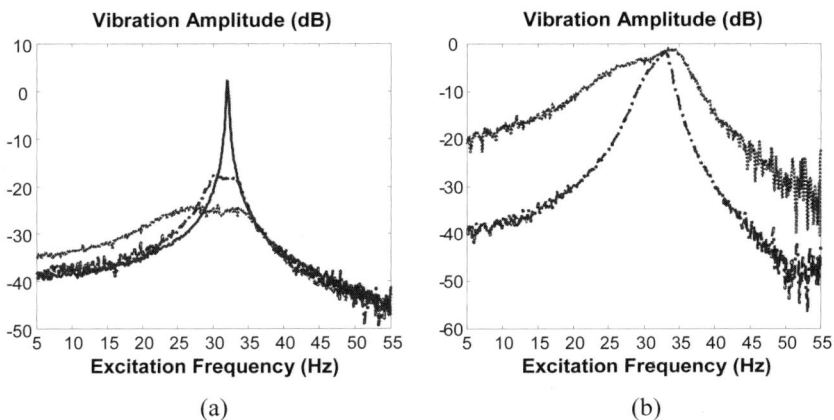

(a)   (b)

Fig. 2.9. Performance comparison: with or without negative capacitance. (a) passive damping (b) voltage driving response.
——————: open circuit response; — · — · — : shunt circuit response (without negative capacitance); ·················· : shunt circuit response with negative capacitance.

2.4 CONCLUDING REMARKS

This chapter presents a nondimensionalized analysis of four basic configurations of active-passive hybrid piezoelectric transducer circuitry; focused on their open-loop passive damping and active authority amplification performance. The investigation is based on that all the circuitry configurations are under optimal passive tuning, which corresponds to the best fail-safe situation. The effect of the generalized electro-mechanical coupling effect is highlighted throughout this study. It is found that under all the electro-mechanical coupling coefficients:

- Properly designed piezoelectric transducer circuitry can possess both passive damping and active authority amplification capabilities.
- The four configurations have virtually the same passive damping and driving source amplification effect around the circuit resonant frequency. This implies that the four configurations would have similar hybrid damping ability.
- Increase of the electro-mechanical coupling in general benefits the system passive damping.
- Increase of the electro-mechanical coupling reduces the driving source amplification effect because larger optimal resistance in the shunt circuit is required.

With the study of electro-mechanical coupling coefficient as basis, this chapter also presents a method of enhancing the voltage-source-based circuitry performance by adding a negative capacitance. The negative capacitance is realized by a negative impedance converter circuit with operational amplifier. It is shown via analysis that:

- Adding a negative capacitance in the piezoelectric circuitry can increase the electro-mechanical coupling of the integrated system.
- The negative capacitance can significantly increase the system passive damping.
- Although adding the negative capacitance leads to larger optimal resistance and reduces the voltage amplification effect around the circuit resonant frequency, the overall control authority is significantly improved (compared to the circuitry without negative capacitance) since the structure can be driven to a higher amplitude in a wider bandwidth given the same level of voltage input.

The analytically predicted enhancements are verified experimentally. The test results illustrate that the negative capacitance increases both the passive damping and active authority performance of a voltage-source-based piezoelectric transducer circuit significantly. This provides an additional design variable for control development.

APPENDIX 2.1 BEAM STRUCTURE INTEGRATED WITH ACTIVE-PASSIVE HYBRID PIEZOELECTRIC CIRCUITRY: EQUATIONS OF MOTION

In general, the equations of motion of a beam integrated with an active-passive hybrid piezoelectric circuitry can be derived using the Hamilton's principle,

$$\int_{t_1}^{t_2} [\delta T - \delta U_b - \delta U_p + \delta W_v] dt = 0 \qquad (A2.1)$$

where T, U_b, U_p and δW_v are, respectively, the kinetic energy of the integrated system, the potential energy of the beam, the elastic and electrical energy of the piezoelectric transducer, and the virtual work term.

Recalling Eq. (2.2), from beam theory we can derive,

$$T = T_b + T_p = \frac{1}{2} m_b \dot{q}^2 + \frac{1}{2} m_p \dot{q}^2 \qquad (A2.2)$$

$$U_b = \frac{1}{2} k_b q^2 \qquad (A2.3)$$

where

$$m_b = \int_0^{l_b} \rho_b A_b \phi^2 dx, \quad m_p = \int_0^{l_b} \rho_p A_p \phi^2 \Delta H dx, \quad k_b = \int_0^{l_b} E_b I_b \phi^2 dx \qquad (A2.4)$$

Here $\Delta H = H(x - x_l) - H(x - x_r)$ and $H(x)$ is the Heaviside step function. All relevant notations are listed in the Nomenclature.

The electrical displacement D of the piezoelectric patch is discretized as

$$D = \psi p \tag{A2.5}$$

In virtue of the linear constitutive relation Eq. (2.1), the elastic and electrical energy variation of the piezoelectric transducer can be obtained as

$$\delta U_p = \int_V (\tau \delta \varepsilon + E \delta D) dV = \int_V (E_p \varepsilon - h_{31} D) \delta \varepsilon dV \tag{A2.6}$$
$$+ \int_V (-h_{31} \varepsilon + \beta_{33} D) \delta D dV = k_p q \delta q + k_{pq} p \delta q + k_{pq} q \delta p + k_{pp} p \delta p$$

where

$$k_p = \int_0^{l_b} E_p I_p \phi''^2 \Delta H dx , \quad k_{pp} = \int_0^{l_b} A_p \beta_{33} \psi^2 \Delta H dx , \tag{A2.7}$$
$$k_{pq} = \int_0^{l_b} F_p h_{31} \phi'' \psi \Delta H dx$$

The virtual work term is due to the voltage across the piezoelectric material, the disturbance, as well as the beam structural damping,

$$\delta W_v = F_e \delta p + F_m \delta q - g \dot{q} \delta q \tag{A2.8}$$

where

$$F_e = \int_0^{l_b} V_a w_p \Delta H \psi dx , \quad F_m = \int_0^{l_b} \hat{F}(x,t) \phi dx , \quad g = \int_0^{l_b} c_b \phi^2 dx \tag{A2.9}$$

Substitution of the above expressions into Eq. (A2.1) leads to the following equations:

$$(m_b + m_p) \ddot{q} + g \dot{q} + (k_b + k_p) q + k_{pq} p = F_m \tag{A2.10}$$
$$k_{pq} q + k_{pp} p = F_e$$

Hereafter we assume that the electrical displacement is independent of the spatial coordinates, *i.e.*, $\psi = 1$ and $D = Q/(w_p l_p)$. Also, we define

$$m = m_b + m_p, \quad k = k_b + k_p, \quad k_1 = \frac{k_{pq}}{w_p l_p}, \quad k_2 = \frac{k_{pp}}{(w_p l_p)^2} \tag{A2.11}$$

We then have

$$m\ddot{q} + g\dot{q} + kq + k_1 Q = F_m \tag{A2.12}$$
$$k_2 Q + k_1 q = V_a$$

For configuration (a), the piezoelectric transducer is connected in series with the *RL* elements and the voltage source. Therefore, the voltage across the piezoelectric transducer is related to the external voltage input in the following manner,

$$V_a = -L\ddot{Q} - R\dot{Q} + V_i \tag{A2.13}$$

The system equations now become

$$m\ddot{q} + g\dot{q} + kq + k_1 Q = F_m \tag{A2.14a,b}$$
$$L\ddot{Q} + R\dot{Q} + k_2 Q + k_1 q = V_i$$

Note here k_2 actually is the inverse of the capacitance of the piezoelectric transducer.

For configuration (b), the *RL* elements are in parallel, and the voltage source is in series with the shunting elements. From Kirchhoff's current and voltage laws, we have, respectively,

$$V_a = V_i - R\dot{Q}_1 \tag{A2.15a}$$

$$R\dot{Q}_1 = L\ddot{Q}_2 \qquad\qquad (A2.15b)$$

$$Q - Q_1 - Q_2 = 0 \qquad\qquad (A2.15c)$$

For configuration (c), the piezoelectric transducer is connected in series with the RL elements, and the charge/current source is in parallel with the shunting elements. From Kirchhoff's current and voltage laws, we have, respectively,

$$Q_i + Q + Q_1 = 0 \qquad\qquad (A2.16a)$$

$$L\ddot{Q}_1 + R\dot{Q}_1 = V_a \qquad\qquad (A2.16b)$$

For configuration (d), the RL elements and the charge source are all in parallel. From Kirchhoff's current and voltage laws, we have, respectively,

$$Q_i + Q + Q_1 + Q_2 = 0 \qquad\qquad (A2.17a)$$

$$L\ddot{Q}_1 = V_a \qquad\qquad (A2.17b)$$

$$R\dot{Q}_2 = V_a \qquad\qquad (A2.17c)$$

APPENDIX 2.2 OPTIMAL TUNINGS FOR PASSIVE DAMPING

There are different ways of finding the optimal tunings for passive damping. Here we derive the optimal passive tuning for the serial and the parallel shunting schemes based on the frequency response function approach, *i.e.*, the tuning parameters are called optimal for passive damping when they minimize the maximum of the frequency response functions given by Eq. (2.5a) or Eq. (2.7a). The approach presented here is analogous to that developed by Den Hartog (1956) for mechanical vibration absorbers. As usual, the structural damping is neglected in the following derivations.

The frequency response function of the series RL configuration is given as Eq. (2.5a). This function intersects that of the baseline system (without the RL shunt) at two invariant points. Clearly, the transfer function will be independent of r if and only if

$$\frac{\delta^2 - \Omega^2}{(1 - \Omega^2)(\delta^2 - \Omega^2) - \delta^2 \xi^2} = -\frac{1}{1 - \Omega^2} \qquad (A2.18)$$

Therefore, the two invariant points are the roots of the following equation (which is derived from the above equation),

$$\Omega^4 - (1 + \delta^2)\Omega^2 + \delta^2 - \frac{1}{2}\delta^2 \xi^2 = 0 \qquad (A2.19)$$

The first step is to determine the tuning ratio such that the frequency responses at these two points are equal. Let r goes to infinity. The responses at these two points are equal if

$$\frac{1}{1 - \Omega_1^2} = -\frac{1}{1 - \Omega_2^2} \qquad (A2.20)$$

or

$$\Omega_1^2 + \Omega_2^2 = 2 \qquad (A2.21)$$

Combining Eqs. (A2.19) and (A2.21), we have the optimal inductance tuning

$$\delta^* = 1 \qquad (A2.22)$$

When $\delta = 1$, the two invariant points can be solved from Eq. (A2.19) as

$$\Omega_{1,2}^2 = 1 \pm \frac{\sqrt{2}}{2}\xi \qquad (A2.23)$$

The optimal resistance tuning can be derived by equating the frequency response at these two invariant points to that at $\Omega = 1$ (resonant frequency of the baseline system),

$$r^* = \sqrt{2}\xi \qquad (A2.24)$$

In the parallel configuration, the frequency response given in Eq. (2.7a) will be independent of r if and only if

$$\frac{\delta^2 - \Omega^2}{(1-\Omega^2)(\delta^2 - \Omega^2) - \delta^2\xi^2} = -\frac{1}{1-\Omega^2 - \xi^2} \qquad (A2.25)$$

or

$$\Omega^4 - (1 + \delta^2 - \frac{1}{2}\xi^2)\Omega^2 + \delta^2(1-\xi^2) = 0 \qquad (A2.26)$$

When r goes to zero, we have

$$\frac{1}{1-\Omega_1^2 - \xi^2} = -\frac{1}{1-\Omega_2^2 - \xi^2} \qquad (A2.27)$$

or,

$$\Omega_1^2 + \Omega_2^2 = 2(1-\xi^2) \qquad (A2.28)$$

Combining Eqs. (A2.26) and (A2.28), we have the optimal inductance tuning for the parallel configuration,

$$\delta^* = \sqrt{1 - \frac{3}{2}\xi^2} \qquad (A2.29)$$

Under the above optimal inductance tuning, the two invariant points can be solved from Eq. (A2.26) as

$$\Omega_{1,2}^2 = 1 - \xi^2 \pm \frac{1}{2}\sqrt{2\xi^2(1-\xi^2)} \qquad (A2.30)$$

Finally, the optimal resistance tuning for the parallel configuration can be solved by equating the frequency response at these two invariant points to that at $\Omega = \delta^*$,

$$r^* = \frac{1}{\sqrt{2}\xi} \qquad (A2.31)$$

3 Closed-Loop Synthesis for Active-Passive Structural Modal Control and Damping

As mentioned in the preceding chapter, considerable amount of work has been performed to utilize the piezoelectric transducer circuitry for active-passive hybrid structural vibration control and damping. This hybrid control concept could have the advantages of both the passive (stable, fail-safe, requiring low power) and active (high performance, feedback or feedforward actions) systems. Several topologies of such active-passive hybrid piezoelectric circuitry configurations are illustrated and studied from open-loop standpoint in Chapter 2. It is observed that with the circuitry treatment, not only the passive damping effect can be significantly enhanced; the modal response of the structure to the input voltage or current signal can also be greatly increased.

With these basic understandings on the respective effects of the circuitry elements to the passive damping and active authority amplification, the next question in line is how to determine the active control and passive parameters to achieve an effective hybrid control for structural vibration damping/suppression. While the tuning method presented in Chapter 2 can minimize the maximum of frequency response for a passive system, such selection of the circuitry elements might not necessarily be good choice for an active-passive hybrid system. From the driving voltage (control input) standpoint, the circuitry inductance value determines the electrical resonant frequency around which the active control authority is amplified, and, although appropriate resistance is necessary to achieve broadband passive damping, resistance in general reduces the voltage amplification effect. To address such trade-off effect, Kahn and Wang (1994, 1995) and Tsai and Wang (1996, 1999) have proposed methods to synthesize the circuitry elements and control parameters and performed analysis to examine the system performance for closed-loop modal damping/control.

K.W. Wang and J. Tang, *Adaptive Structural Systems with Piezoelectric Transducer Circuitry*, doi: 10.1007/978-0-387-78751-0_3, © Springer Science + Business Media, LLC 2008

3.1 PROBLEM STATEMENT AND OBJECTIVE

The main objective of this chapter is to present a methodology for the design and synthesis of closed loop modal damping and control using the active-passive hybrid piezoelectric transducer circuitry. As discussed in Chapter 2, a variety of basic circuitry configurations exist for such application. Without loss of generality, in this chapter we focus on the serial configuration with voltage source (*i.e.*, configuration (a) in Fig. 2.1), which was referred to as Active-Passive Piezoelectric Network (APPN) in the literature (Kahn and Wang, 1994, 1995; Tsai and Wang, 1996, 1999). Throughout the methodology development to be presented in the following sections, some fundamental issues are highlighted:

(a) What are the interactions between the active and passive components in the APPN configuration-based closed-loop control? Is an integrated configuration better than a design with separated active and passive components?

(b) How do we choose the active control parameters and passive elements? Should we design the active and passive parameters simultaneously or sequentially?

   The performance enhancement of the APPN compared with purely passive and purely active systems is demonstrated. In order for the presentation to be self-contained, the complete analysis of APPN design that includes model development, open-loop evaluation, control design and system synthesis, and closed-loop performance evaluation is presented in this chapter.

3.2 SYSTEM MODEL

For the purpose of illustration, a cantilevered beam partially covered with a pair of piezoelectric patch transducers is employed in our analysis. A schematic of the configuration is shown in Fig. 3.1. The top piezoelectric transducer is used as an actuator for vibration control, which is connected in series to an external voltage source and a resistance-inductance (*RL*) circuit. The bottom transducer, which is mounted to the beam at the same axial location, is used as the sensor. The system equation is derived based on the following assumptions:

(a) The poling direction of the piezoelectric transducers is in the positive *w*-direction;

(b) The rotational inertia is negligible;
(c) Only uni-axial loading of the piezoelectric transducers in the beam longitudinal direction is considered;
(d) The piezoelectric transducers are thin and short compared to the beam;
(e) The voltage applied is uniform.

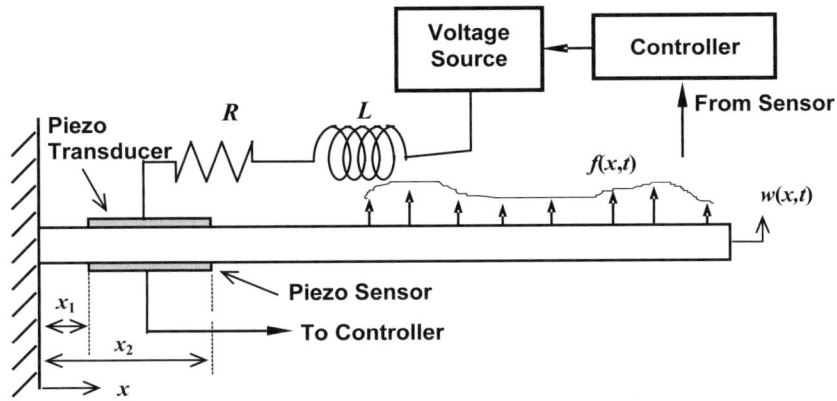

Fig. 3.1. Schematic of a cantilevered beam with APPN.

Using Hamilton's principle, one can construct

$$\int_{t_1}^{t_2} [\delta T_b - \delta U_b - \delta U_c + \delta W_v]\,\mathrm{d}t = 0 \tag{3.1}$$

where T_b is the beam kinetic energy, U_b is the beam strain energy, U_c is the mechanical and electrical energy of the piezoelectric material, and δW_v is the virtual work term.

For one-dimensional structures with uni-axial loading, the constitutive relation of the piezoelectric transducers (IEEE, 1988) can be written as

$$\begin{bmatrix} \tau \\ E_v \end{bmatrix} = \begin{bmatrix} C_{11}^D & -h_{31} \\ -h_{31} & \beta_{33} \end{bmatrix} \begin{bmatrix} \varepsilon \\ D \end{bmatrix} \tag{3.2}$$

where D is the electrical displacement (charge/area in the transverse direction), E_v is the electric field (volts/length along the transverse direction), ε is the mechanical strain in the x-direction, and τ is the stress in the x-direction. C_{11}^D is the open circuit elastic modulus of the piezoelectric transducer, β_{33} is the dielectric constant, and h_{31} is the piezoelectric constant of the transducer. Please note that in order to be consistent with the relevant literature (Tsai and Wang, 1999), the notations used here are slightly different from what are used in Chapter 2. Based on the above constitutive equation, and assuming D is constant along the piezoelectric transducer thickness for thin materials, one can derive

$$U_c = \frac{1}{2}\int_V (\tau\varepsilon + E_{va}D_a)\,dV + \frac{1}{2}\int_V (\tau\varepsilon + E_{vs}D_s)\,dV \tag{3.3}$$

$$= \frac{1}{2}\int_0^{L_b} \begin{aligned}[t] &[2C_{11}^D I_c(\frac{\partial^2 w}{\partial x^2})^2 + 2h_{31}J_c(\frac{\partial^2 w}{\partial x^2})D_a + A_c\beta_{33}D_a^2 \\ &+2h_{31}J_c(\frac{\partial^2 w}{\partial x^2})D_s + A_c\beta_{33}D_s^2](H(x-x_1)-H(x-x_2))\,dx \end{aligned}$$

The other functions are presented as follows

$$T_b = \frac{1}{2}\int_0^{L_b} \rho_b A_b (\frac{\partial w}{\partial t})^2 \, dx \tag{3.4}$$

$$U_b = \frac{1}{2}\int_0^{L_b} E_b I_b (\frac{\partial^2 w}{\partial x^2})^2 \, dx \tag{3.5}$$

$$\delta W_v = V_a(t)\int_0^{L_b} b_s \delta D_a (H(x-x_1)-H(x-x_2))\,dx \tag{3.6}$$

$$+V_s(t)\int_0^{L_b} b_s \delta D_s (H(x-x_1)-H(x-x_2))\,dx$$

$$+\int_0^{L_b} (f(x,t)-c_b\frac{\partial w(x,t)}{\partial t})\delta w(x,t)\,dx$$

Note that the voltage is related to the external circuit in the following manner,

$$V_a = -L\frac{d^2 Q_a}{dt^2} - R\frac{dQ_a}{dt} - V_c \tag{3.7}$$

$$V_a = E_{va}(h_s - h_b), \quad V_s = E_{vs}(h_s - h_b) \tag{3.8}$$

$$Q_a = b_s(x_2 - x_1)D_a, \quad Q_s = b_s(x_2 - x_1)D_s \tag{3.9}$$

In the above expressions, $w(x,t)$ is the transverse displacement of the beam, E_b is the beam elastic modulus, L_b is the beam length, ρ_b is the beam density, b_s is the width of the beam and the piezoelectric transducer, h_b is the distances from the beam neutral axis to the top surface of the beam, h_s is the distance from the beam neutral axis to the top surface of the piezo-electric transducer, A_b, A_c are the cross sectional area of the beam and pie-zoelectric transducer, respectively. c_b is the uniform damping constant, and I_b and I_c are the beam and piezoelectric transducer moments of inertia, respectively. Also, $J_s = b_s(h_s^2 - h_b^2)/2$, and $(x_2 - x_1)$ is the length of the piezo-electric transducer, R is the resistance, L is the inductance, V_c is the exter-nal voltage for control, V_s is the sensor voltage, D_a, D_s are the electric displacement of the actuator and sensor, respectively. Q_a, Q_s are the elec-trical charge of the piezoelectric actuator and sensor, E_{va}, E_{vs} are the elec-trical field of the actuator and sensor, respectively. $f(x,t)$ is the external load distribution, and H is the Heaviside step function. The other parame-ters are defined in Nomenclature.

Substituting the above equations into Eq. (3.1) and further assuming that the field within and the electrical displacement on the surface are both uni-form for the piezoelectric transducer, one can derive the system model. The structure equation is

$$\rho_b A_b \frac{\partial^2 w}{\partial t^2} + c_b \frac{\partial w}{\partial t} + E_b I_b \frac{\partial^4 w}{\partial x^4} \tag{3.10}$$

$$+(2C_{11}^D I_c \frac{\partial^4 w}{\partial x^4} + h_{31}J_c \frac{\partial^2 D_a}{\partial x^2})[H(x - x_1) - H(x - x_2)]$$

$$+(4C_{11}^D I_c \frac{\partial^3 w}{\partial x^3} + 2h_{31}J_c \frac{\partial D_a}{\partial x})[\delta(x - x_1) - \delta(x - x_2)]$$

$$+(2C_{11}^D I_c \frac{\partial^2 w}{\partial x^2} + h_{31}J_c D_a)[\delta'(x - x_1) - \delta'(x - x_2)] = f(x,t)$$

where δ is the Dirac Delta function and $0 < x < L_b$. The boundary conditions are

$$w(0,t) = \frac{\partial w(0,t)}{\partial x} = \frac{\partial^2 w(L_b,t)}{\partial x^2} = \frac{\partial^3 w(L_b,t)}{\partial x^3} = 0 \qquad (3.11)$$

The actuator circuit equation is

$$(V_c + L\frac{\mathrm{d}^2 Q_a}{\mathrm{d}t^2} + R\frac{\mathrm{d}Q_a}{\mathrm{d}t} + \frac{\beta_{33}(h_s - h_b)}{b_s(x_2 - x_1)}Q_a + \frac{h_{31}J_c}{b_s}\frac{\partial^2 w}{\partial x^2}) \qquad (3.12)$$
$$\times[(H(x - x_1) - H(x - x_2)] = 0$$

Assume an open circuit ($D_s = 0$) for the sensor. The sensor equation is

$$(V_s + \frac{h_{31}J_c}{b_s}\frac{\partial^2 w}{\partial x^2})[(H(x - x_1) - H(x - x_2)] = 0 \qquad (3.13)$$

Galerkin's method can be used to discretize Eqs. (3.10) – (3.13) into a set of ordinary differential equations. For the beam structure, one can obtain

$$\mathbf{M}_b\ddot{\mathbf{q}} + \mathbf{C}_b\dot{\mathbf{q}} + \mathbf{K}_b\mathbf{q} + \frac{h_{31}(h_s^2 - h_b^2)}{2(x_2 - x_1)}[\boldsymbol{\varphi}'(x_2) - \boldsymbol{\varphi}'(x_1)]Q_a = \hat{\mathbf{f}} \qquad (3.14)$$

For the actuator and sensor circuits,

$$L\ddot{Q}_a + R\dot{Q}_a + \frac{\beta_{33}(h_s - h_b)}{b_s(x_2 - x_1)}Q_a + \frac{h_{31}(h_s^2 - h_b^2)}{2(x_2 - x_1)}[\boldsymbol{\varphi}'(x_2) - \boldsymbol{\varphi}'(x_1)]^{\mathrm{T}}\mathbf{q} \qquad (3.15)$$
$$= -V_c$$

$$\frac{h_{31}(h_s^2 - h_b^2)}{2(x_2 - x_1)}[\boldsymbol{\varphi}'(x_2) - \boldsymbol{\varphi}'(x_1)]^{\mathrm{T}}\mathbf{q} = -V_s \qquad (3.16)$$

where $\mathbf{q}$, $\dot{\mathbf{q}}$ and $\ddot{\mathbf{q}}$ are vectors of the generalized displacement, velocity, and acceleration. $(\dot{\ })$ and $(')$ are derivatives with respective to time and x, respectively. $\hat{\mathbf{f}}$ is the external disturbance vector. The vector $\boldsymbol{\varphi}$ contains the comparison functions, which are chosen to be the eigenfunctions of a uniform fixed-free beam without circuit.

The discretized adaptive structure model, Eqs. (3.14) and (3.15), can be expressed in a standard state-space form:

$$\dot{\mathbf{y}} = \mathbf{A}(R,L)\mathbf{y} + \mathbf{B}_1\mathbf{f} + \mathbf{B}_2(R,L)\mathbf{u} \qquad (3.17)$$

$$\mathbf{z} = \mathbf{C}_1\mathbf{y}$$

$$\mathbf{y} = [\mathbf{q}^\mathrm{T} \quad Q_a \quad \dot{\mathbf{q}}^\mathrm{T} \quad \dot{Q}_a \quad]^\mathrm{T}, \qquad \mathbf{u} = V_c$$

Where $\mathbf{y}$ is the state vector, $\mathbf{u}$ is the control input, $\mathbf{f}$ is the external disturbance vector, and $\mathbf{z}$ is the measurement output. The system matrix, $\mathbf{A}$, and control matrix, $\mathbf{B}_2$ are functions of the passive resistance and inductance.

The system described above has $(N+1)$ modes, where N is the number of terms in the Galerkin expansion. The $(N+1)$th mode is due to the circuitry dynamics. Because the comparison functions in the expansion are chosen to be the eigenfunctions of a fixed-free beam, the ith generalized coordinate will closely resemble the ith structural modal coordinate (i=1,2,3,....,N). This model is used for system analysis and controller development, as discussed in the following sections. The system parameters used for the investigation reported in this chapter are shown in Table 3.1 unless stated otherwise.

Table 3.1 System parameters used in modal control/damping study

t_b = 3.175 mm	b_s = 19.05 mm	c_b = 0.12 N-s/m^2
E_b = 7.1 × 10^{10} Pa	C_{11}^D = 7.4x10^{10} Pa	h_s = 0.0016 m
β_{33} = 7.331x10^7 V m/C	h_{31} = 7.664x10^8 N/C	t_c = 0.25 mm
x_1 = 0.0127m	x_2 = 0.0762m	L_b = 0.1524m
ρ_b = 2700 kg/m^3	ρ_c = 7600 kg/m^3	

3.3 OPEN-LOOP SYSTEM ANALYSIS of APPN

In this section, the system model is used to investigate the open-loop effects that the circuitry parameters (resistance and inductance) have on the passive damping ability and active control authority of the APPN. Also, the APPN is compared to the configuration with separated active and passive elements. Note that although some of these issues are also presented in Chapter 2, the discussions here help to provide a self-contained explanation of the comparison results.

3.3.1 Effects of Circuitry Elements on Passive Damping and Active Authority

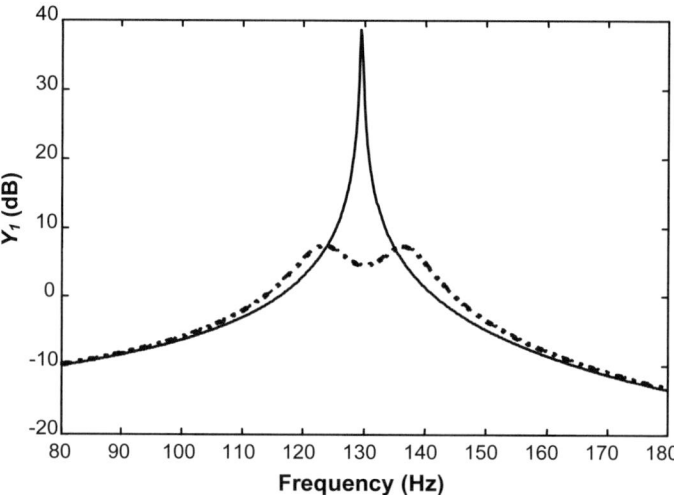

Fig. 3.2. Passive damping index Y_1:——— : short circuit, -- · -- ·· : APPN with $V_c=0$.

For a linear system described in the previous section, the overall structural response is the summation of the response contributed from the excitation force f and the response contributed from the control voltage V_c. We define Y_1 to be the magnitude of the transfer function w_1/f and Y_2 to be the magnitude of the transfer function w_2/V_c. Here, w_1 and w_2 are the responses close to the beam tip ($x=0.92L_b$) caused by the point excitation force at $x=0.95L_b$ and the control voltage, respectively. With $V_c=0$, Y_1 represents the structural vibration amplitude without active control. That is, Y_1 is an index of the system's passive damping ability (the smaller the better). On the other hand, Y_2 represents the structural vibration amplitude created by the active actuator. Larger Y_2 indicates that the actuator has more authority to excite the host structure for the given voltage input. Thus, Y_2 is used as an index for the system's active control authority (the larger the better).

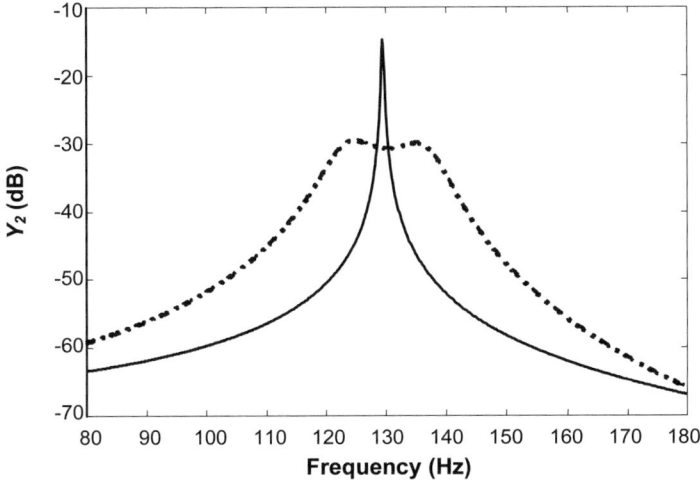

Fig. 3.3. Active control authority index Y_2 for the purely active and APPN systems. ——— : purely active, – · – ·· : APPN.

To illustrate the basic concepts and observations, we focus on the first modal response in the analysis presented in this section, thus only a one-term Galerkin expansion is used. We first compare the APPN with the purely active system. The R and L values are chosen to be the optimal values for the passive system following the procedure described in Appendix A3.1. Figs. 3.2 and 3.3 show the passive damping index Y_1 and the active authority index Y_2 for both systems. The Y_1 plot reconfirms that the passive shunt circuit behaves like a tuned resonant damper. In other words, the RL circuit will enhance the passive damping ability around the first resonant frequency. Fig. 3.3 (Y_2 plot) illustrates that the shunt circuit is broadening the actuator active authority bandwidth around the tuned frequency, while reducing the peak value. However, since the R and L values here are chosen to optimize the passive system, it is not obvious that they will be the best for maximizing the active action. For example, while the resistor is designed to dissipate the structure vibration energy, it could be dissipating the control power from the active element as well. This can be observed in Fig. 3.4, where Y_2 is shown to be decreasing with increasing R. In other words, the resistance reduces the active authority of the actuator.

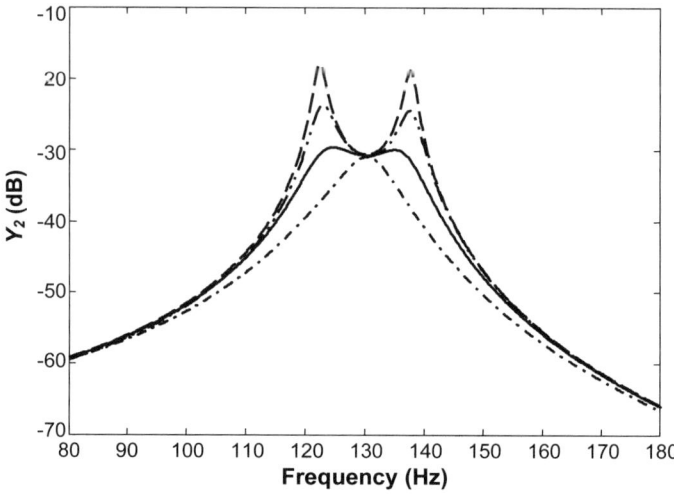

Fig. 3.4. Active control authority index Y_2 with different resistance values.
$---$ $R = 500\ \Omega$, $-\cdot\cdot-$ $R = 1.0\ K\Omega$, $\underline{\quad\quad}$ $R = 2.3\ K\Omega$, $-\cdot--\cdot$ $R = 5.0\ K\Omega$.

3.3.2 Integrated versus Separated Active and Passive Components

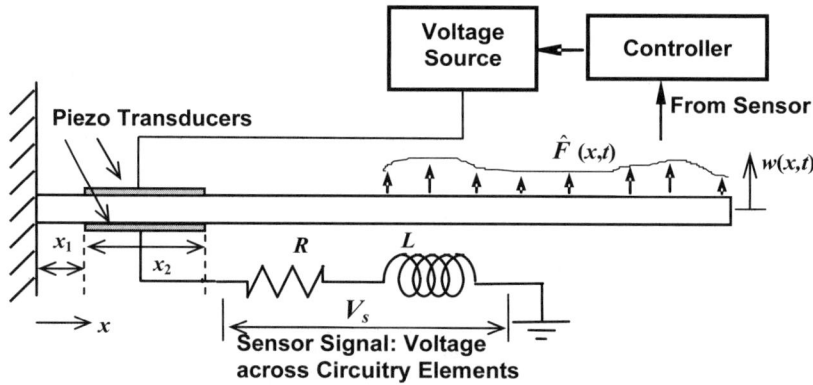

Fig. 3.5. Schematic of a cantilevered beam with separated active and passive elements.

One simple-minded approach to resolve the problem discussed above is to separate the passive shunt circuit from the active source (Fig. 3.5). While this is still an active-passive hybrid configuration, the active and passive elements do not interact directly anymore. That is, we are simply applying active control on an optimized (tuned circuit) passive system. The structure equation of motion for such a separated configuration is

$$\rho_b A_b \frac{\partial^2 w}{\partial t^2} + c_b \frac{\partial w}{\partial t} + E_b I_b \frac{\partial^4 w}{\partial x^4} + (2C_{11}^D I_c \frac{\partial^4 w}{\partial x^4} + h_{31} J_c \frac{\partial^2 D_a}{\partial x^2} \tag{3.18}$$

$$+ h_{31} J_c \frac{\partial^2 D_s}{\partial x^2})[H(x-x_1) - H(x-x_2)] + (4C_{11}^D I_c \frac{\partial^3 w}{\partial x^3} + 2h_{31} J_c \frac{\partial D_a}{\partial x}$$

$$+ 2h_{31} J_c \frac{\partial D_s}{\partial x})[\delta(x-x_1) - \delta(x-x_2)] + (2C_{11}^D I_c \frac{\partial^2 w}{\partial x^2} + h_{31} J_c D_a$$

$$+ h_{31} J_c D_s)[\delta'(x-x_1) - \delta'(x-x_2)] = f(x,t)$$

The circuit and actuator equations are

$$(L \frac{d^2 Q_s}{dt^2} + R \frac{dQ_s}{dt} + \frac{\beta_{33}(h_s - h_b)}{b_s(x_2 - x_1)} Q_s + \frac{h_{31} J_c}{b_s} \frac{\partial^2 w}{\partial x^2}) \tag{3.19}$$

$$\times [H(x-x_1) - H(x-x_2)] = 0$$

$$(V_a + \frac{\beta_{33}(h_s - h_b)}{b_s(x_2 - x_1)} Q_a + \frac{h_{31} J_c}{b_s} \frac{\partial^2 w}{\partial x^2})[H(x-x_1) - H(x-x_2)] = 0 \tag{3.20}$$

The sensor equation is

$$(V_s + \frac{\beta_{33}(h_s - h_b)}{b_s(x_2 - x_1)} Q_s + \frac{h_{31} J_c}{b_s} \frac{\partial^2 w}{\partial x^2})[H(x-x_1) - H(x-x_2)] = 0 \tag{3.21}$$

Here, the sensor voltage V_s is the voltage across the shunt circuit. Since the two configurations (integrated versus separated) are the same without the active source, the passive damping index Y_1 plot is the same for the two (Fig. 3.6). This implies that the two configurations have the same passive damping ability. However, the active authority index Y_2 plot shows that the APPN can drive the host structure much more effectively than the separated treatment (Fig. 3.7).

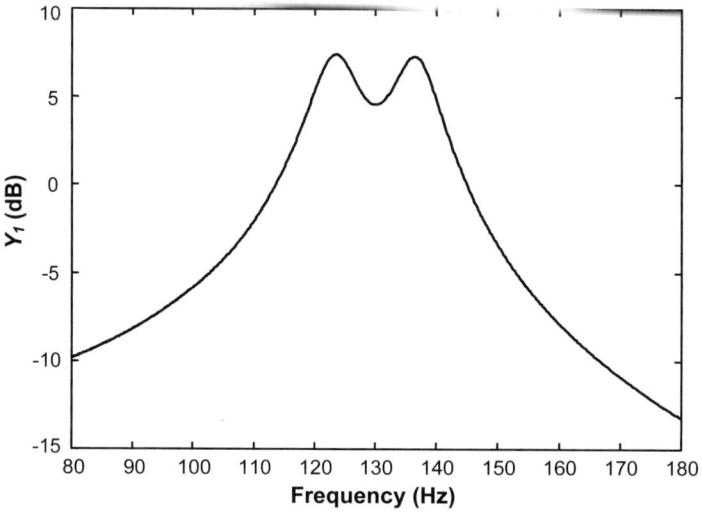

Fig. 3.6. Passive damping index Y_1 for the APPN and separated systems.

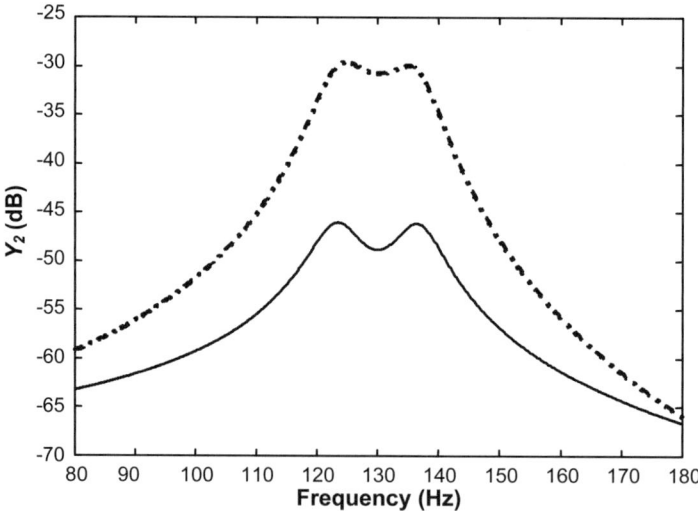

Fig. 3.7. Active control authority index Y_2 for the APPN and separated systems. ——— : separated, – · – · · : APPN.

The reason behind the active authority enhancement is that the circuit can amplify the voltage input into the piezoelectric transducer around the resonant frequency, even when the *RL* parameters are not optimized for enhancing the active action. Experimental investigations are also performed to examine the circuitry's passive damping ability as well as its active authority enhancement ability (Fig. 3.8), using a piezoelectric transducer treated beam structure similar to that used in the analysis. Through exciting the structure with the piezoelectric transducer, the open-loop structural response of the integrated APPN and the configuration with separated passive resonant shunt and active piezoelectric actuator are measured. While the two configurations have the same passive damping ability, the APPN configuration can drive the host structure much more effectively than the separated treatment can (Fig. 3.8), which clearly demonstrated the merit (high active authority) of the integrated APPN design. With these observations, a sensible approach is to use the coupled APPN configuration with better selected *R* and *L* values (instead of the *R* and *L* values optimized for the purely passive system). To achieve such a purpose, a concurrent design/control method is developed and presented in the next section.

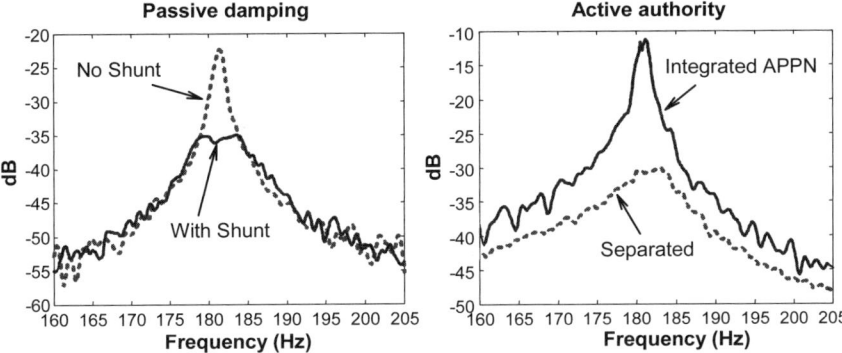

Fig. 3.8. Exerimental results on system passive damping and active authority. Left: structrual response under disturbance. Right: structural response under actuator input.

3.4 ACTIVE-PASSIVE CONTROL SYNTHESIS

In this section, a methodology is presented to *concurrently* design the passive elements and the active control action. This approach ensures that the active and passive elements are configured in a systematic and integrated manner.

The strategy developed is to combine the optimal control theory with an optimization process and to determine the active control gains together with the values of the passive system parameters (the resistance and inductance). The procedure contains three major steps as outlined in the following subsections.

3.4.1 Determining Active Gains

In order to examine the system response under different bandwidth excitations, the disturbance is modeled as the result of passing a Gaussian, white noise process through a second order low pass Butterworth filter. The bandwidth of the filter is defined to be the radian cutoff frequency at which the filter output has a 3-dB attenuation of its value at zero frequency. The filter bandwidth is in fact the bandwidth of the external disturbance. The equations that describe such a process are

$$\dot{\mathbf{s}} = \mathbf{A}_f \mathbf{s} + \mathbf{B}_f \mathbf{d} \tag{3.22}$$
$$\mathbf{f} = \mathbf{C}_f \mathbf{s}$$

$$\dot{\mathbf{y}} = \mathbf{A}(R,L)\mathbf{y} + \mathbf{B}_1 \mathbf{f} + \mathbf{B}_2(R,L)\mathbf{u} \tag{3.23}$$

The inputs in $\mathbf{d}$ are Gaussian and white. Here, the mean and spectral density of $\mathbf{d}$ are given by $E[\mathbf{d}(t)]=\mathbf{0}$ and $E[\mathbf{d}(t)\mathbf{d}^T(\tau)]=\mathbf{D}(t)\delta(t-\tau)$. Here, $E[\ ]$ is the expectation operator. By defining an augmented state as $\mathbf{x}=[\mathbf{y}\quad \mathbf{s}]^T$, the overall system state equations become

$$\dot{\mathbf{x}} = \begin{bmatrix} \mathbf{A} & \mathbf{B}_1\mathbf{C}_f \\ \mathbf{0} & \mathbf{A}_f \end{bmatrix}\mathbf{x} + \begin{bmatrix} \mathbf{0} \\ \mathbf{B}_f \end{bmatrix}\mathbf{d} + \begin{bmatrix} \mathbf{B}_2 \\ \mathbf{0} \end{bmatrix}\mathbf{u} = \mathbf{A}_a\mathbf{x} + \mathbf{B}_{1a}\mathbf{d} + \mathbf{B}_{2a}(R,L)\mathbf{u} \tag{3.24}$$

With a given set of passive parameters (R and L), the optimal control theory is used to determine the value of the active gains. The cost function is

$$J_e = \lim_{t \to \infty} E[\mathbf{x}^T\mathbf{Q}_e\mathbf{x} + \mathbf{u}^T\mathbf{S}\mathbf{u}] \tag{3.25}$$

$\mathbf{Q}_e$ is a positive-semi-definite weighting matrix chosen to be

$$Q_e = \begin{bmatrix} K_b & & \\ & 0 & 0 \\ & M_b & \\ & 0 & 0 \\ & & & 0 \end{bmatrix} \qquad (3.26)$$

Here, $x^T Q_e x$ represents the overall structure energy. While S is the positive-definite weighting matrix on the control inputs (Note that for the single input example system, S becomes a scalar S). Since the purpose of this chapter is to discuss the characteristics of the APPN actuator, the sensor equation is not used here and full state feedback is assumed for demonstration.

With this stochastic regulator control problem, the optimal control gain is given as

$$K^C = S^{-1} B_{2a}{}^T P_r \qquad (3.27)$$

where P_r satisfies the Riccati equation

$$A_a{}^T P_r + P_r A_a - P_r B_{2a} S^{-1} B_{2a}{}^T P_r + Q_e = 0 \qquad (3.28)$$

The closed-loop system thus becomes

$$\frac{d}{dt} x = (A_a - B_{2a} K^C) x + B_{1a} d = A_{cl} x + v \qquad (3.29)$$

Hereafter we define $v = B_{1a} d$ for notation simplification. v is Gaussian and white. Here, the mean and spectral density of v are given as $E[v(t)] = 0$ and $E[v(t)v^T(\tau)] = V(t)\delta(t-\tau)$, respectively.

3.4.2 Determining Objective Function

The objective function is selected to be the minimized cost function of the stochastic regulator problem (Kwakernaak and Sivan, 1972)

$$J = \operatorname{Min} J_e = \operatorname{trace}(\mathbf{P}_1 \mathbf{V}) \qquad (3.30)$$

$\mathbf{P}_1$ is the solution to the following Lyapunov equation,

$$\mathbf{A}_{cl}\mathbf{P}_1 + \mathbf{P}_1\mathbf{A}_{cl}^{T} + \mathbf{Q}_e = 0 \qquad (3.31)$$

3.4.3 Iteration on Active Gains and Passive Parameters to Minimize *J*

Note that for each set of the passive control parameters R and L, there exists an optimal control with the corresponding minimized cost function and control gains. That is, J is a function of R and L. Utilizing a sequential quadratic programming algorithm (Arora, 1989), one can determine the resistance and inductance which further minimize J. As the passive parameters vary, the active gain will be updated as well. In other words, by varying the values of the active gains and passive parameters simultaneously, the "optimized" optimal control can be obtained.

3.5 ANALYSIS OF THE OPTIMIZED CLOSED-LOOP SYSTEM

3.5.1 Effect of Weighting

As stated earlier, the optimal resistance and inductance for the active-passive hybrid system may not be the same as the optimal R and L for a purely passive arrangement. In order to evaluate the differences, the optimal R and L for different weightings on control effort are shown in Fig. 3.9. Here, a three term Galerkin's expansion is used and the excitation bandwidth (bandwidth of the Butterworth filter described in Section 3.4.1) is set to be 192 Hz for us to focus on the first mode.

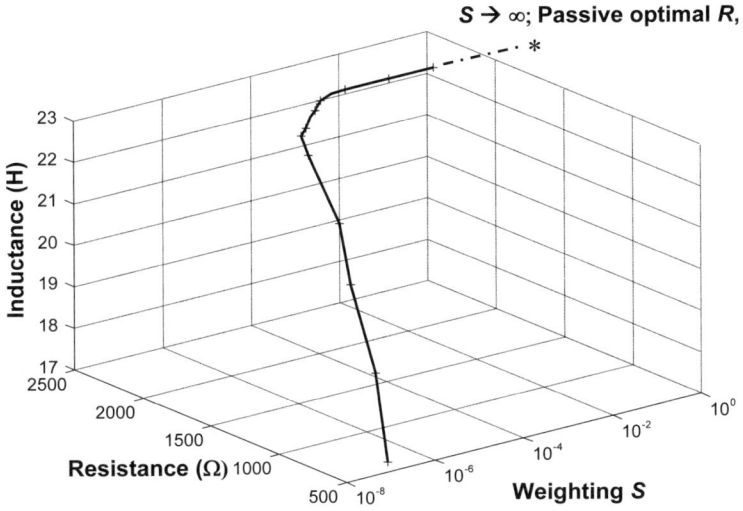

Fig. 3.9. Optimal resistance and inductance for different weighting S.

The optimal R and L for the passive case shown in Fig. 3.9 are calculated using the passive optimization process stated in Appendix 3.1. Note that when the weighting on control effort S increases (*i.e.*, higher penalty on control effort), the optimal R and L values using the concurrent design approach those derived from the passive optimization procedure. This is because the input control effort would be very small under the condition of large S and passive damping is dominating. However, we see that the optimal R, L values for the hybrid system depart quite significantly from the R, L values for the purely passive system as the demand on performance increases (i.e., S decreases). This indicates that the simultaneous controller/circuit optimization process is necessary, especially when high system performance is required.

It is also shown in Fig. 3.9 that the higher the S, the smaller the resistance and inductance. In order to understand this trend and further illustrate the effect of S on the system characteristics, another set of indexes for passive damping (I_p) and active authority (I_a) are defined as follows:

$$I_p = (J_{p\_pa} - J_{p\_ap})/J_{p\_pa} \qquad (3.32)$$

$$I_a = (J_{a\_ap} - J_{a\_pa})/J_{a\_pa} \tag{3.33}$$

where J_{p_pa} and J_{p_ap} represent the total structural energy (potential and kinetic energy) of the purely active and APPN systems under the same level of external disturbances, respectively. Higher I_p indicates more passive damping ability. J_{a_pa} and J_{a_ap} represent the total structural energy of the purely active and APPN systems when the structure is driven by the same level of control voltage. Larger I_a indicates more control authority. J_{p_pa}, J_{p_ap}, $J_{\_pa}$ and J_{a_ap} can be calculated from Eq. (A3.6) in Appendix 3.1. Both the external force and the control voltage are modeled as white noise passing through a low pass Butterworth filter with bandwidth of 192 Hz. Figs 3.10 and 3.11 show that I_p decreases while I_a increases as the weighting S decreases. This implies that when the demand on the performance becomes higher (lower S), the optimal R and L values will vary in a direction to increase active authority while reducing passive damping. We therefore examine the frequency response of Y_2 (active authority index) in the cases of high and low weightings S. It is clear from Fig. 3.12 that when the demand on performance is high, R becomes smaller to enhance the active authority peak magnitude, and L reduces to push the actuator frequency response up to cover a wider frequency bandwidth.

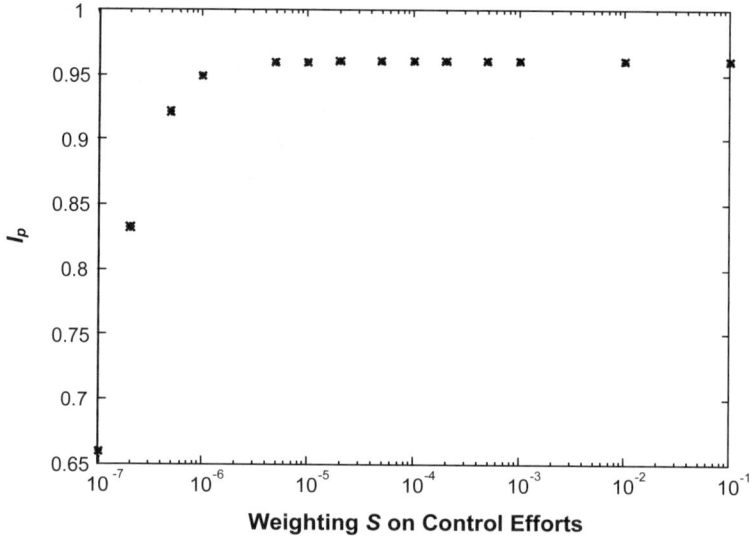

Fig. 3.10. Passive damping I_p for different weighing S.

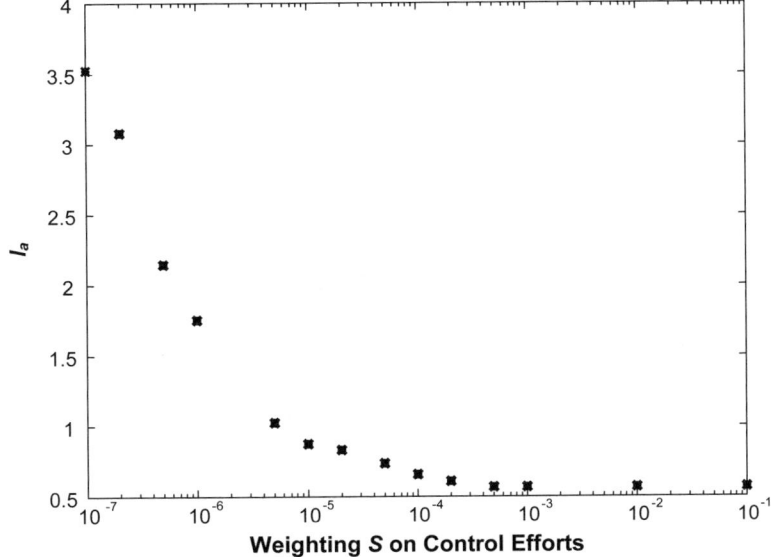

Fig. 3.11. Active authority I_a for different weighting S.

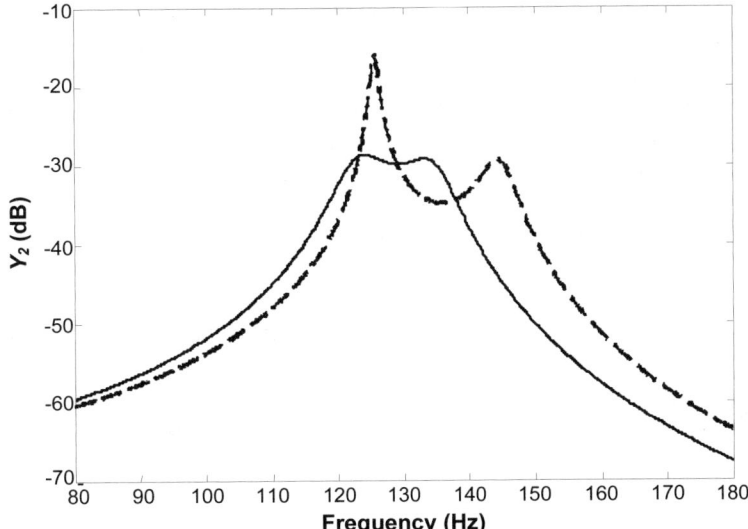

Fig. 3.12. Active control authority index Y_2 for different weighting S.
——— : $S=1\times10^{-1}$, – – – – : $S=2\times10^{-7}$.

3.5.2 Effect of Excitation Bandwidth

In order to examine the APPN system under broadband excitations, a three-mode expansion is used which covers modal frequencies up to 1.97 kHz. The APPN system is designed based on the algorithm presented in Section 3.4. The weighting on control efforts S is chosen to be 4×10^{-7}. To evaluate the effects of excitation frequency bandwidth (bandwidth of the Butterworth filter described in Section 3.4.1) on the optimal values of R and L, the contour of the objective function J_e for different bandwidth (190 Hz, 560 Hz and 640 Hz) are plotted in Figs 3.13 to 3.15. Fig. 3.13 shows that there is only one global minimum for the objective function J_e when the excitation frequency bandwidth is 190 Hz. When the bandwidth increases to 560 Hz, the objective function J_e will have two local minima with the right one as the global minimum. As the bandwidth increases to 640 Hz, the global minimum will jump from the right to the left region. To explain the phenomenon of this jump, the active authority index Y_2 for bandwidth 192 Hz and 640 Hz are shown in Figs 3.16 and 3.17. When the excitation bandwidth is narrow (192 Hz), the optimal R and L are tuned to enhance the active authority around the first mode. Although the Y_2 value of the hybrid system is about 20 dB lower than that of the purely active system for responses above 195 Hz, the system performance will not be affected. This is because the response of the system at higher frequencies is very small when the excitation bandwidth is low and narrow (192 Hz). However, when the excitation frequency increases to 640 Hz, the large response at higher frequencies will affect the objective function significantly and the optimal L and R will jump to the second mode region.

Fig. 3.18 shows the value of J versus excitation bandwidth for the purely active and APPN cases. A different set of optimal APPN parameters is designed for each given bandwidth. It is clear that the hybrid system outperforms the purely active structure significantly when the bandwidth is low and narrow. However, as the bandwidth increases, the APPN starts to approach the purely active case. This shows that since the circuitry is used to enhance the active action around a certain frequency, the results will become less effective when the number of contributing modes greatly exceeds the number of actuators. Nevertheless, the APPN will always outperform the purely active system, as illustrated in Fig. 3.18.

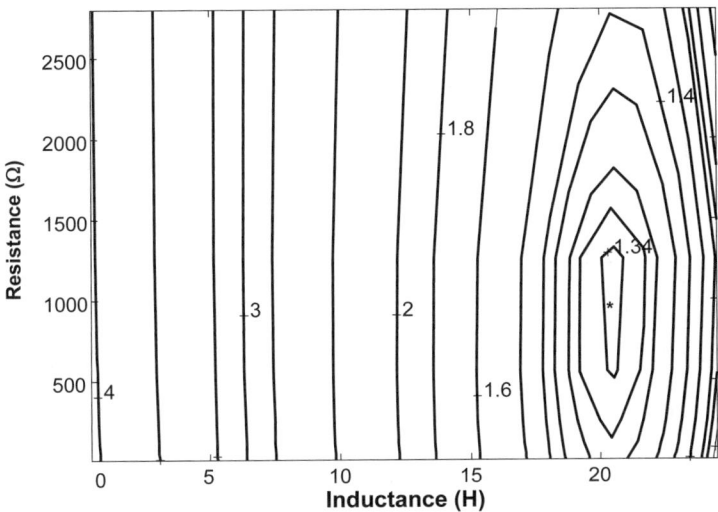

Fig. 3.13. Contour of the cost function J_e under the excitation bandwidth 190 Hz, * : optimal resistance and inductance, J=1.335.

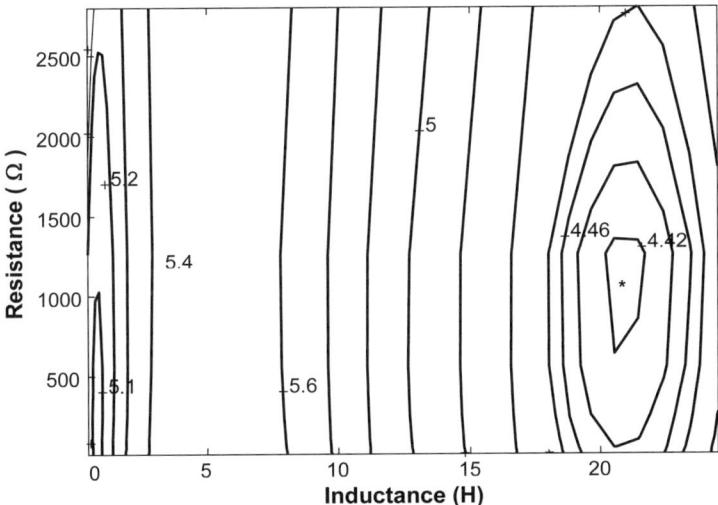

Fig. 3.14. Contour of the cost function J_e under the excitation bandwidth 560 Hz, * : optimal resistance and inductance, J= 4.414.

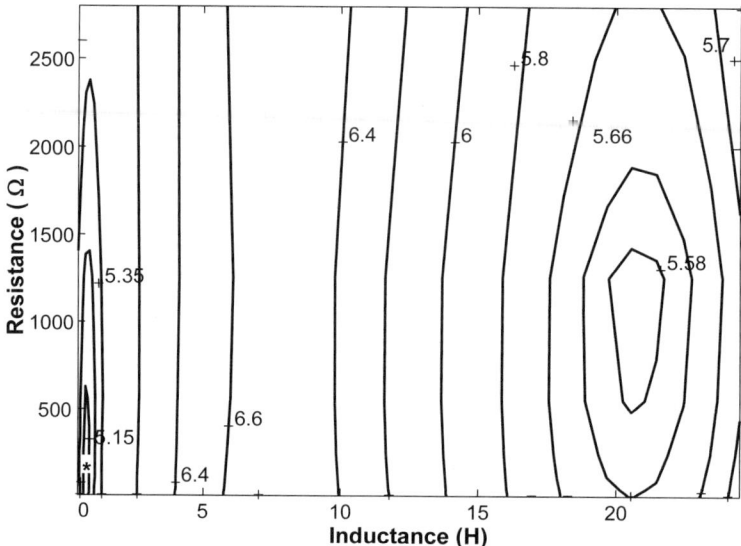

Fig. 3.15. Contour of the cost function J_e under the excitation bandwidth 640 Hz,
* : optimal resistance and inductance, J= 5.119.

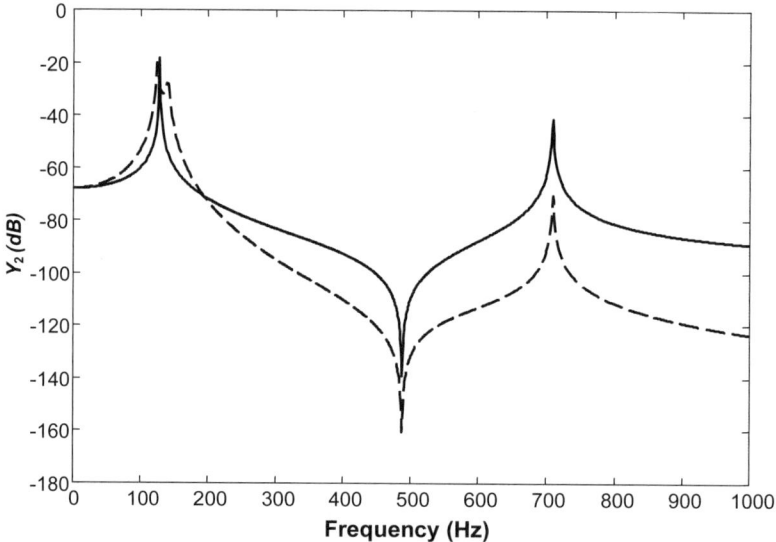

Fig. 3.16. The active authority Y_2 under excitation bandwidth 190 Hz.
————— : purely active, – – – – ·: APPN.

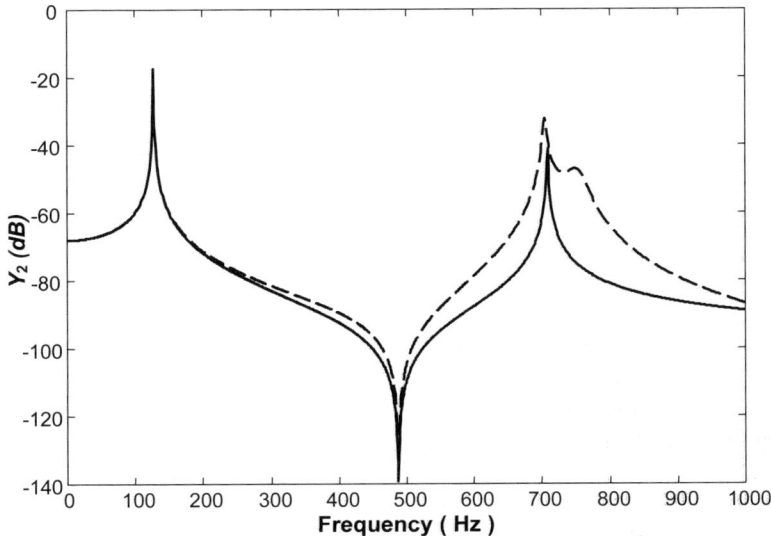

Fig. 3.17. Active authority Y_2 under excitation bandwidth 640 Hz.
————: purely active, – – – – · : APPN.

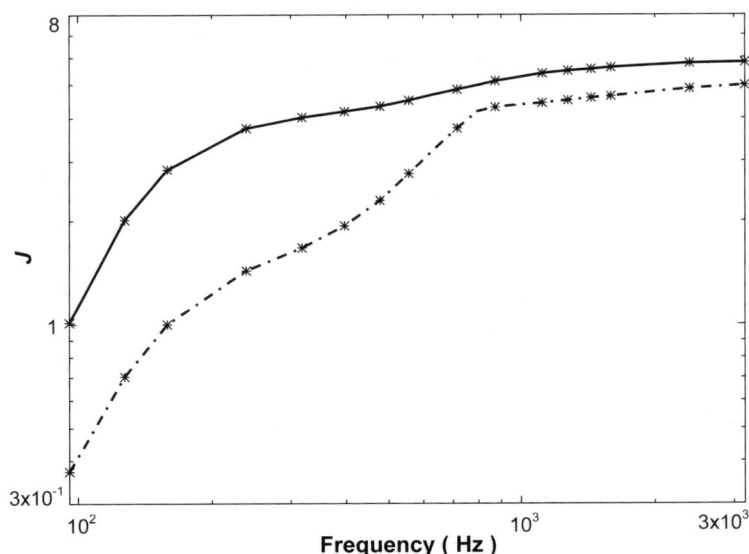

Fig. 3.18. Value of objective function J versus excitation frequency bandwidth.
————: purely active, – – – – · : APPN.

3.6 CONCLUDING REMARKS

This chapter presents a comprehensive analysis of vibration damping/control using a representative piezoelectric circuitry, the active-passive hybrid piezoelectric network (APPN). It is shown that the circuitry not only can provide passive damping, they can also enhance the active action authority around the tuned frequency. Therefore, the integrated APPN design is more effective than a system with separated active and passive elements. However, it is also clear that the active authority will degrade if the inductance is mistuned or if the resistance is too high. Some of the phenomena predicted in the open-loop analysis have also been observed experimentally. Therefore, a systematic design/control method is developed to ensure that the passive and active actions are optimally synthesized. This chapter also addresses the effects of weighting and frequency bandwidth on the APPN configuration. It is shown that the optimal resistance and inductance values for the hybrid system could be quite different from that of the passive system, especially when the demand on performance is high. It is concluded that the difference between the APPN and purely active cases becomes smaller as the excitation frequency bandwidth increases. One possible method to enhance the system's broadband performance is to increase the number of actuators. Another approach is to integrate the APPN design with other broadband damping elements, such as active constrained layer damping layers (Tsai and Wang, 1997).

APPENDIX 3.1 AN ALTERNATIVE METHOD FOR OPTIMAL PASSIVE TUNING

The selection of optimal passive resistance and inductance for vibration absorbing was originally derived by Hagood and von Flotow (1991). Chapter 2 has outlined such a classical procedure. For easy comparison with the active-passive hybrid system presented in this chapter, in what follows we present an alternative method for deciding optimal passive tuning. The concept and results, nevertheless, are generally similar.

The system equations can be expressed by the state space form

$$\dot{\mathbf{y}} = \mathbf{A}(R, L)\mathbf{y} + \mathbf{B}_1\mathbf{f} \qquad (A3.1)$$

Here, the control action **u** is not included since this system is a passive system. The disturbance **f** can modeled as the result of passing a Gaussian, white noise process through a second order low pass Butterworth filter. The bandwidth of the filter is defined to be the radian cutoff frequency at which the filter output has a 3-dB attenuation of its value at zero frequency. The equations that describe such a process are,

$$\dot{\mathbf{s}} = \mathbf{A}_f \mathbf{s} + \mathbf{B}_f \mathbf{d} \tag{A3.2}$$
$$\mathbf{f} = \mathbf{C}_f \mathbf{s}$$

The inputs in **d** are Gaussian and white. Here, the mean and spectral density of **d** is given by $E[\,\mathbf{d}(t)\,]=\mathbf{0}$ and $E[\,\mathbf{d}(t)\mathbf{d}^{\mathrm{T}}(\tau)\,]=\mathbf{D}(t)\delta(t-\tau)$. Here, $E[\ ]$ is the expectation operator.

By defining an augmented state vector as $\mathbf{x}=[\mathbf{y}\quad \mathbf{s}]^{\mathrm{T}}$, the overall system state equations become,

$$\dot{\mathbf{x}} = \begin{bmatrix} \mathbf{A} & \mathbf{B}_1\mathbf{C}_f \\ \mathbf{0} & \mathbf{A}_f \end{bmatrix}\mathbf{x} + \begin{bmatrix} \mathbf{0} \\ \mathbf{B}_f \end{bmatrix}\mathbf{d} = \mathbf{A}_a\mathbf{x} + \mathbf{B}_{1a}\mathbf{d} = \mathbf{A}_a\mathbf{x} + \mathbf{v} \tag{A3.3}$$

Here we define $\mathbf{v} = \mathbf{B}_{1a}\mathbf{d}$ for notation simplification. The mean and spectral density of **v** are given as $E[\,\mathbf{v}(t)\,]=\mathbf{0}$ and $E[\,\mathbf{v}(t)\mathbf{v}^{\mathrm{T}}(\tau)\,]=\mathbf{V}(t)\delta(t-\tau)$, respectively. The cost function is defined as

$$J_e = \lim_{t\to\infty} E[\mathbf{x}^{\mathrm{T}}\mathbf{Q}_e\mathbf{x}] \tag{A3.4}$$

Q$_e$ is a positive-semi-definite weighting matrix chosen to be

$$\mathbf{Q}_e = \begin{bmatrix} \mathbf{K}_b & & \\ & \mathbf{0} & & \mathbf{0} \\ & & \mathbf{M}_b & \\ & \mathbf{0} & & \mathbf{0} \\ & & & \mathbf{0} \end{bmatrix} \tag{A3.5}$$

Here, $\mathbf{x}^T\mathbf{Q}_e\mathbf{x}$ represents the overall structure energy. With a given set of passive parameters (R and L), the cost function J_e is $\mathrm{trace}(\mathbf{P}_1\mathbf{V})$.

The system response will consist of a state vector with zero mean and a variance given by the solution ($\mathbf{P}_1$) to the Lyapunov equation,

$$\mathbf{A_a}^T\mathbf{P_1} + \mathbf{P_1}\mathbf{A_a} + \mathbf{Q_e} = \mathbf{0} \tag{A3.6}$$

Note that the cost function is a function of R and L. Utilizing a sequential quadratic programming algorithm (Arora, 1989), one can determine the optimal passive R and L to minimize J_e.

4 Active-Passive Adaptive Design for Frequency-Varying Narrowband Disturbance Rejection

The preceding two chapters illustrate the usage of piezoelectric transducer circuitry for vibration suppression. In particular, the closed-loop design discussed in Chapter 3 features the active-passive hybrid action for modal damping and control. Such an approach, on the other hand, is not the most effective one when the disturbance is narrow-banded and off-resonance. For the simplest case of such excitation, *i.e.*, harmonic excitation at a given frequency, one can recognize that an undamped absorber (*e.g.*, a piezoelectric transducer with undamped resonant shunt) can in theory be tuned to the excitation frequency and completely suppress the vibration. The principal drawback of such tonally tuned absorbers, however, is that they require the absorber damping to be as low as possible to achieve good performance at the designated frequency. This low damping, however, causes the effective bandwidth of the absorber to be quite small (von Flotow *et al.*, 1994). In such a situation, the system performance is extremely sensitive to tuning errors and the device may in fact increase the structural vibration if mistuned. For this reason, passive undamped absorbers with fixed design are not generally useful in off-resonance situations, especially when the excitation frequency is unsteady or changing.

The concept of semi-active piezoelectric absorber has been explored to suppress harmonic excitations with time-varying frequencies. An example of this type of excitation is rotating unbalance in machinery. The frequency variation could be a slow drift due to changes in operating conditions or a rapid spin-up/spin-down when the machine is turned on/off. The implementation of these semi-active absorbers requires either a variable inductor or a variable capacitor element, but both these methods have some inherent limitations. For instance, the variable capacitor method (Davis and Lesieutre, 2000) limits the tuning of the piezoelectric absorber to a relatively small frequency range. The variable inductor approach (Hollkamp and Starchville, 1994) typically requires the using of a synthetic inductance circuit. This device adds a parasitic resistance to the circuit, which is often significant enough to deteriorate the vibration absorbing

K.W. Wang and J. Tang, *Adaptive Structural Systems with Piezoelectric Transducer Circuitry*, doi: 10.1007/978-0-387-78751-0_4,
© Springer Science + Business Media, LLC 2008

performance. In both cases, it is difficult to tune the variable elements rapidly with high accuracy.

Morgan and Wang (2002a, 2002b) proposed an alternative approach to deal with narrowband excitations with time-varying frequencies, based on the collective design of a piezoelectric circuitry and an active controller. Such approach thus also falls into the general category of active-passive piezoelectric circuitry. Their basic idea is to tune the passive inductor in the piezoelectric circuitry to the nominal or steady-state excitation frequency, and then use active feedback to adapt the circuitry to varying frequency and also enhance system performance. Such active inductance tuning allows fast and accurate adjustment of the tuned frequency with no parasitic resistance. In another study (Morgan and Wang, 2002c), they further extended this adaptive active-passive piezoelectric absorber configuration so that it can track and suppress multiple harmonic excitations.

4.1 PROBLEM STATEMENT AND OBJECTIVE

The objective of this chapter is to provide a detailed discussion on the development of active-passive piezoelectric circuitry for narrowband disturbance rejection with time-varying frequencies. The system explored by Morgan and Wang (2002a) has the same configuration as the APPN (see Chapter 3), which consists of a piezoelectric transducer integrated with an active voltage source in series with an electric circuitry. However, the control strategy is completely different. For frequency-varying narrowband disturbance rejection, the passive inductance is tuned to the nominal excitation frequency whereas the design centers around the active control actions. The main issues that need to be addressed are the system adaptability and robustness with respect to the frequency variation and the control performance. The foci of the discussion in this chapter are:

(a) How to design active control that can allow simultaneously the adaptive tracking of the varying frequency, the reduction of the inherent resistance, and the enhancement of the overall vibration suppression performance and robustness?
(b) How to develop control actions that can lead to effective vibration suppression under multiple harmonic excitations?

The active control law discussed in this chapter consists of three parts. The first part is designed to imitate a variable inductor so that the absorber can be adaptively tuned to the correct frequency. The advantages of the active inductance tuning include fast and accurate adjustment and no parasitic resistance. The second control action is the negative resistance, which is used to reduce the absorber damping (*i.e.*, the inherent resistance in the circuit). Finally, an active coupling enhancement action is employed to increase the robustness and overall performance of the absorber. To ensure that the active inductance is properly tuned, an expression for the optimal tuning on a general multiple-degrees-of-freedom (MDOF) structure is derived. The closed-loop inductance is achieved using this optimal tuning law in conjunction with an algorithm that estimates the fundamental frequency of the measured excitation. This idea is then extended to multiple harmonic excitation applications. Numerical simulation and experimental investigation efforts are presented in this chapter to demonstrate the new idea and provide guidelines for implementation.

4.2 ACTIVE INDUCTANCE AND RESISTANCE TUNING

While the configuration considered in this study is identical to the active-passive hybrid piezoelectric circuitry network (APPN) configuration discussed in the previous chapters, the application and the control algorithm used here are completely different. Without loss of generality, we assume that the local vibration in the structure can be measured using a piezoelectric sensor. We also assume that the model of the system can be obtained, either analytically or experimentally, in the form shown as

$$\mathbf{M}\ddot{\mathbf{q}} + \mathbf{C_d}\dot{\mathbf{q}} + \mathbf{K}^{\mathbf{D}}\mathbf{q} + \mathbf{K_C}Q = \hat{\mathbf{F}} \cdot f(t) \tag{4.1}$$

$$L_p\ddot{Q} + R_p\dot{Q} + \frac{1}{C_p^S}Q + \mathbf{K_C^T}\mathbf{q} = V_c$$

Here, $\mathbf{M}$, $\mathbf{C_d}$, and $\mathbf{K}^{\mathbf{D}}$ are the mass, damping, and open-circuit stiffness matrices, $\mathbf{q}$ is a vector of the generalized coordinates of the structure, L_p and R_p are the passive inductance and resistance of the circuit, Q is the charge on the piezoelectric transducer, C_p^S is the capacitance of the piezoelectric transducer under constant strain, and V_c is the control voltage. The coupling vector $\mathbf{K_C}$ characterizes the electro-mechanical coupling effect. The passive inductance of the circuit, L_p, is selected so that the absorber is tuned to the nominal or expected excitation frequency. No resistance is intentionally added to the circuit. However, the passive inductor may have non-negligible internal resistance, which is represented by R_p.

4.2.1 Tuning Law

The purpose of the active tuning law is to simulate an ideal variable inductor. The variable inductance is realized as shown below

$$\left(V_C\right)_{inductor} = -L_a(t) \cdot \ddot{Q} \tag{4.2}$$

The total inductance of the closed-loop system is the sum of the fixed passive inductance L_p and the variable active inductance L_a, which can be positive or negative. To find the desired active inductance, the total inductance needed to properly tune the absorber to the excitation frequency must be derived. An optimal tuning for this purpose is developed (see Section 4.2.2). Since the passive inductance is tuned to the nominal excitation frequency, the active inductance necessary to keep the absorber tuned is significantly low, as compared to a purely active inductor.

For a tonally tuned absorber to be effective, it is usually desired to have very low damping (resistance) in the piezoelectric absorber. Therefore, it may be desirable to partially cancel the internal resistance of the passive inductor using an active negative resistance action. This negative resistance can be implemented using the simple control law shown in Eq. (4.3). The circuit equation for the closed-loop system with the active tuning and negative resistance actions is shown in Eq. (4.4), while the structure equation remains the same as in Eq. (4.1).

$$\left(V_C\right)_{resistor} = R_a\dot{Q} \tag{4.3}$$

$$\left(L_p + L_a(t)\right)\ddot{Q} + \left(R_p - R_a\right)\dot{Q} + \frac{1}{C_p^S}Q + \mathbf{K}_c^\mathsf{T}\mathbf{q} = 0 \tag{4.4}$$

This control law can be implemented by simply measuring the voltage across the passive inductor. Using this approach along with a quasi-steady-state assumption (i.e., slowly varying excitation frequency), the active tuning control law can be implemented using the transfer function shown in Eq. (4.5), where V_L is the voltage across the inductor and L_a is the variable active inductance,

$$\frac{V_C}{V_L} = \frac{-L_a s^2 + R_a s}{L_p s^2 + R_p s} \tag{4.5}$$

Unlike the semi-active piezoelectric absorbers that use a variable capacitor or variable inductor element, this active-passive tuning mechanism can be tuned quickly and accurately. Compared to a variable capacitor semi-active absorber, the active inductor tuning method can cover a much wider range of frequencies. When combined with the negative resistance action, the proposed method offers better performance than a variable inductor semi-active device, due to its ability to achieve very low resistance in the absorber circuit. It should be noted here that, in principle, the negative resistance concept could also be used to reduce the resistance in the synthetic inductor circuit of a variable inductor semi-active absorber. However, the parasitic resistance of the synthetic inductor is frequency dependent, so it cannot be effectively cancelled without a considerably more complicated control law.

4.2.2 Optimal Tuning Frequency

To implement the control law given in Eq. (4.5), we need to develop a tuning law that provides the optimal absorber inductance for a given excitation frequency. A quasi-steady-state assumption is used for this optimization, which means that the excitation frequency is assumed to be changing very slowly, so that the transient effects due to frequency variations are negligible. This assumption allows the optimization to be performed using the system steady-state frequency response rather than the much more complicated transient dynamics of the time-varying system. To further simplify the derivation of the optimal tuning law, it is assumed that the damping in both the structure and the absorber are small and can be neglected. The assumption of small structural damping is valid for many practical applications, and the negligible absorber damping can be assured by using the negative resistance control. The first step in the derivation is to transform Eq. (4.1) into modal coordinates (see Eq. (4.6)) using the transformation $\mathbf{q}=\mathbf{U}\boldsymbol{\eta}$, where $\mathbf{U}$ is a matrix of mass-normalized eigenvectors of the original system. Note that $L_t = L_p + L_a$ is the total inductance, $\mathbf{I}$ is an $n \times n$ identity matrix, where n is the number of modes in the model, and $[\omega_n^2]$ is a diagonal matrix of the open-circuit modal frequencies.

$$\mathbf{I}\ddot{\boldsymbol{\eta}}+\left[\omega_n^2\right]\boldsymbol{\eta}+\tilde{\mathbf{K}}_c Q = \tilde{\mathbf{F}} \cdot f(t) \tag{4.6}$$

$$L_t\ddot{Q}+\frac{1}{C_p^S}Q+\tilde{\mathbf{K}}_c^{\mathrm{T}}\boldsymbol{\eta}=0$$

The response of each mode can be solved directly from Eq. (4.6). The response of the jth mode at the excitation frequency ω_e is given by Eq. (4.7). The response is expressed as a function of the absorber tuning ratio δ, which is a parameter to be optimized. The tuning ratio is defined in Eq. (4.8), where ω_a is the natural frequency of the circuit.

$$\frac{\eta_j}{F}(\omega_e)=M_j\left[\frac{a_j}{\dfrac{1}{C_p^S}\left(1-\dfrac{1}{\delta^2}\right)-b}+\tilde{F}_j\right] \tag{4.7}$$

where,

$$\delta=\frac{\omega_a}{\omega_e} \quad \text{with} \quad \omega_a=\frac{1}{\sqrt{\left(L_p+L_a\right)C_p^S}} \tag{4.8}$$

and

$$M_j=\frac{1}{\omega_{n_j}^2-\omega_e^2}\quad,\quad a_j=\tilde{K}_{c_j}\sum_{i=1}^{n}\frac{\tilde{K}_{c_i}\tilde{F}_i}{\omega_{n_i}^2-\omega_e^2}\quad,\quad b=\sum_{i=1}^{n}\frac{\tilde{K}_{c_i}^2}{\omega_{n_i}^2-\omega_e^2}$$

Along with this result, the other item that is needed to optimize the tuning ratio is an expression that relates the displacement at the point of interest to the modal displacements, which is given as

$$w_d(t)=\mathbf{W}^{\mathrm{T}}\boldsymbol{\eta}(t) \tag{4.9}$$

The derivation of the modal transformation vector $\mathbf{W}$ depends on how the original model is derived. For a model derived using a discretization method such as the Galerkin's method, the transformation vector is a product of the comparison functions evaluated at the point of interest and the eigenvector matrix. Similarly, for a finite element model the vector $\mathbf{W}$ would contain the interpolation functions evaluated at the point of interest along with elements from the eigenvector matrix. The objective function to be minimized is the squared magnitude of the transfer function of vibration amplitude versus excitation force,

$$J = \left| \frac{w_d}{F}(\omega_e) \right|^2 = \left[\sum_{j=1}^{n} W_j \frac{\eta_j}{F}(\omega_e) \right]^2 \tag{4.10}$$

After substituting Eq. (4.7) into Eq. (4.10), the partial derivative of J with respect to the tuning ratio δ is set to zero and the possible minimizing values of δ are analytically solved. This procedure gives two possible solutions for the optimal tuning ratio as follows,

$$\delta_{opt} = \left[1 + C_p^S \frac{\sum_{j=1}^{n} W_j M_j \left(a_j - \tilde{F}_j b \right)}{\sum_{j=1}^{n} W_j M_j \tilde{F}_j} \right]^{-\frac{1}{2}} \tag{4.11}$$

$$\text{or} \quad \delta_{opt} \to \infty$$

In the absence of damping in the structure and the absorber, it is generally possible for the absorber to force the response of the structure to zero at the point of interest. Thus, instead of performing the optimization procedure, the objective function J could be directly equated to zero. The solution for δ obtained in this way is identical to the first solution given in Eq. (4.11), which confirms that this expression gives the optimal tuning ratio when it is possible to eliminate the response. The special case where this is not possible and the significance of the second solution shown in Eq. (4.11) are discussed in Section 4.3.4.

The expression for δ_{opt} can be calculated online by the controller or calculated offline and stored in a look-up table. Note that for a SDOF (single-degree-of-freedom) system, the solution given by Eq. (4.11) reduces to an optimal tuning ratio of unity at all frequencies. This is consistent with the classical result for a vibration absorber on an undamped SDOF system.

The desired active inductance gain can be solved directly from Eq. (4.8) if both the optimal tuning ratio and the excitation frequency are known. In some systems, a direct measurement of the excitation frequency may be available. For instance, if the excitation source is a rotating unbalance, a tachometer measuring the rotational speed will directly give the excitation frequency. In other cases, it is necessary to estimate the excitation frequency from an additional sensor such as a force transducer or accelerometer. For this purpose, an algorithm proposed by Handel and Tichavsky (1994) is used here. This scheme is based on the discounted least squares identification of a harmonic signal and the measurement of phase differences to estimate the fundamental frequency of the signal. This method has been shown to have excellent convergence speed and noise rejection characteristics. Ideally, a force transducer located at the disturbance input point could be used. An accelerometer or other sensor could also be used, but it should not be placed near the point where the response is to be minimized to ensure that the signal-to-noise ratio is sufficient.

4.3 ACTIVE COUPLING ENHANCEMENT

For tonally tuned passive undamped absorbers with fixed design, if the excitation frequency or the system parameters change even slightly, the device effectiveness could be greatly reduced. The mistuning due to a varying excitation frequency can be addressed by making the absorber adaptive, as discussed above. However, there may be other sources of mistuning such as system parameters fluctuations, changes in operating conditions, and the effects of signal noise. In addition, the above optimal tuning law is derived using a quasi-steady-state assumption, which may lead to mistuning if the excitation frequency is changing rapidly. Increasing the robustness of the absorber would help minimize the performance loss due to all these possible errors. From a steady-state point of view, the robustness of an absorber can be measured in the frequency domain by the width of the notch produced near the tuned frequency. For a piezoelectric absorber, this notch width is directly dependent on the level of the coupling between the structure and the electrical circuit. In the following section, we will demonstrate that the effective coupling of the piezoelectric absorber can be increased using active action.

4.3.1 Active Coupling Feedback

The electro-mechanical coupling of the piezoelectric element is represented in Eq. (4.1) by the terms $\mathbf{K_C}$ and $\mathbf{K_C}^T$. The control voltage V_c cannot modify the coupling term in the structure equation, but it can increase the effective coupling in the circuit equation by using the following control law,

$$V_c = -\left(G_{ac} - 1\right)\mathbf{K_C^T}\mathbf{q} , \quad G_{ac} \geq 1 \qquad (4.12)$$

By applying this action, the circuit equation for the closed-loop system (see Eq. (4.13)) will have an apparent coupling vector that is increased by a factor of G_{ac}, compared to the open-loop system.

$$L_p\ddot{Q} + R_p\dot{Q} + \frac{1}{C_p^S}Q + G_{ac}\mathbf{K_C^T}\mathbf{q} = 0 \qquad (4.13)$$

In general, this control law requires the measurement or estimation of the generalized coordinates in the vector $\mathbf{q}$, as well as an accurate model of the coupling vector $\mathbf{K_C}$. However, if a collocated sensor and actuator pair can be used, this control law becomes extremely simple to implement. Consider the equation for a piezoelectric sensor that is attached at the same location as the piezoelectric actuator. If the sensor is attached to a high impedance measurement device, it is valid to treat the sensor connection as an open circuit. In this case, the circuit equation from Eq. (4.1) reduces to the form shown in Eq. (4.14), and the active coupling control law becomes a proportional feedback of the piezoelectric sensor signal.

$$V_S = \mathbf{K_C^T}\mathbf{q} \qquad (4.14)$$

Collocated piezoelectric patches are feasible in many structures subjected to pure bending loads, such as beams, plates, and shells. In these structures, the sensor and actuator are generally attached on opposite sides of the neutral axis of the structure. Piezoelectric stack actuators are also available with integrated strain gauges, which allow collocated actuation and sensing. In the case where a collocated sensor and actuator pair cannot be achieved or an accurate system model is not available, an alternate form of coupling enhancement can be used. The use of a negative capacitance element to increase the effective coupling of a piezoelectric absorber has also been demonstrated (Morgan *et al.*, 2000; also see Chapter 2).

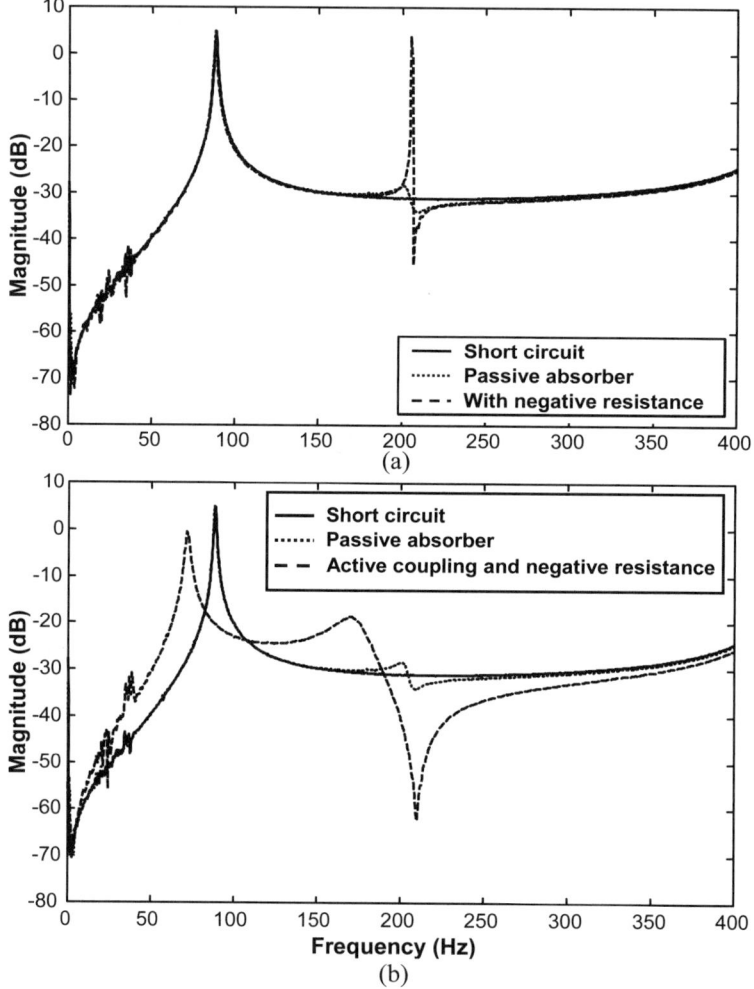

Fig. 4.1. Experimental verification of active coupling feedback. (a) Negative resistance effects, (b) Active coupling and negative resistance effects.

The effectiveness of the active coupling feedback can be seen from the steady-state experimental data (frequency response function between structural vibration amplitude and input force) shown in Fig. 4.1. Fig. 4.1a shows the effect of the negative resistance action. From this plot, it is clear that the internal resistance of the passive inductor is too large to yield the deep notch that is desired (here the target frequency is at 205 Hz) from the passive absorber. The notch created by the absorber is much deeper with the addition of the negative resistance, thus the maximum performance is increased significantly. However, the notch is still very narrow, which implies poor robustness. Fig. 4.1b shows the same system with the addition of the active coupling feedback. This additional active action dramatically increases both the width and depth of the notch produced by the absorber, so the steady-state performance and robustness properties are improved. The transient effects of the active coupling enhancement are further discussed in the subsequent sections.

4.3.2 Effective Coupling Coefficient

In this section, we examine how the active coupling feedback increases the effective coupling of the piezoelectric absorber. First, we introduce the generalized coupling coefficient, K_{ij}, which is defined by Hagood and von Flotow (1991) and analyzed in Chapter 2,

$$K_{ij}^2 = \frac{\left(\omega_n^D\right)^2 - \left(\omega_n^E\right)^2}{\left(\omega_n^E\right)^2} \tag{4.15}$$

Here ω_n^D and ω_n^E are the open-circuit and short-circuit natural frequencies of the system, respectively. The subscript ij denotes the mode of operation where the applied electrical field is in the i direction and the mechanical stresses and strains act in the j direction. We define the *effective* coupling coefficient as the apparent generalized coupling coefficient that is observed when the active coupling feedback is considered to be part of the piezoelectric device. To derive this effective coupling coefficient the active coupling control law shown in Eq. (4.12) is substituted into a SDOF system model in the form of Eq. (4.1) and the open-circuit and short-circuit natural frequencies of the closed-loop system are derived. The definition in Eq. (4.15) is then used to find the effective generalized coupling coefficient, which can be expressed in terms of the original K_{ij}^2 and the active coupling gain, as shown below,

$$\left(K_{ij}^2\right)_{eff} = \frac{G_{ac}K_{ij}^2}{1+\left(1-G_{ac}\right)K_{ij}^2} \tag{4.16}$$

A representative plot of the effective coupling coefficient versus the coupling gain G_{ac} is shown in Fig. 4.2. This plot shows that the square of the coupling coefficient increases almost linearly with the active coupling gain G_{ac} for small gains. When the gain becomes larger, the effective coupling coefficient increases much faster and actually goes to infinity at the critical gain shown. Beyond this critical gain, the square of the effective coupling coefficient is negative, which implies that the generalized coupling coefficient is imaginary (not physically meaningful). In fact, it can be shown that the system is unstable in this case by applying the Routh-Hurwitz criterion (Kuo, 1995) to the transfer function of the closed-loop system. The resulting stability criterion for the active coupling gain is given as follows,

$$G_{ac} < 1 + \frac{1}{K_{ij}^2} \tag{4.17}$$

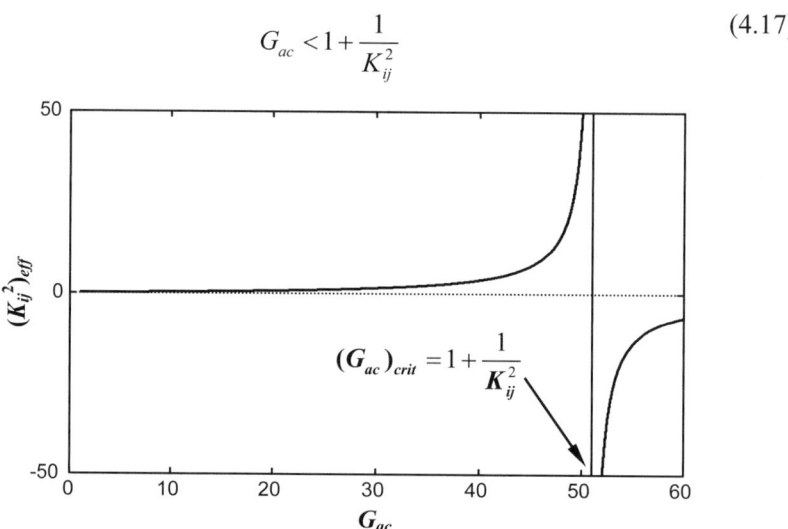

Fig. 4.2. Effective coupling coefficient versus active coupling gain ($K_{ij}^2 = .02$).

4.3.3 Optimal Inductance Tuning for Systems with Active Coupling

With the introduction of the active coupling feedback, the optimal tuning law derived previously needs to be modified. Starting with Eq. (4.6), the coupling term in the circuit equation is multiplied by the active coupling gain G_{ac}. By repeating the previous procedure, the resulting optimal tuning law can be shown to be identical to Eq. (4.11), but with the parameters a_j and b modified as:

$$a_j = G_{ac} \tilde{K}_{c_j} \sum_{i=1}^{n} \frac{\tilde{K}_{c_i} \tilde{F}_i}{\omega_{n_i}^2 - \omega_e^2} \qquad b = G_{ac} \sum_{i=1}^{n} \frac{\tilde{K}_{c_i}^2}{\omega_{n_i}^2 - \omega_e^2} \qquad (4.18)$$

The control law implementation can now be summarized as follows. First, the controller estimates the instantaneous excitation frequency using the estimation algorithm described in Section 4.2.2. Next, the optimal tuning ratio for the current excitation frequency is calculated online using Eqs. (4.11) and (4.18), or obtained from a look-up table of previously calculated optimal tuning ratios. The optimal active inductance gain L_a is then calculated using Eq. (4.8). The total control voltage is then calculated in the Laplace domain as shown in Eq. (4.19), where s is the Laplace operator, V_S is the piezoelectric sensor voltage and V_L is the voltage measured across the passive inductor.

$$V_C = -(G_{ac} - 1)V_S + \frac{-L_a s^2 + R_a s}{L_p s^2 + R_p s} V_L \qquad (4.19)$$

4.3.4 Discussion of Typical Optimal Tuning Curves

In this section, we examine some typical optimal tuning curves. The frequency response functions and the corresponding optimal tuning curves are generated using an analytical model of a cantilevered aluminum beam. In the first case, we consider the displacement of the beam tip for a force applied at the tip. The frequency response function (FRF) for this example is given in Fig. 4.3a. Since this is a driving point FRF, it contains alternating resonant and anti-resonant frequencies. The optimal tuning ratio predicted by Eq. (4.11) is given in Fig. 4.3b. These optimal tuning curves are generated using the first solution given in Eq. (4.11), except at frequencies where this solution is imaginary. At these points, the optimal solution is an infinite tuning ratio, which corresponds to a short circuit (no absorber).

From Fig. 4.3b it is apparent that the optimal tuning ratio is close to unity at all frequencies except near the anti-resonance. The anti-resonant frequency is the only point where the first solution in Eq. (4.11) does not exist, thus the optimal tuning ratio goes to infinity. The optimal tuning ratio also approaches zero just beyond the anti-resonant frequency. The physical significance of these observations is that the optimal tuning for frequencies near the anti-resonance is far away from the anti-resonance. In other words, the absorber cannot further reduce the response at an anti-resonant point; it can only increase it. Indeed, tuning the absorber to the anti-resonant frequency produces a resonance at this frequency and a notch on either side of this frequency. This effect is exactly the opposite of what is observed when an absorber is tuned to a resonant frequency.

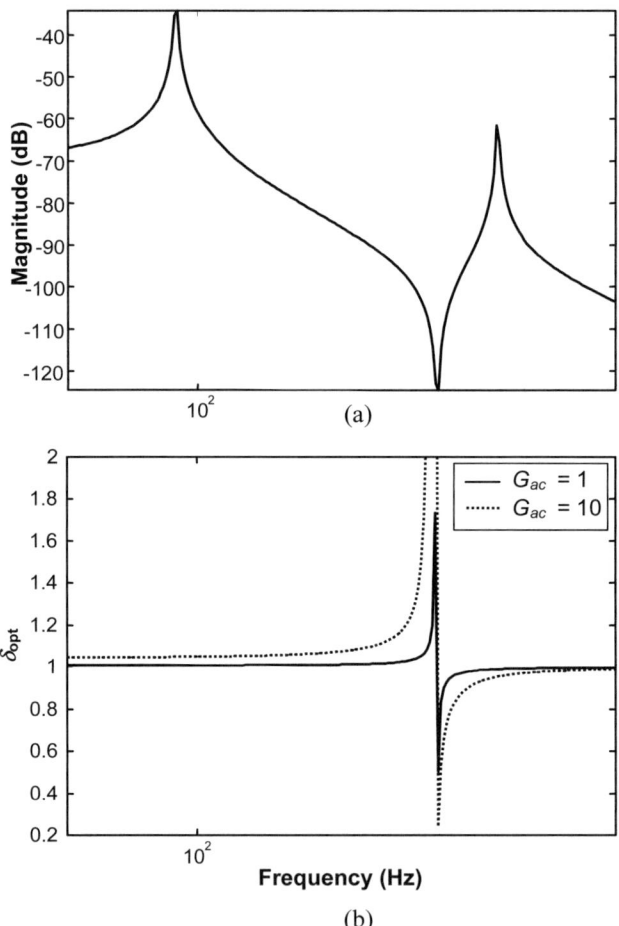

Fig. 4.3. (a) Frequency response function (anti-resonant case), (b) optimal tuning curve (anti-resonance case).

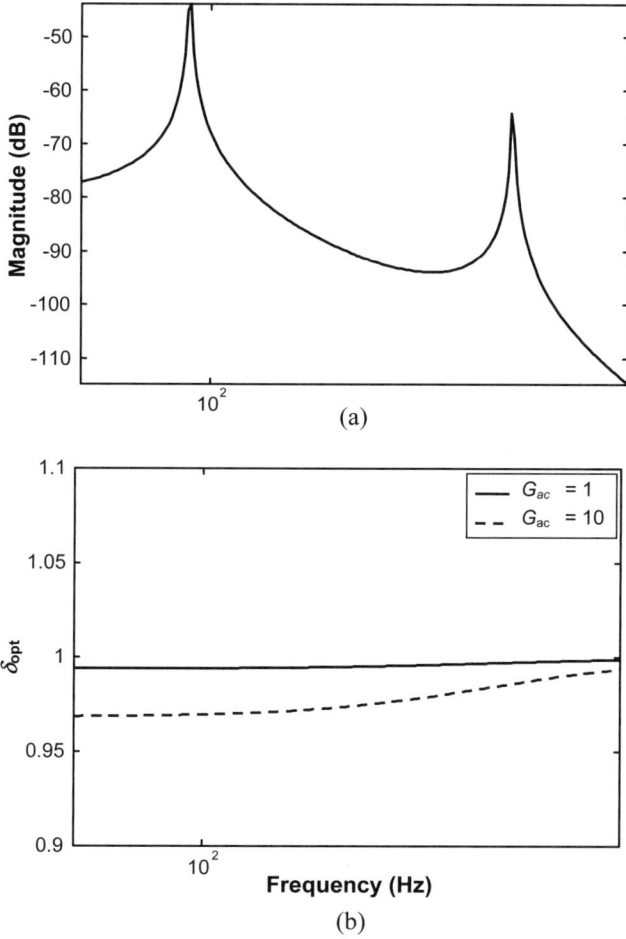

Fig. 4.4. (a) Frequency response function (no anti-resonance), (b) optimal tuning curve (no anti-resonance).

In the second case, we consider the transfer function for the displacement at the midpoint of the beam versus the same tip load, which is shown in Fig. 4.4a. Since this transfer function does not contain any structural anti-resonant frequencies the tuning curve becomes much simpler, as illustrated in Fig. 4.4b. In this case, the optimal tuning ratio is relatively constant over the entire frequency range. Although the optimal tuning ratio is close to unity (the optimal tuning ratio for a SDOF system), the error in assuming a tuning ratio of unity can be significant enough to result in severe performance degradation. Since all flexible structures in theory have infinite degrees of freedom, and the effects of the higher order modes could be significant, the optimal tuning law of Eq. (4.11) is preferred over the classical absorber theory that assumes a tuning ratio of unity.

Using this optimal tuning law requires an accurate system model of the form shown in Eq. (4.1), which in some cases may be difficult to obtain. Fortunately, the qualitative information gained from the above examples can be combined with experimental data to reconstruct the optimal tuning curves. In most systems, getting the frequency response function due to the excitation force is a relatively simple matter. Using steady-state FRF data similar to Fig. 4.1, the optimal tuning ratio for a single frequency can easily be obtained experimentally. This is accomplished by first measuring the circuit inductance and the capacitance of the piezoelectric material and using Eq. (4.8) to calculate the natural frequency of the absorber. The notch frequency corresponding to this absorber tuning is then obtained from the experimental data. The optimal tuning ratio at this frequency is found by dividing the natural frequency of the absorber by the measured notch frequency. In this manner, the optimal tuning ratio for a few points within the frequency bandwidth of interest can be obtained. Since the optimal tuning ratio is relatively constant in the frequency bands between anti-resonances, an optimal tuning curve can be constructed (approximated) by using a simple linear interpolation for the frequencies between the measured points. If the optimal tuning can be determined experimentally in this way, it is not necessary to obtain a model of the system in order to implement the proposed absorber design.

4.4 STABILITY ANALYSIS

The stability condition derived in Section 4.3.2 is the maximum coupling gain for an active-passive piezoelectric absorber with fixed gains. In this section, the stability analysis is extended to the adaptive absorber, which is a linear time-varying system. Lyapunov's stability theory is used to derive conditions under which the system is guaranteed to be stable.

To begin the derivation, the homogeneous dynamic equations for the closed-loop system are written in matrix form, as shown in Eq. (4.20), where L_t is the total (time-varying) inductance and R_t is the net resistance $R_p - R_a$.

$$\begin{bmatrix} \mathbf{M} & 0 \\ 0 & G_{ac}L_t(t) \end{bmatrix}\ddot{\mathbf{z}} + \begin{bmatrix} \mathbf{C}_d & 0 \\ 0 & G_{ac}R_t \end{bmatrix}\dot{\mathbf{z}} + \begin{bmatrix} \mathbf{K}^\mathbf{D} & G_{ac}\mathbf{K}_\mathbf{C} \\ G_{ac}\mathbf{K}_\mathbf{C}^\mathbf{T} & \dfrac{G_{ac}}{G_{nc}}C_p^S \end{bmatrix}\mathbf{z} = \begin{Bmatrix} \mathbf{0} \\ 0 \end{Bmatrix} \qquad (4.20)$$

$$\underbrace{\phantom{\begin{bmatrix} \mathbf{M} & 0 \\ 0 & G_{ac}L_t(t) \end{bmatrix}}}_{\bar{\mathbf{M}}} \qquad \underbrace{\phantom{\begin{bmatrix} \mathbf{C}_d & 0 \\ 0 & G_{ac}R_t \end{bmatrix}}}_{\bar{\mathbf{C}}} \qquad \underbrace{\phantom{\begin{bmatrix} \mathbf{K}^\mathbf{D} & \mathbf{K} \\ \mathbf{K} & \mathbf{K} \end{bmatrix}}}_{\bar{\mathbf{K}}}$$

where $\mathbf{z} = \begin{Bmatrix} \mathbf{q} \\ \tilde{Q} \end{Bmatrix}$ and $\tilde{Q} = G_{ac}Q$. The controller is assumed to have an active coupling gain G_{ac} greater than or equal to one (a coupling gain of less
than unity would decrease the effective coupling of the system). The new
vector $\mathbf{z}$ contains the n generalized coordinates from the original structure
equation as well as a scaled version of the charge on the piezoelectric actuator. The charge Q is scaled by the active coupling gain G_{ac} so that the
resulting system stiffness matrix ($\bar{\mathbf{K}}$) is symmetric.

The Lyapunov function candidate is selected to be the total energy of
the system,

$$V(\mathbf{z}) = \frac{1}{2}\mathbf{z}^\mathrm{T}\bar{\mathbf{K}}\mathbf{z} + \frac{1}{2}\dot{\mathbf{z}}^\mathrm{T}\bar{\mathbf{M}}\dot{\mathbf{z}} \qquad (4.21)$$

If the system mass and stiffness matrices are positive definite, the total system energy is positive definite by definition. Clearly, the system mass matrix is positive definite in this case since the structural mass matrix must be
positive definite and the total inductance is always positive. For now, we
assume that the system stiffness matrix is positive definite as well, which
implies that Eq. (4.21) is a valid Lyapunov function. Therefore, the system
is stable when the time derivative of V is at least negative semi-definite.
Since the system mass and stiffness matrices are symmetric, the derivative
of V can be found as follows,

$$\dot{V}(\mathbf{z}) = -\dot{\mathbf{z}}^\mathrm{T}\left(\bar{\mathbf{C}} - \frac{1}{2}\dot{\bar{\mathbf{M}}}\right)\dot{\mathbf{z}} \qquad (4.22)$$

The condition under which the energy rate is negative semi-definite is
given as

$$R_t - \frac{1}{2}\dot{L}_t(t) \geq 0 \tag{4.23}$$

Using Eq. (4.8) to relate the total inductance to the excitation frequency and using the Invariance Principle (Kahlil, 1996), the stability condition can alternately be expressed in terms of the frequency rate of change,

$$-R_t C_p^S \delta^2 \omega^3 \leq \dot{\omega} \tag{4.24}$$

Since the left side of this inequality is negative, there is no upper bound on the frequency rate. The reason for this is that as the excitation frequency increases, the inductance decreases and thus the total system energy decreases. However, if the excitation frequency is decreasing rapidly the system may not be stable if the circuit resistance is too low to compensate for the additional inductance (electrical inertia) being added to the system. Numerical examples can be used to show that this stability condition is not very restrictive, even in the worst-case scenario where the frequency and resistance are both low. For example, for a capacitance of 40 nF, a nominal excitation frequency of 10 Hz, and a closed-loop damping ratio of only 0.1%, the minimum frequency rate for stable operation is −1.26 Hz/s, which is quite fast relative to the nominal frequency.

In deriving the stability condition given in Eq. (4.24), it was assumed that the system stiffness matrix $\bar{K}$ is positive definite. The conditions under which this assumption is valid are now examined. Sylvester's criterion for positive definiteness of a matrix requires that the determinant of the matrix be positive and the principal minors of this determinant are positive as well. Since $\mathbf{K^D}$ is positive definite, all of the principal minors of the determinant of $\bar{K}$ are guaranteed to be positive. The matrix identity shown in Eq. (4.25) can be used to rewrite the determinant of the system stiffness matrix, where $\mathbf{A}$, $\mathbf{B}$, $\mathbf{C}$, and $\mathbf{D}$ are $n \times n$, $n \times m$, $m \times n$, and $m \times m$ matrices, respectively.

$$\begin{vmatrix} \mathbf{A} & \mathbf{B} \\ \mathbf{C} & \mathbf{D} \end{vmatrix} = |\mathbf{D}||\mathbf{A} - \mathbf{BD^{-1}C}| \quad \text{if} \quad |\mathbf{D}| \neq 0 \tag{4.25}$$

The condition for positive definiteness of the system stiffness matrix is given as

$$\left| \mathbf{K^D} - G_{ac} C_p^S \mathbf{K_C} \mathbf{K_C^T} \right| > 0 \qquad (4.26)$$

For a SDOF mechanical system, Eq. (4.26) is identical to Eq. (4.17), which also defines the maximum active coupling gain for stability. From the Lyapunov stability analysis, the closed-loop system is guaranteed to be stable when the conditions given in Eqs. (4.26) and (4.24) are both satisfied. Physically, this means that the system will be stable as long as the frequency is not decreasing very rapidly and the active coupling gain is not excessively high. It is also important to remember that the conditions resulting from Lyapunov's method are sufficient, but not necessary, to guarantee stability.

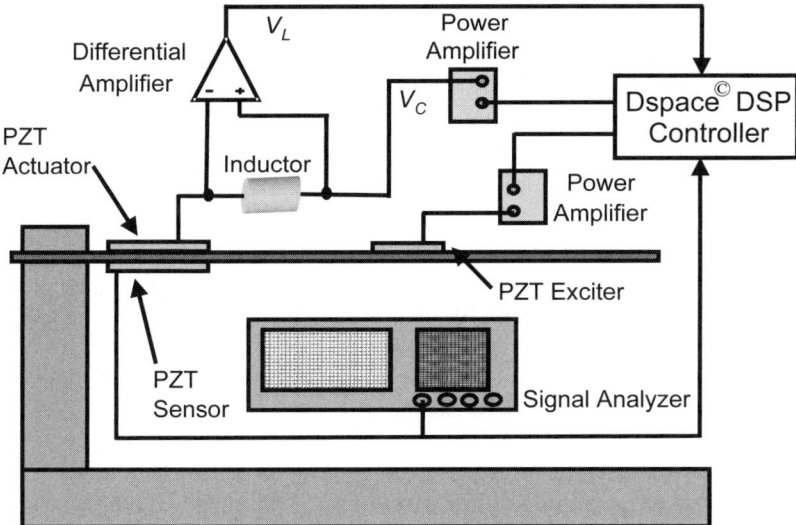

Fig. 4.5. Experimental setup.

4.5 EXPERIMENTAL VALIDATION

The test setup is shown in Fig. 4.5. The system consists of a cantilevered beam with two identical surface-bonded piezoelectric ceramic (PZT) patches at the root of the beam, which serve as the sensor and actuator.

The dimensions and parameters of the test system are given in Table 4.1. These same parameters are used in the analytical model for the simulations in this chapter. The beam is excited using a third piezoelectric patch located at the midpoint of the beam. The voltage from the piezoelectric sensor is used to measure the beam vibration level.

Table 4.1. System parameters

Length of beam	196 mm
Thickness of beam	3.2 mm
Width of beam and PZT	25.4 mm
Length of PZT	58.8 mm
Thickness of PZT	0.5 mm
Piezoelectric constant (d_{31})	-175e-12 m/V
Modulus of PZT ($1/s_{11}^E$)	65 GPa
Dielectric constant of PZT (ε_{33}^T)	1.5e-8 F/m
Generalized coupling coeff. (K_{31})	0.141

Two test cases are considered: the first case is for a non-resonant excitation (in this book chapter, non-resonant excitation means that the excitation frequencies are sufficiently far from the system resonant frequencies), and the second case is for an excitation near a resonant frequency of the structure. The experimental baseline system for the near-resonant excitation case is an optimally damped passive piezoelectric absorber (see Chapters 2 and 3). That is, the absorber is tuned to the first resonant frequency and sufficient damping (resistance) is added to give a flat frequency response around that frequency. In the non-resonant case, a passive absorber with fixed design would not be effective for an excitation with varying frequency. Therefore, the baseline is selected to be the structural response with the piezoelectric transducer shorted.

Since the beam structure used here has a pair of collocated actuator and sensor, the active coupling implementation requires only a proportional feedback of the sensor voltage. The optimal tuning curve is obtained experimentally using only two frequency points, since the frequency response of the structure does not contain any anti-resonant points in the frequency band of interest. The inputs to the controller are the sensor voltage, the voltage across the passive inductor, and the excitation signal. The controller is implemented using a Simulink© model, which is compiled to run on a dSPACE© DSP board. The controller also contains the frequency estimation algorithm (Handel and Tichavsky, 1994), which uses the measured excitation signal to continually estimate the excitation frequency.

The purpose of the experiment is to study the performance of the system when subject to a harmonic excitation with varying frequency. The simplest such excitation is a linear chirp signal, which is a sinusoid of linearly increasing frequency. The three parameters that characterize the chirp signal are the nominal frequency f_o, the bandwidth of the frequency variation Δ, and the frequency rate of change $\dot{f}$ (Hz/s). For the linear chirp used here the frequency starts at $(1-\Delta)f_o$ at time t_s and increases at a rate of $\dot{f}$ until reaching a maximum frequency of $(1+\Delta)f_o$ at time t_f. In this experiment, the nominal excitation frequency and bandwidth are constant in each case and the frequency rate of change is varied. Four tests are carried out for both the near-resonant and non-resonant cases, with the frequency rate of change varying from 2 to 8 Hz per second. The excitation is applied at time $t = 0$, but the data acquisition system is set to have a trigger delay of t_s seconds. The purpose of this delay is to discard the large transient response caused by initially applying the excitation, which gives a better estimate of the performance of the system under extended operating conditions. The controller and excitation parameters used in both experiments are given in Table 4.2.

Table 4.2. Experimental parameters

Non-Resonant Case	
$L_p = 16.1$ H	$G_{ac} = 12$
$f_o = 205$ Hz	$\Delta = .10$
$R_p = 900\ \Omega$	$t_s = 1$ sec
$R_a = 800\ \Omega$	$\dot{f} = 2, 4, 6,$ and 8 Hz/sec
Resonant Case	
$L_p = 79.0$ H	$G_{ac} = 10$
$f_o = 92$ Hz	$\Delta = .10$
$R_p = 1530\ \Omega$	$t_s = 1$ sec
$R_a = 1000\ \Omega$	$\dot{f} = 2, 4, 6,$ and 8 Hz/sec

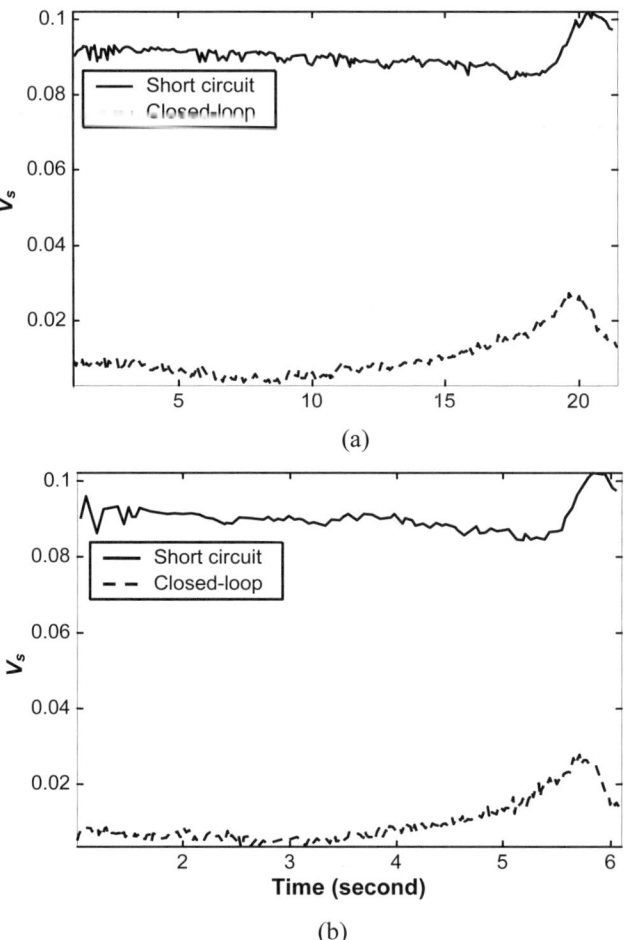

(a)

(b)

Fig. 4.6. Experimental response envelops (non-resonant case); (a) $\dot{f}$ = 2 Hz/s, (b) 8 Hz/s.

Table 4.3. Summary of experimental results for non-resonant excitation case

$\dot{f}$ (Hz/s)	% Reduction Response		Control Voltage
	RMS	Peak	(RMS volts)
2	89.2	73.4	.097
4	89.4	71.8	.092
6	89.6	73.8	.083
8	89.1	72.8	.079

Table 4.4. Summary of experimental results for near-resonant excitation case

$\dot{f}$ (Hz/s)	% Reduction Response		Control Voltage (RMS volts)
	RMS	Peak	
2	79.1	48.9	1.22
4	80.1	48.1	1.11
6	83.0	52.9	.95
8	81.2	48.5	1.07

Experimental data for the non-resonant case is shown in Fig. 4.6. The excitation bandwidth used for the non-resonant case is +/-10% of the nominal frequency of 205 Hz, which corresponds to approximately 40 Hz. These plots show the response envelopes for the cases $\dot{f}$ = 2 Hz/s and $\dot{f}$ = 8 Hz/s. Note that the spin-up response magnitudes are very similar for the slowest and fastest cases. This is not surprising for the passive baseline system, since there are no parameters varying in response to the changing excitation frequency. However, it is interesting that the closed-loop response for the fastest excitation case is similar to the slowest case, since the optimal tuning law is derived using a quasi-steady-state (QSS) assumption, which should only be valid for very slowly changing excitations. The experimental results for the non-resonant case are summarized in Table 4.3. This table shows percentage reduction in RMS and peak response, compared to the short circuit baseline system, as well as the required control voltage for each case. These results also show that the performance of the active-passive absorber is relatively unaffected by the frequency rate of change. The conclusion is that the added robustness due to the active coupling enhancement allows the QSS-tuned adaptive absorber to achieve good performance even for rapidly varying excitations (e.g., the spin-up of a rotating machine when it is turned on). The combination of a rapidly changing excitation frequency and a very wide frequency bandwidth is a difficult problem for a semi-active device. However, the active-passive piezoelectric absorber presented here may have the performance and robustness necessary for these rapid spin-up applications.

Next we consider the case where the excitation frequency is close to a resonance of the structure. The test results for the near-resonant case tests are summarized in Table 4.4. Results for two of the test cases are shown in Fig. 4.7. Once again we see that the performance of the active-passive absorber is relatively unaffected by the frequency rate of change. Although the performance of the optimal passive absorber baseline is already much better than the original system (no absorber), the adaptive active-passive absorber still can outperform the passive baseline system significantly. Compared to the system with no absorber, the performance of the active-passive absorber is much better in the near-resonant case than in the non-resonant case. The reason for this is that the size of the notch produced by the absorber is considerably larger near the resonant frequencies of the structure.

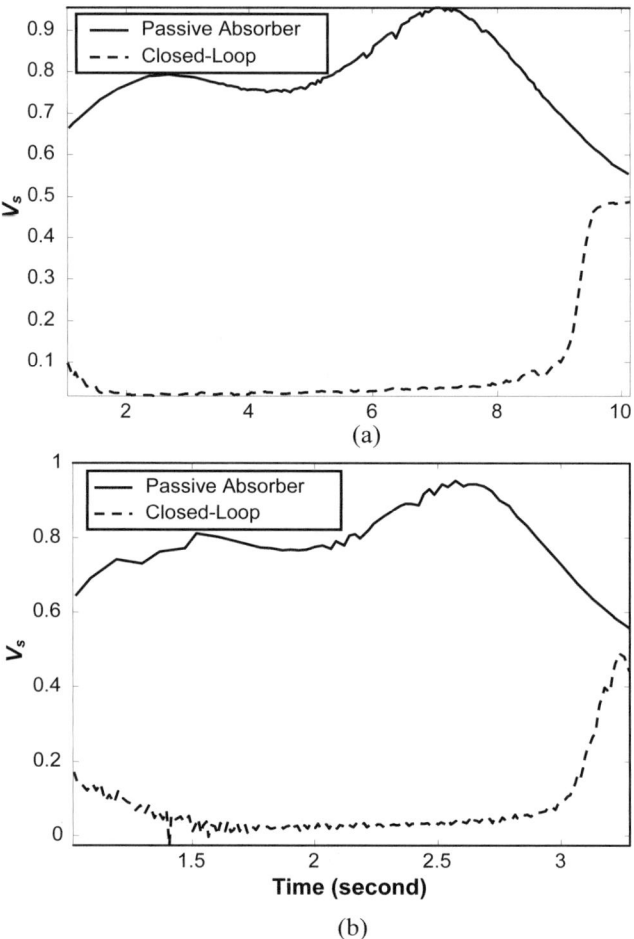

Fig. 4.7. Experimental response envelops (near-resonant case); (a) $\dot{f} = 2$ Hz/s, (b) $\dot{f} = 8$ Hz/s.

4.6 COMPARISON STUDIES

4.6.1 Comparison to Adaptive Feedforward Control

The effect of the proposed adaptive active-passive piezoelectric absorber is similar in concept to an adaptive feedforward cancellation of a harmonic disturbance. However, the control laws presented here are obtained using physical insight of vibration absorber characteristics instead of traditional control theory. Therefore, it would be interesting to compare the performance and efficiency of the adaptive active-passive absorber system to a state-of-the-art active control approach. For the purpose of suppressing a harmonic excitation with varying frequency, one logical choice would be an adaptive feedforward control law. One of the most widely used adaptive feedforward schemes is the filtered-x algorithm (Fuller *et al.*, 1996). In general, such controllers generate the control signal by filtering the reference signal x with an adaptive FIR filter $H(z)$, where z is the discrete-time z-transform operator. In the filtered-x algorithm, the updating law for the ith coefficient of the adaptive FIR filter is the difference equation shown in Eq. (4.27), where α is the adaptation gain, n is the current sample number, $e(n)$ is the observed error, and $r(n)$ is the reference signal x filtered by a model of the plant. The order of the filter $H(z)$, and thus the number of filter coefficients h_i, depend on the complexity of the disturbance to be modeled. For a harmonic disturbance, a second order filter is sufficient.

$$h_i(n+1) = h_i(n) - \alpha e(n) r(n-i) \qquad (4.27)$$

It has also been shown by Sievers and von Flotow (1992) that in the case of a SISO (single-input-single-output) system and where the reference signal x represents a harmonic disturbance, the filtered-x feedforward algorithm is equivalent to using the second order transfer function shown in Eq. (4.28), which gives the controller output $u(n)$ for a given error signal $e(n)$. Here, T is the sample time and ω_e is the excitation frequency.

$$H(z) = \frac{U(z)}{E(z)} = -\alpha \hat{A} \frac{\cos\left(\omega_e T - \hat{\phi}\right) \cdot z - \cos\hat{\phi}}{z^2 - 2\cos\left(\omega_e T\right) \cdot z + 1} \qquad (4.28)$$

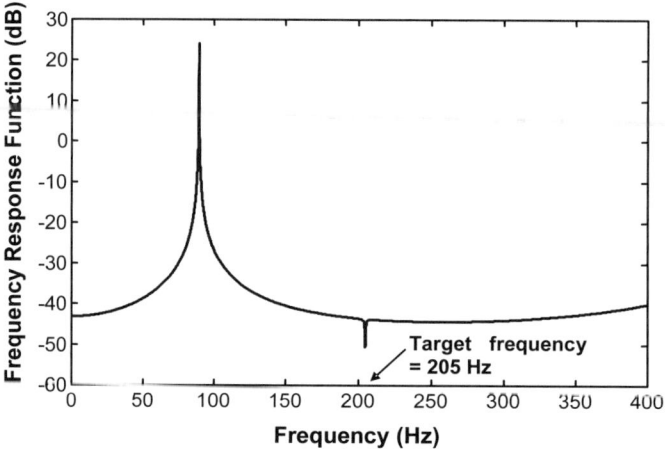

Fig. 4.8. Steady-state performance example of filtered-x controller.

In order to implement this transfer function, it is necessary to know the excitation frequency ω_e as well as the estimated amplitude and phase response of the plant at this frequency, which are represented by $\hat{A}$ and $\hat{\phi}$, respectively. To perform a fair comparison between the filtered-x baseline and the active-passive piezoelectric absorber, the same frequency estimation algorithm is used for both controllers. Fig. 4.8 shows an example of the steady-state performance of the filtered-x controller. The controller introduces a very sharp notch at the target frequency of 205 Hz, which can provide effective attenuation of the harmonic excitation, provided the controller is properly tuned. Comparing this figure to Fig. 4.1b, which shows the same structure with a piezoelectric absorber, it can be seen that the notch produced by the active-passive piezoelectric absorber is much wider than that of the filtered-x controller, even when no active coupling gain is used. For this reason, we may expect the filtered-x controller to be less robust to mistuning errors, and thus not as effective for suppressing excitations with rapidly varying frequency.

In what follows, we use the filtered-x adaptive feedforward control law as a baseline for comparison with the adaptive active-passive piezoelectric absorber. The example used here is the same cantilever beam system specified in the experimental study. The analytical model is constructed by first deriving the equations of motion using Hamilton's principle, then discretizing this model using Galerkin's method. The linear chirp signal is used as the excitation signal for all of the simulations performed.

There are five parameters considered in this study: Δ, $\dot{f}$, f_o, G_{ac} and ζ_{cl}. The excitation parameters are the bandwidth Δ, the rate of frequency change $\dot{f}$, and the nominal excitation frequency f_o. In this section, the nominal excitation frequency is fixed at $f_o = 205$ Hz, which is between the first two resonant frequencies of the system. The case where the nominal excitation frequency is near resonance will be considered separately in the next section. The parameters for the active-passive piezoelectric absorber are the active coupling gain G_{ac} and the closed-loop damping ratio of the absorber circuit, ζ_{cl}, which includes the effect of the negative resistance action. For the filtered-x baseline controller, the only adjustable parameter is the adaptation gain α, which is optimized for each case considered.

For now, we consider the case where the excitation bandwidth is fixed at $\Delta=0.05$. That is, the excitation frequency is always within 5% of the nominal frequency. First, the effects of the adaptation gain and the frequency rate of change $\dot{f}$ are examined. Fig. 4.9 shows the RMS vibration response versus the adaptation gain for several values of the normalized frequency rate of change (normalized with respect to the nominal frequency, f_o). The physical interpretation of the normalized frequency rates shown here is that the excitation frequency changes at a rate of 0.25%/sec in the slowest case and 2.5%/sec in the fastest case. From Fig. 4.4, it can be seen that the performance of the filtered-x baseline system is much better for slowly varying excitations. In addition, there is clearly an optimal adaptation gain for each case. For the parameters shown here, the optimal adaptation gain ranges from 2.2 for the slowest excitation to 1.8 for the fastest case. It is also observed that the control voltage requirement increases only slightly as α increases, so the gains from Fig. 4.9 that give the best performance can be considered to be optimal.

Next we investigate the effects of the absorber damping ratio and the active coupling gain on the performance of the active-passive absorber. The performance index is the percentage reduction in the RMS vibration response compared to the optimized filtered-x baseline system. First the damping ratio is varied while the active coupling gain is kept fixed at $G_{ac} = 5$. The results, which are illustrated in Fig. 4.10, show that the absorber optimal damping ratio increases slightly in the cases where the excitation frequency is changing faster. This trend is not surprising since the increased damping ratio helps increase the effective bandwidth of the absorber, which improves performance for rapidly changing excitations.

This effect is not very significant, however, and the optimal closed-loop damping ratio is very close to $\zeta_{cl} = 0.0004$ for all cases.

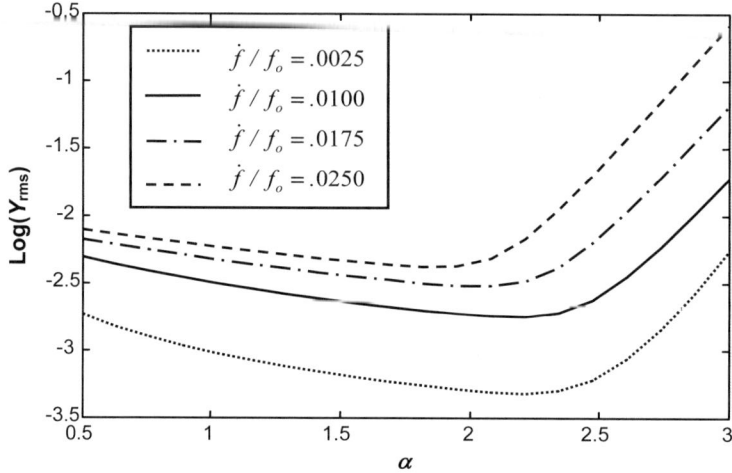

Fig. 4.9. Performance of filtered-x controller versus adaptation gain α ($\Delta = .05$).

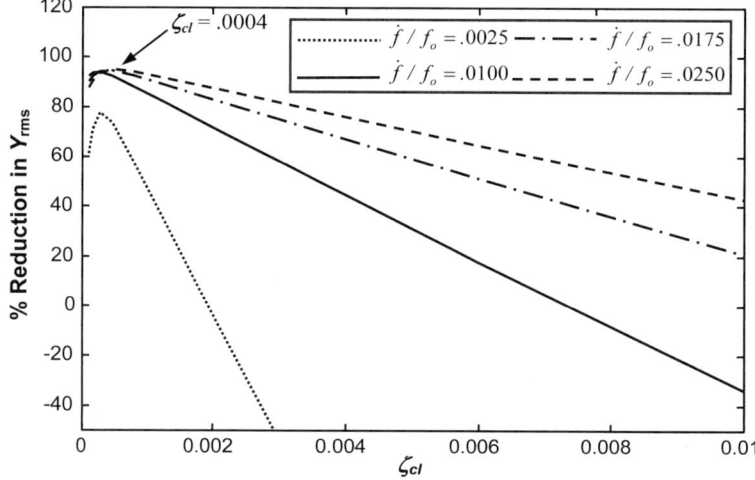

Fig. 4.10. Effects of absorber damping ratio ($G_{ac} = 5, \Delta = .05$).

Next, the effects of the active coupling gain are examined while fixing the absorber damping ratio. The percentage reduction in RMS vibration response compared to the baseline filter-x system is shown as a function of the active coupling gain in Fig. 4.11a. It is apparent that the active-passive absorber can outperform the filtered-x baseline, especially when the excitation frequency is changing rapidly. The reason for this is that the filtered-x controller is much less robust and is not effective for rapidly changing excitations. The poor robustness characteristics of the filtered-x controller are directly related to the narrow notch that the controller produces, which is illustrated in Fig. 4.8.

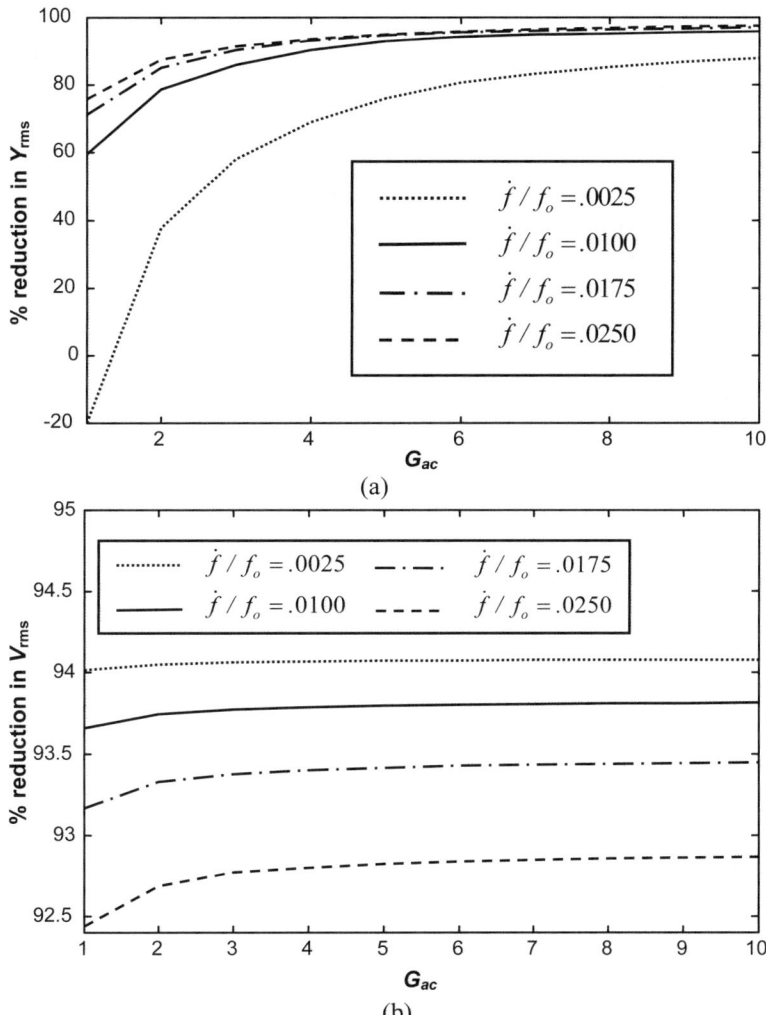

Fig. 4.11. Effects of active coupling gain ($\zeta_{cl} = .0004$, $\Delta = .05$); (a) relative performance, (b) relative control voltage.

It is also interesting to compare the corresponding control efforts of the active-passive absorber and the filtered-x controller. The percentage reduction in RMS control voltage, which is plotted in Fig. 4.11b, shows that the active-passive absorber requires much less control effort than the filtered-x controller to achieve superior performance. The increased efficiency of the adaptive absorber is largely due to its effective use of passive circuit elements compared to the filtered-x algorithm, which is a purely active method. Although the performance increases significantly as the active coupling gain is increased, the voltage requirement is increased only slightly. This suggests that the majority of the control voltage requirement for the active-passive absorber is due to the active inductor tuning.

Now that the effects of the active-passive absorber parameters have been investigated, the effects of the excitation parameters are studied. First, the excitation bandwidth is changed from $\Delta = 0.05$ to $\Delta = 0.15$. The adaptation gain of the filtered-x baseline is first re-optimized for the new bandwidth. It is observed that the relative performance is affected very little by the wider excitation bandwidth, and the resulting plots are nearly identical to Figs. 4.10 and 4.11a. However, the relative control voltage for the wider bandwidth case, which is shown in Fig. 4.12, does show differences. The control voltage requirement of the active-passive absorber is larger for the wider bandwidth case, although it is still significantly less than that of the filtered-x controller. The reason for the increased voltage requirement is that the active inductor tuning requires more voltage when the excitation frequency is farther away from the nominal frequency.

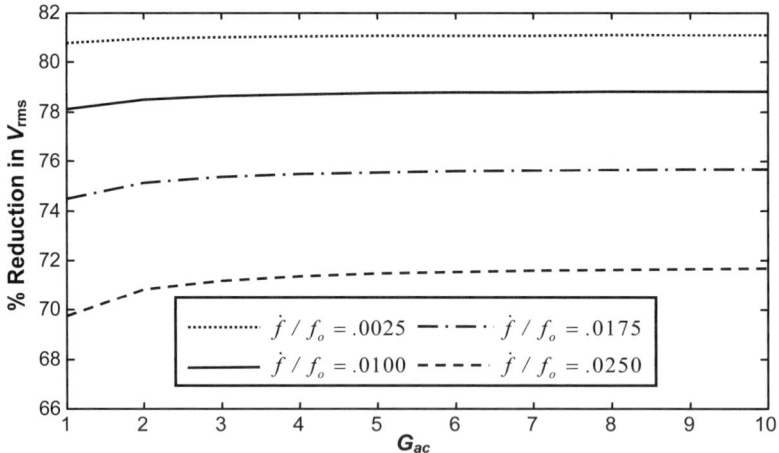

Fig. 4.12. Effects of active coupling gain on relative control voltage ($\zeta_{cl} = .0004$, $\Delta = .15$).

The only parameter that has not yet been varied is the nominal excitation frequency f_o. In this section we have considered a case where the nominal excitation frequency is between two system natural frequencies. In the next section we consider the case where the excitation frequency is near a resonant frequency of the structure.

4.6.2. Comparison to Concurrently Designed Active-Passive Piezoelectric Damping

The purpose of this section is to present the case when the nominal excitation frequency is close to a structural resonant frequency. In this situation, adaptive feedforward control may not be a good choice. It has been shown that the maximum stable adaptation gain for the filtered-x algorithm decreases as the magnitude of the filtered reference signal $r(t)$ increases, and $r(t)$ becomes very large when the excitation frequency is near resonance (Fuller *et al.*, 1996). The result is that the maximum stable gain for the filtered-x controller is quite small and it cannot provide very effective suppression near resonance. Thus, another method must be chosen as a baseline for the resonant case parameter studies. Fortunately, in this case there are many other options available since any method that adds damping to the system will be effective at suppressing excitations near resonance. For example, an optimal passive piezoelectric absorber could be used as a baseline, as was shown in the experimental validation effort. A more effective choice would be a concurrently designed active-passive hybrid piezoelectric network (APPN) configuration (See Chapter 3). The APPN is synthesized by a simultaneous optimal control - optimization algorithm that provides the best combination of the passive RL values and active gain for a given bandwidth. In this case, the desired bandwidth is around the first natural frequency of the structure. The principal difference between the concurrent APPN approach and the adaptive active-passive absorber presented in this chapter is that the former scheme uses an optimal but fixed configuration to achieve good performance within the frequency range of interest but does not on-line adapt to frequency variations.

A series of parametric studies similar to the ones shown in Section 4.6.1 are performed for the case where the nominal excitation frequency is at the first resonance of the structure, $f_o = 92$ Hz. The excitation bandwidth is again fixed at $\Delta = 0.05$ initially. First the relative performance is plotted versus the closed-loop damping ratio of the absorber (as in Fig. 4.10) and it is observed that the optimal damping is approximately $\zeta_{cl} = 0.002$ for all cases. Next, the relative performance and control voltages are plotted versus the active coupling gain, as shown in Fig. 4.13. Fig. 4.13a shows that the adaptive piezoelectric absorber is capable of reducing the vibration response by 98% compared to the fixed APPN system. It is also interesting to note that there exists an optimal active coupling gain of approximately G_{ac} = 5.5 for each case, whereas in the non-resonant situation the performance increases continuously as the coupling gain is increased. One possible reason for this is that the notch width of the absorber near resonance is already quite large, so a large coupling gain is not needed. It is also reasonable that there may be some undesirable effects in the transient dynamics of the system if the active coupling gain is made too large. As the coupling gain approaches its maximum stable value, which in this case is approximately $(G_{ac})_{max}$ = 50, two or more of the closed-loop poles become less damped and eventually become unstable. Therefore, there must be some tradeoff between better steady-state performance/ robustness (higher G_{ac}) and better transient dynamics (lower G_{ac}).

Fig. 4.13b shows the relative control voltage versus active coupling gain. Once again it is observed that increasing the active coupling gain has little effect on the total control voltage requirement of the adaptive piezoelectric absorber. It is also evident that the adaptive absorber system requires significantly less control voltage than the concurrent APPN system while achieving better performance. Although the fixed APPN design is an active-passive configuration, it requires much more control effort because it must be designed for a much larger bandwidth, since it does not have the ability to adapt to the changing excitation frequency. While the optimal control weightings for the fixed APPN system could be modified so that the control effort requirement becomes smaller, this would only further degrade the performance as compared to the adaptive active-passive absorber.

Next the effects of a wider excitation bandwidth are investigated by selecting Δ = .15. The relative performance and control voltage are shown in Fig. 4.14. Fig. 4.14a shows that the percentage reduction in vibration response is even higher than that in the Δ = .05 case. The reason for this is that the APPN system is mainly effective for frequencies close to resonance, so its effectiveness is reduced as the excitation bandwidth is increased. The other interesting observation is that there is no optimal active coupling gain within the range of gains shown in Fig. 4.14a. The reason for this is that the wider bandwidth case is more like the non-resonant case presented in Section 4.6.1. The width of the absorber notch decreases significantly when the notch frequency is further away from the resonant frequency, thus more active coupling gain is needed to achieve the optimal notch width. Fig. 4.14b shows that the adaptive absorber once again requires more control voltage in the wider bandwidth case (compared to Fig. 4.13b), but it still requires less voltage than the fixed APPN system.

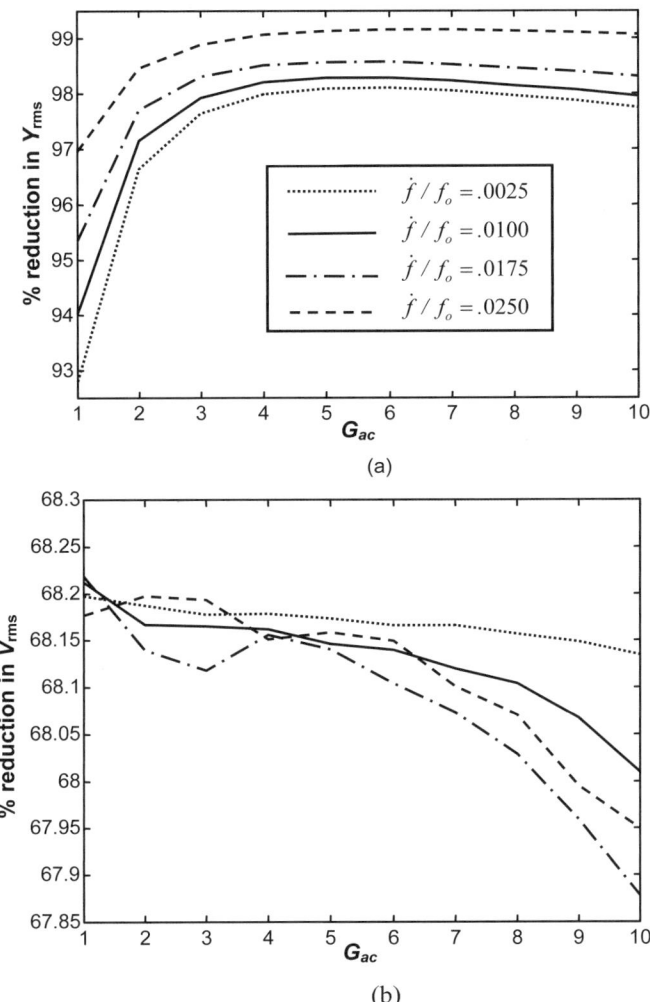

(a)

(b)

Fig. 4.13. Effects of active coupling gain ($\zeta_{cl} = .002$, $\Delta = .05$); (a) relative performance, (b) relative control voltage.

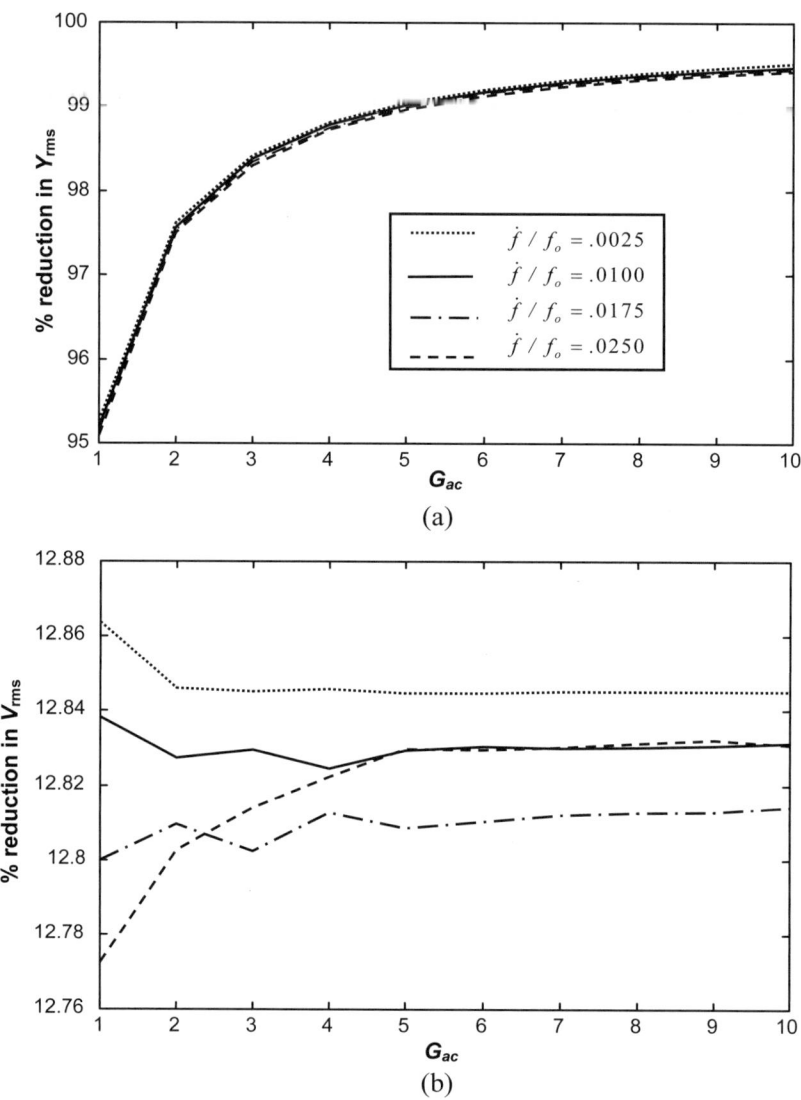

Fig. 4.14. Effects of active coupling gain $(\zeta_{cl} = .002, \Delta = .15)$; (a) relative performance, (b) relative control voltage.

4.7 EXTENSION TO MULTIPLE-FREQUENCY EXCITATION – BASIC IDEA

The previous sections have focused on discussions of structures that are under single variable frequency excitations. Hereafter we extend the investigation to mechanical systems subject to time-varying multiple-frequency excitations. We begin by considering multi-frequency circuit designs for passive (fixed) piezoelectric vibration absorbers that have been developed by previous researchers. One possible approach is to add resistance-inductance-capacitance (RLC) circuits in parallel with the original RL circuit, thus introducing additional resonances that can be tuned to other frequencies (Hollkamp, 1994). The shortcoming of this approach is that the dynamics of the individual circuit branches are strongly coupled, so a numerical optimization routine is needed to determine the circuit parameters that will achieve the desired tunings. For this reason, this approach is not applicable to an adaptive piezoelectric absorber. Another approach is to introduce resonant LC blocking filters to effectively decouple the individual circuit branches (Wu, 1998). A schematic for a two-mode passive piezoelectric absorber using this approach is given in Fig. 4.15. The purpose of such circuit, as originally proposed, is to provide modal damping around the resonant frequencies ω_1 and ω_2, where it is assumed that $\omega_1 < \omega_2$. The parallel combination of L_{21} and C_{21} has a resonant frequency, which is tuned to ω_1. The subscript 21 denotes the blocking filter in branch 2 that is tuned to ω_1. The decoupling effect of this blocking filter can be seen in the electrical impedances of the circuit branches, as shown in Fig. 4.16. Note that the impedance of branch 1 is higher than that of branch 2 (without the blocking filter) because the lower frequency ω_1 requires a larger inductance L_1. For frequencies near ω_1, most of the current from the piezoelectric transducer would flow into the second branch, causing the first branch to be ineffective. The purpose of the blocking filter is to increase the impedance of the second circuit branch for frequencies near ω_1. At these frequencies, the second branch acts like an open circuit due to the large impedance of the blocking filter, which causes the first branch to function like a single-frequency absorber. The blocking filter is, in general, not necessary for the second branch of the circuit to function because the impedance of the higher frequency branch is already significantly lower than that of the first branch (Wu, 1998). Ideally, we would like the blocking filter to have very little effect on the impedance of the second branch for frequencies near ω_2. Although the impedance of the blocking filter must be sufficiently large to provide the decoupling effect that enables the first branch to function, a large blocking impedance can have a significant effect on the second circuit branch as well. For example, in Fig. 4.16 it can be seen that at the frequency ω_2 the impedance of the second branch with blocking filter is slightly lower than that of the original circuit consisting only of R_2 and L_2. Physically, this means that the blocking filter impedance is partially canceling the impedance of the inductance L_2, which can decrease the effectiveness of the second branch. The effects of the blocking filter impedance on the performance of the multi-frequency absorber will be further considered in later sections.

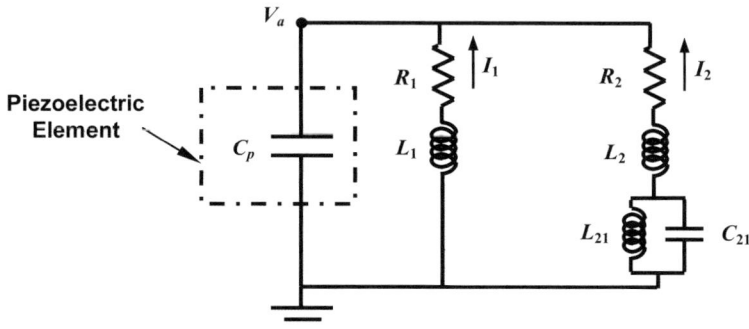

Fig. 4.15. Two-mode passive piezoelectric absorber.

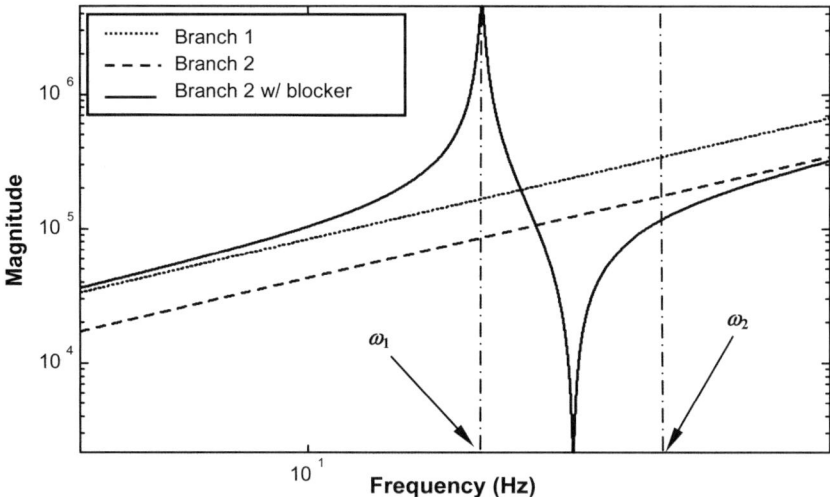

Fig. 4.16. Electrical impedances of circuit branches. ($C_{21} = C_p^s$, $\omega_1 = 20$ Hz, $\omega_2 = 40$ Hz, $\zeta_i = .02$, $\zeta_{21} = .0005$).

Using the configuration shown in Fig. 4.15 as a starting point, the concept is modified so that it can be applied to an adaptive active-passive piezoelectric absorber design. First, we would like to be able to independently vary the tuning of each circuit branch. This requires the use of separate control inputs for each circuit branch. Secondly, we cannot use a passive LC blocking filter if the circuit natural frequencies are changing. For the first branch to be effectively decoupled from the second, the resonant frequency of the blocking filter must always be tuned to ω_1. This problem can be overcome by implementing an adaptive blocking filter. Thus, the circuit design for the active-passive multi-frequency piezoelectric absorber reduces to the configuration shown in Fig. 4.17.

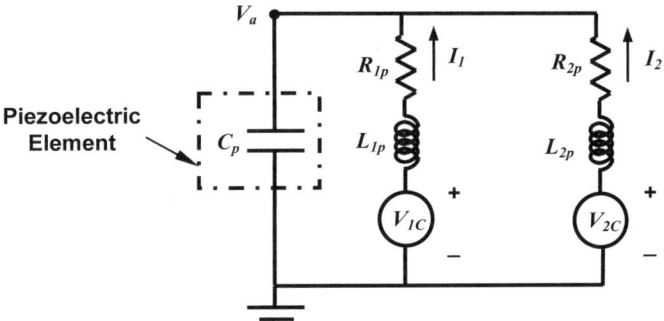

Fig. 4.17. Active-passive multi-frequency piezoelectric absorber configuration.

The passive inductances L_{1p} and L_{2p} are used to tune the circuit to be resonant at the nominal or steady-state excitation frequencies. We begin designing the control law for the multi-frequency adaptive absorber by incorporating the same three components as the single-frequency case. The first part of the control law is designed to imitate a variable inductor so that the absorber can be adaptively tuned to the correct frequency. An active negative resistance action is then used to reduce the effective resistances in the absorber circuit. Finally, the effective coupling in the circuit equation is enhanced by using the third part of the control law, which is the active coupling enhancement action. This action significantly improves the performance and robustness of the absorber.

The control law for the first circuit branch, which is shown in Eq. (4.29), is actually identical to that of the single-frequency piezoelectric absorber, where L_{1a} is the active variable inductance, R_{1a} is the negative resistance, and G_{ac} is the active coupling gain. In addition to these three elements, the control input for the second branch must also provide the voltage for the adaptive blocking filter. To calculate this voltage, the impedance of the blocking filter (in the Laplace domain) is first expressed as shown in Eq. (4.30). The damping term ζ_{21} has been added to the original LC design to improve the stability of the blocking filter. The resulting control voltage for the second branch of the circuit is given in Eq. (4.31).

$$V_{1C} = -L_{1a}(t)\cdot\dot{I}_1 + R_{1a}I_1 - (G_{ac}-1)\mathbf{K}_C^T\mathbf{q} \tag{4.29}$$

$$Z_{21}(s) = \frac{-V_{21}}{I_2} = \frac{\beta_{21}}{C_p^S}\frac{s+2\zeta_{21}\omega_1}{s^2+2\zeta_{21}\omega_1 s+\omega_1^2} \quad \text{where} \quad \beta_{21} = \frac{C_p^S}{C_{21}} \tag{4.30}$$

$$V_{2C} = -L_{2a}(t)\cdot\dot{I}_2 + R_{2a}I_2 - (G_{ac}-1)\mathbf{K}_C^T\mathbf{q} + V_{21} \tag{4.31}$$

4.8 OPTIMAL TUNING FOR MULTI-FREQUENCY EXCITATION

To ensure that the active inductance is properly tuned, an expression for the optimal tuning on a general MDOF structure is derived. The tuning ratio δ_i is defined in Eq. (4.32), where ω_{ia} is the ith circuit frequency and ω_{ie} is the estimate of the ith excitation frequency. The frequency estimates are obtained on-line using an algorithm based on the recursive least squares method (Handel and Tichavsky 1994).

$$\delta_i = \frac{\omega_{ia}}{\omega_{ie}} \quad \text{where} \quad \omega_{ia} = \frac{1}{\sqrt{(L_{ip}+L_{ia})C_p^S}} \tag{4.32}$$

To begin the derivation of the optimal tuning law, the system model is first transformed into modal space using the transformation $\mathbf{q} = \mathbf{U}\eta$, where $\mathbf{U}$ is a matrix containing the eigenvectors of the structure model. The displacement of the structure at a desired point is chosen as the objective function, which is then expressed as a weighted sum of the modal responses, as shown in Eq. (4.33), where $\mathbf{W}$ is the modal weighting vector. The optimal tuning ratio for the single-frequency absorber is given in terms of the system modal parameters by Eq. (4.34), where ω_n is a vector of the modal frequencies of the structure. The ~ notation denotes the system parameters after the modal transformation, and the subscripts i and j denote the ith and jth elements of a vector.

$$w_d(t) = \mathbf{W}^T \mathbf{\eta}(t) \tag{4.33}$$

$$\delta_{opt} = \left[1 + G_{ac} C_p^S \frac{\sum_{j=1}^{n} W_j M_j \left(a_j - \tilde{F}_j b \right)}{\sum_{j=1}^{n} W_j M_j \tilde{F}_j} \right]^{-\frac{1}{2}} \tag{4.34}$$

where

$$M_j = \frac{1}{\omega_{n_j}^2 - \omega_e^2}, \quad a_j = \tilde{K}_{c_j} \sum_{i=1}^{n} \frac{\tilde{K}_{c_i} \tilde{F}_i}{\omega_{n_i}^2 - \omega_e^2}, \quad b = \sum_{i=1}^{n} \frac{\tilde{K}_{c_i}^2}{\omega_{n_i}^2 - \omega_e^2}$$

The key assumptions in this derivation are that the excitation frequency is changing relatively slowly compared to the system dynamics (*i.e.*, quasi steady-state) and that the structural and circuit damping are negligible.

The main challenge in the design of the multi-frequency absorber is how to determine the optimal tunings for each branch of the circuit. As in the single frequency case, we would like to find an expression for the optimal tuning ratio in terms of the modal parameters of the system. Recall that the blocking filter configuration was chosen to facilitate the determination of the optimal circuit tunings. While the decoupling effect of the blocking filter is indeed helpful in this respect, the dynamics of the blocking filter can also cause the optimal circuit tunings for the multi-frequency absorber to be considerably different from those of the corresponding single frequency absorbers.

We begin by considering the optimal tuning of the first circuit branch for the excitation frequency ω_1. Since the blocking filter 21 is resonant at this frequency, the second branch acts like an open circuit because of the large impedance. For this reason, the first branch can be treated as a single frequency absorber, and Eq. (4.34) can be used to solve for the optimal tuning ratio $(\delta_1)_{opt}$. Next we consider the optimal tuning of the second branch for the given excitation frequencies ω_1 and ω_2. There are two reasons that the optimal branch inductance L_2 is dependent on ω_1. First, the blocking filter contains a capacitance as well as a variable inductance that depends on ω_1, which changes the effective inductance in the second branch of the circuit. Secondly, the first circuit branch and the blocking filter both add additional dynamics to the system. These additional modes can have a significant effect on the optimal tuning ratio δ_2, just like higher-order structural modes. These problems can be solved by lumping the known circuitry dynamics with the structural model, as shown in Eqs. (4.35) and (4.36). This augmented system model treats the states Q_1 and Q_{21}, which are defined as the integrals of the currents in the inductors L_1 and L_{21}, similarly as the structural coordinates. Note that the augmented system is in the same form as Eqs. (4.1) and (4.2), except that the damping terms are absent and the effective capacitance is different.

$$
\underbrace{\begin{bmatrix} \mathbf{M} & 0 & 0 \\ 0 & G_{ac}L_1 & 0 \\ 0 & 0 & \dfrac{G_{ac}\beta_{21}}{\omega_1^2 C_p^S} \end{bmatrix}}_{\mathbf{M_2}} \underbrace{\left\{ \begin{array}{c} \ddot{\mathbf{q}} \\ \dfrac{1}{G_{ac}}\ddot{Q}_1 \\ \dfrac{1}{G_{ac}}\ddot{Q}_{21} \end{array} \right\}}_{\ddot{\mathbf{q}}_2} + \underbrace{\begin{bmatrix} \mathbf{K^D} & G_{ac}\mathbf{K_C} & 0 \\ G_{ac}\mathbf{K_C^T} & \dfrac{G_{ac}}{C_p^S} & 0 \\ 0 & 0 & \dfrac{G_{ac}\beta_{21}}{C_p^S} \end{bmatrix}}_{\mathbf{K_2}} \underbrace{\left\{ \begin{array}{c} \mathbf{q} \\ \dfrac{1}{G_{ac}}Q_1 \\ \dfrac{1}{G_{ac}}Q_{21} \end{array} \right\}}_{\mathbf{q}_2}
$$

$$
+ \underbrace{\begin{bmatrix} \mathbf{K_C} \\ \dfrac{1}{C_p^S} \\ \dfrac{-\beta_{21}}{C_p^S} \end{bmatrix}}_{\mathbf{K_{C2}}} Q_2 = \underbrace{\begin{bmatrix} \hat{\mathbf{F}} \\ 0 \\ 0 \end{bmatrix}}_{\hat{\mathbf{F}}_2} f(t)
$$

$$
(4.35)
$$

$$
L_2 \ddot{Q}_2 + \underbrace{\frac{1+\beta_{21}}{C_p^S}}_{\frac{1}{C_{p2}^S}} Q_2 + G_{ac} \underbrace{\left[\mathbf{K}_C^T \quad \frac{1}{C_p^S} \quad \frac{-\beta_{21}}{C_p^S} \right]}_{\mathbf{K}_{C2}^T} \mathbf{q}_2 = 0 \tag{4.36}
$$

The previously derived optimal tuning law (Eq. (4.34)) can now be applied to this augmented system model. First, the augmented system model is transformed to modal space using the mass-normalized eigenvector matrix, as shown in Eq. (4.37). The modal frequencies of the augmented system ω_{n2} are obtained from the solution of the eigenvalue problem, and the remaining parameters needed in the optimal tuning law are given in Eq. (4.38).

$$
\mathbf{q}_2 = \mathbf{U}_2 \boldsymbol{\eta}_2 \quad \text{where} \quad \mathbf{K}_2 \mathbf{U}_2 = \mathbf{M}_2 \mathbf{U}_2 \left[\omega_{n2}^2 \right] \quad \text{and} \quad \mathbf{U}_2^T \mathbf{M}_2 \mathbf{U}_2 = \mathbf{I} \tag{4.37}
$$

$$
\tilde{\mathbf{K}}_c = \mathbf{U}_2^T \mathbf{K}_{C2} \quad , \quad \tilde{\mathbf{F}} = \mathbf{U}_2^T \hat{\mathbf{F}}_2 \quad , \quad C_{p2}^S = \frac{C_p^S}{1 + \beta_{21}} \tag{4.38}
$$

We now examine a typical optimal tuning curve predicted by the procedure specified above. First, we introduce the example system used to generate these results. The system used in this specific study is again a cantilevered beam with a collocated piezoelectric actuator and sensor pair attached near the root of the beam. The parameters of the example system are given in Table 4.1. In all subsequent results, the response of interest is the tip displacement of the beam.

In some situations, the optimal tuning law for the multi-frequency piezoelectric absorber gives results that are similar to the single frequency absorber. If the frequency ratio $r = \omega_2 / \omega_1$ is relatively constant, the optimal tuning ratios will be only slightly frequency dependent, much like in the single frequency absorber case. However, a varying frequency ratio can cause the optimal tuning ratio for the second branch to be strongly frequency dependent, as shown in Fig. 4.18. In the case shown here, both excitation frequencies are changing at a normalized rate of $|\dot{\omega} / \omega_o| = .02$ (the rate of change is 2% of the nominal excitation frequency per second) but ω_1 is increasing while ω_2 is decreasing. Because the excitation frequencies are moving in opposite directions, the frequency ratio r decreases from

3.33 to 1.2 in this case, which causes the optimal tuning ratio δ_2 to decrease sharply.

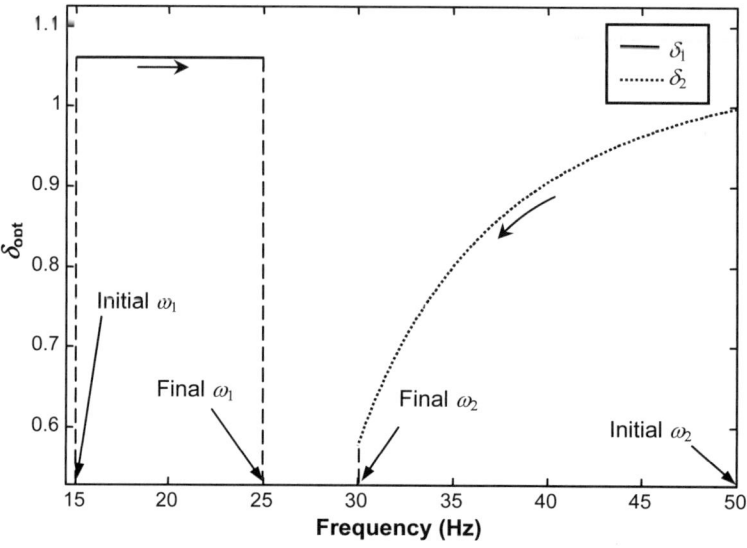

Fig. 4.18. Example optimal tuning curves: optimal δ_1 and optimal δ_2.
$\left(\omega_{1_o} = 20 \text{ Hz}, \ \omega_{2_o} = 40 \text{ Hz}, \ \Delta_1 = \Delta_2 = .25, \ \dot{\omega}_1 / \omega_{1_o} = .02, \ \dot{\omega}_2 / \omega_{2_o} = -.02, \ G_{ac} = 14 \right)$.

This phenomenon can be explained qualitatively by considering the impedance matching criterion, which states that the impedance of the multi-frequency circuit should be matched to that of a single frequency absorber (R_2 and L_2 only) at the frequency ω_2 (Wu, 1998). As the frequency ratio r approaches unity, the impedance of the blocking filter at ω_2 approaches negative infinity. To counteract this and allow the impedance matching condition to be satisfied, the inductance L_2 must become infinitely large, which causes the optimal tuning ratio δ_2 to approach zero. Although the impedance matching criterion can be used to explain some of the trends shown in Fig. 4.18, it neglects the effects of the additional electrical dynamics and higher-order structural modes on the optimal tuning ratios. The results shown in Fig. 4.18 are generated using the optimal tuning procedure outlined above, which gives more accurate results than the impedance matching criteria since it takes into account all the dynamics of the system.

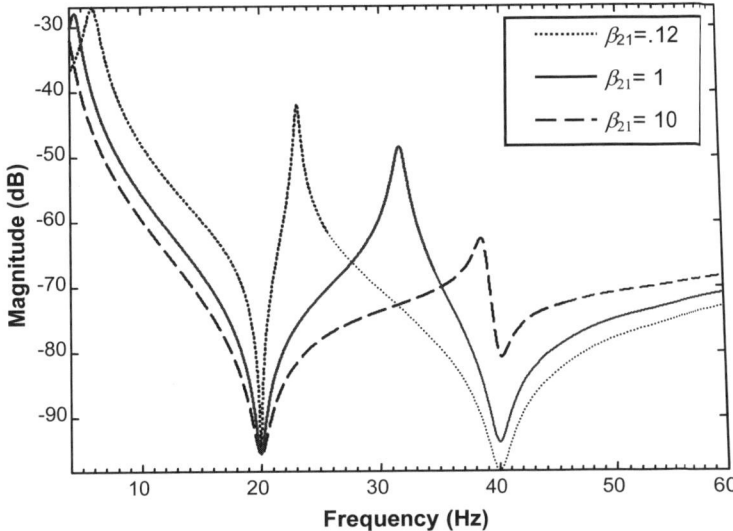

Fig. 4.19. Steady-state performance comparison. $\omega_{1_o} = 20$ Hz , $\omega_{2_o} = 40$ Hz , G_{ac} = 14, $\zeta_i = .02$, $\zeta_{21} = .0005$.

Note that the use of only one blocking filter provides decoupling of the two circuit branches only at the frequency ω_1. This partial decoupling is sufficient to allow the optimal tuning ratios to be solved sequentially, as described above. We originally assumed that the circuit branches are partially decoupled at the frequency ω_2 because the impedance of the lower frequency branch should be significantly higher due to the larger inductance L_1. This assumption is invalid when the frequencies are close together $(r \approx 1)$ because the impedances of the two branches will be similar in magnitude. However, because the dynamics of the first branch are included when solving for the optimal tuning ratio δ_2, the above procedure still provides very accurate tuning in these situations.

Next we examine the steady-state performance and consider the effect of β_{21}, which is essentially the gain parameter for the adaptive blocking filter. The steady-state frequency responses of the system for several different values of β_{21} are shown in Fig. 4.19. As discussed earlier, there is a compromise in the performance at the two excitation frequencies that depends on how large the impedance of the blocking filter is. Note that for the $\beta_{21} = 1$ case the response notches produced at ω_1 and ω_2 are approxi-

mately equal in size. For large values of β_{21} (larger blocking filter impedances), the decoupling action of the blocking filter is more effective so the
performance is increased slightly near ω_1. However, this excessively large
impedance of the blocking filter also degrades the performance near ω_2
considerably.   For smaller values of β_{21} (smaller blocking filter impedances), the decoupling action becomes less effective, which decreases the
performance near ω_1 but increases the performance at ω_2 slightly. If our
goal is to produce equally large notches at both excitation frequencies, then
$\beta_{21} = 1$ appears to be the best in this case. However, it will be shown that
the optimal selection of β_{21} for transient excitation suppression may be
considerably different than these steady-state results suggest.

These steady-state results also show that the active coupling gain causes
the performance to be quite robust with respect to frequency variations
even without using the adaptive tuning mechanism, $i.e.$, either excitation
frequency can vary as much as 15% from nominal (for the $\beta_{21} = 1$ case)
and the performance will still be better than the original system (without
the absorber). Without the active coupling gain, however, the performance
can be worse than the original system for even a 2% change from the
nominal frequency, unless the adaptive tuning is used to compensate for
this frequency change. A general observation is that when the excitation
frequency is changing even slightly, the active tuning action is extremely
helpful in achieving good performance. This is true even in the case where
a large active coupling gain is used to increase the robustness of the absorber, since the vibration attenuation provided by a fixed-frequency absorber decreases rapidly as the excitation frequency varies from nominal.

4.9 STABILITY ANALYSIS FOR MULTI-FREQUENCY DISTURBANCE REJECTION

In section 4.4, the stability criteria for the single frequency adaptive piezoelectric absorber are derived using Lyapunov's method. A similar analysis
is now performed for the multi-frequency absorber. The total system energy is selected to be the Lyapunov function candidate. The system mass,
damping and stiffness matrices are derived similarly to those shown in Eq.
(4.35), except that the state Q_{21} is replaced by the voltage across the blocking filter (V_{b2}), which eliminates the need for the coupling vector in the
system equation. In order to simplify the derivation of the Lyapunov stability criteria, it is desired that the system mass and stiffness matrices be
symmetric. Initially, these matrices are not symmetric, but they can be

$$2\zeta_1\omega_1\delta_1 + \frac{\dot{\omega}_1}{\omega_1} \geq 0 \quad \text{and} \quad 2\zeta_2\omega_2\delta_2 + \frac{\dot{\omega}_2}{\omega_2} \geq 0 \tag{4.44}$$

$$\frac{L_2}{\omega_1^2}\left(2\zeta_2\omega_2\delta_2 + \frac{\dot{\omega}_2}{\omega_2}\right)\left(2\zeta_b\omega_1 + \frac{\dot{\omega}_1}{\omega_1}\right) + \frac{\dot{\omega}_2}{\omega_2}\left(2\zeta_b\omega_1 - 2\zeta_2\omega_2\delta_2 - \frac{\dot{\omega}_2}{\omega_2}\right) \geq 0 \tag{4.45}$$

$$R_i = R_{ip} - R_{ia} = \frac{2\zeta_i}{C_p^S\omega_{ia}} \tag{4.46}$$

The closed-loop system can then be guaranteed to be stable if the conditions given in these equations, as well as Eq. (4.42), are satisfied. Note that the failure to meet these conditions does not imply instability, as Lyapunov's method only provides sufficient conditions for stability.

The first two conditions are similar to the previously derived stability conditions for the single frequency piezoelectric absorber in that they give a lower bound on the frequency rate of change. By examining these conditions, it can be seen that they will not be violated unless the damping in the absorber circuit is extremely small and the excitation frequencies are decreasing very rapidly. The third and final stability condition, which is given in Eq. (4.45), also gives bounds on the frequency rates of change that are dependent on both the circuit damping and the damping in the blocking filter. Except in the case where the damping terms are extremely small and the excitation frequencies are decreasing very rapidly, the first half of the condition shown in Eq. (4.45) will be positive. The sign of the second term is primarily dependent on whether the second excitation frequency is increasing or decreasing, and the relative magnitude of the terms containing the damping ratios for branch two and the blocking filter. This condition can then be used to safely select the damping quantities in the absorber design given the expected frequency change rates.

The above equations give the conditions under which the transformed system can be guaranteed to be stable. The definition of stability for nonautonomous systems states that for every $\varepsilon > 0$ and $t_o \geq 0$ there exists a $\bar{\delta}(\varepsilon, t_o) > 0$ such that $\|\bar{z}(t_o)\| < \bar{\delta}$ implies $\|\bar{z}(t)\| < \varepsilon$ for every $t \geq t_o$ (Kahlil, 1996). Since the coefficients of the linear coordinate transformation given in Eq. (4.39) are finite, we can always find a $\delta(\varepsilon, t_o) > 0$ such that $\|z(t_o)\| < \delta$ implies $\|z(t)\| < \varepsilon$ for every $t \geq t_o$. In other words, the only task necessary to prove stability of the original system from the above results is to choose a different bound on the initial conditions of the system. Note that the above Lyapunov stability analysis alone does not guarantee asymptotic stability, since the time-derivative of $V(\overline{z})$ is only negative semi-definite even when $\mathbf{C}_{eff}$ is positive definite. This is because $V(\overline{z}) = 0$ any time $\dot{\overline{z}} = \mathbf{0}$, even if $\overline{z} \neq \mathbf{0}$. However, it can be shown using La Salle's theorem that the above conditions do indeed guarantee asymptotic stability (Kahlil, 1996). Notice from Eq. (4.43) that the only way in which the system can maintain the $\dot{V}(\overline{z}) = 0$ condition is if $\dot{\overline{z}} = \ddot{\overline{z}} = \mathbf{0}$. From the system equation, Eq. (4.41), it can be seen that this is possible only when $\overline{z} = \mathbf{0}$. Thus for positive definite $\mathbf{C}_{eff}$, $\dot{V}(\overline{z})$ is negative definite everywhere, except at the origin $\overline{z} = \mathbf{0}$, so the system is asymptotically stable. Proving asymptotic stability of the transformed system also proves it for the original system because $\overline{z} = \mathbf{0}$ implies that $\mathbf{z} = \mathbf{0}$ as well (from Eq. (4.39)).

4.10 TRANSIENT PERFORMANCE OF MULTIPLE-FREQUENCY DISTURBANCE REJECTION

In this section, we evaluate the effectiveness of the multi-frequency adaptive piezoelectric absorber for suppressing harmonic excitations with time-varying frequencies. For the purpose of comparison, the filtered-x adaptive feedforward control law (Fuller et al., 1996) is again selected as a baseline. The excitation signal used in this section is a summation of two linear chirp signals, with nominal frequencies of $\omega_{1o} = 20$ Hz and $\omega_{2o} = 40$ Hz. The bandwidth of both chirp signals is fixed at $\Delta = .05$, which means that the excitation frequencies increase linearly from $.95\omega_{1o}$ to $1.05\omega_{1o}$ and from $.95\omega_{2o}$ to $1.05\omega_{2o}$. The frequency rates of change are varied over a range of $\dot{\omega}/\omega_o = .005$ to $.05$. For a given bandwidth and frequency rate of change, the required time span can be calculated. The performance index used is the root-mean-square (RMS) of the response at the point of interest. The control effort index is the RMS control power, which is calculated using the apparent power (the product of the RMS current and RMS voltage) of the source.

We begin by fixing the adaptation gain α of the filtered-x controller and optimizing the parameters of the adaptive blocking filter for the active-passive absorber. For simplicity, the closed-loop damping ratios of the made symmetric by applying the coordinate transformations shown in Eq. (4.39), where the overbar denotes the new (transformed) coordinates.

$$\mathbf{q} = \frac{1}{G_{ac}}\overline{\mathbf{q}}, \quad Q_1 = \overline{Q}_1, \quad Q_2 = \left(1+\beta_{21}\right)\overline{Q}_2 + \frac{C_p^S}{\beta_{21}L_2}\overline{V}_{b2}, \quad V_{b2} = \frac{1}{\omega_1^2}\overline{V}_{b2} \tag{4.39}$$

$$V(\overline{\mathbf{z}}) = \overline{\mathbf{z}}^T\mathbf{K}_{sys}\overline{\mathbf{z}} + \dot{\overline{\mathbf{z}}}^T\mathbf{M}_{sys}\dot{\overline{\mathbf{z}}} \tag{4.40}$$

$$\tag{4.41}$$

$$
\underbrace{\begin{bmatrix} \dfrac{1}{G_{ac}}\mathbf{M} & 0 & 0 & 0 \\[2mm] 0 & L_1 & 0 & 0 \\[2mm] 0 & 0 & \left(1+\beta_{21}\right)L_2 & \dfrac{C_p^S}{\beta_{21}} \\[2mm] 0 & 0 & \dfrac{C_p^S}{\beta_{21}} & \dfrac{1}{L_2}+\dfrac{1}{\omega_1^2} \end{bmatrix}}_{\mathbf{M}_{sys}}
\underbrace{\begin{Bmatrix} \ddot{\overline{\mathbf{q}}} \\[1mm] \ddot{\overline{Q}}_1 \\[1mm] \ddot{\overline{Q}}_2 \\[1mm] \ddot{\overline{V}}_{b2} \end{Bmatrix}}_{\ddot{\overline{\mathbf{z}}}}
$$

$$
+\underbrace{\begin{bmatrix} \dfrac{1}{G_{ac}}\mathbf{C} & 0 & 0 & 0 \\[2mm] 0 & R_1 & 0 & 0 \\[2mm] 0 & 0 & \left(1+\beta_{21}\right)R_2 & \dfrac{R_2 C_p^S}{L_2\beta_{21}} \\[2mm] 0 & 0 & \dfrac{2\zeta_{21}\omega_1}{C_p^S}\beta_{21}\left(1+\beta_{21}\right) & 2\zeta_{21}\omega_1\left(\dfrac{1}{L_2}+\dfrac{1}{\omega_1^2}\right) \end{bmatrix}}_{\mathbf{C}_{sys}}
\underbrace{\begin{Bmatrix} \dot{\overline{\mathbf{q}}} \\[1mm] \dot{\overline{Q}}_1 \\[1mm] \dot{\overline{Q}}_2 \\[1mm] \dot{\overline{V}}_{b2} \end{Bmatrix}}_{\dot{\overline{\mathbf{z}}}}
$$

$$
+\underbrace{\begin{bmatrix} \dfrac{1}{G_{ac}}\mathbf{K}^{\mathbf{D}} & K_C & K_C & 0 \\[2mm] K_C^T & C_p^S & C_p^S & 0 \\[2mm] K_C^T & C_p^S & \dfrac{1+\beta_{21}}{C_p^S} & 0 \\[2mm] 0 & 0 & 0 & 1 \end{bmatrix}}_{\mathbf{K}_{sys}}
\underbrace{\begin{Bmatrix} \overline{\mathbf{q}} \\[1mm] \overline{Q}_1 \\[1mm] \overline{Q}_2 \\[1mm] \overline{V}_{b2} \end{Bmatrix}}_{\overline{\mathbf{z}}} = \mathbf{0}
$$

The resulting system equation is shown as Eq. (4.41). The Lyapunov function candidate is expressed in terms of the transformed system model, as shown in Eq. (4.40). Next, the sufficient conditions under which these system mass and stiffness matrices are positive definite are derived, so that Eq. (4.40) is indeed a valid Lyapunov function. The condition that results from requiring $\mathbf{K}_{sys}$ to be positive definite is such that,

$$\left| \frac{1}{G_{ac}} \mathbf{K}^D - C_p^S \mathbf{K}_C \mathbf{K}_C^T \right| > 0 \tag{4.42}$$

This condition, which gives an upper bound on the active coupling gain, is identical to the one obtained in the stability analysis of the single-frequency piezoelectric absorber. It can be shown that this condition is a necessary condition for stability as well. It has also been shown that in the case of a SDOF system, the maximum active coupling gain can be expressed quite simply in terms of K_{ij}, the generalized coupling coefficient of the system.

Having found the conditions under which the Lyapunov function candidate is valid, we now consider its time derivative,

$$\dot{V}(\overline{z}) = -\dot{\overline{z}}^T \underbrace{\left(\mathbf{C}_{sys} - \frac{1}{2}\dot{\mathbf{M}}_{sys} \right)}_{\mathbf{C}_{eff}} \dot{\overline{z}} + \frac{1}{2}\overline{z}^T \dot{\mathbf{K}}_{sys} \overline{z} \tag{4.43}$$

With the coordinate transformation that was chosen above, the time-derivative of the system stiffness matrix $\mathbf{K}_{sys}$ is identically zero, which greatly simplifies the Lyapunov stability criteria. The time-derivative of $V(\overline{z})$ can then be said to be at least negative semi-definite if the matrix $\mathbf{C}_{eff}$ is positive semi-definite. The conditions under which this can be guaranteed are derived using Sylvester's criteria (Chen, 1984), and the results are shown in Eqs. (4.44) and (4.45). Note that the effective resistances have been expressed in terms of the closed-loop damping ratios ζ_i, as shown in Eq. (4.46).

$$2\zeta_1\omega_1\delta_1 + \frac{\dot{\omega}_1}{\omega_1} \geq 0 \quad \text{and} \quad 2\zeta_2\omega_2\delta_2 + \frac{\dot{\omega}_2}{\omega_2} \geq 0 \tag{4.44}$$

$$\frac{L_2}{\omega_1^2}\left(2\zeta_2\omega_2\delta_2 + \frac{\dot{\omega}_2}{\omega_2}\right)\left(2\zeta_b\omega_1 + \frac{\dot{\omega}_1}{\omega_1}\right) + \frac{\dot{\omega}_2}{\omega_2}\left(2\zeta_b\omega_1 - 2\zeta_2\omega_2\delta_2 - \frac{\dot{\omega}_2}{\omega_2}\right) \geq 0 \tag{4.45}$$

$$R_i = R_{ip} - R_{ia} = \frac{2\zeta_i}{C_p^S\omega_{ia}} \tag{4.46}$$

The closed-loop system can then be guaranteed to be stable if the conditions given in these equations, as well as Eq. (4.42), are satisfied. Note that the failure to meet these conditions does not imply instability, as Lyapunov's method only provides sufficient conditions for stability.

The first two conditions are similar to the previously derived stability conditions for the single frequency piezoelectric absorber in that they give a lower bound on the frequency rate of change. By examining these conditions, it can be seen that they will not be violated unless the damping in the absorber circuit is extremely small and the excitation frequencies are decreasing very rapidly. The third and final stability condition, which is given in Eq. (4.45), also gives bounds on the frequency rates of change that are dependent on both the circuit damping and the damping in the blocking filter. Except in the case where the damping terms are extremely small and the excitation frequencies are decreasing very rapidly, the first half of the condition shown in Eq. (4.45) will be positive. The sign of the second term is primarily dependent on whether the second excitation frequency is increasing or decreasing, and the relative magnitude of the terms containing the damping ratios for branch two and the blocking filter. This condition can then be used to safely select the damping quantities in the absorber design given the expected frequency change rates.

The above equations give the conditions under which the transformed system can be guaranteed to be stable. The definition of stability for nonautonomous systems states that for every $\varepsilon > 0$ and $t_o \geq 0$ there exists a $\bar{\delta}(\varepsilon, t_o) > 0$ such that $\|\bar{z}(t_o)\| < \bar{\delta}$ implies $\|\bar{z}(t)\| < \varepsilon$ for every $t \geq t_o$ (Kahlil, 1996). Since the coefficients of the linear coordinate transformation given in Eq. (4.39) are finite, we can always find a $\delta(\varepsilon, t_o) > 0$ such that $\|z(t_o)\| < \delta$ implies $\|z(t)\| < \varepsilon$ for every $t \geq t_o$. In other words, the only task necessary to prove stability of the original system from the above results is to choose a different bound on the initial conditions of the system. Note that the above Lyapunov stability analysis alone does not guarantee asymptotic stability, since the time-derivative of $V(\bar{z})$ is only negative semi-definite even when $\mathbf{C}_{eff}$ is positive definite. This is because $V(\bar{z}) = 0$ any time $\dot{\bar{z}} = \mathbf{0}$, even if $\bar{z} \neq \mathbf{0}$. However, it can be shown using La Salle's theorem that the above conditions do indeed guarantee asymptotic stability (Kahlil, 1996). Notice from Eq. (4.43) that the only way in which the system can maintain the $\dot{V}(\bar{z}) = 0$ condition is if $\dot{\bar{z}} = \ddot{\bar{z}} = \mathbf{0}$. From the system equation, Eq. (4.41), it can be seen that this is possible only when $\bar{z} = \mathbf{0}$. Thus for positive definite $\mathbf{C}_{eff}$, $\dot{V}(\bar{z})$ is negative definite everywhere, except at the origin $\bar{z} = \mathbf{0}$, so the system is asymptotically stable. Proving asymptotic stability of the transformed system also proves it for the original system because $\bar{z} = \mathbf{0}$ implies that $\mathbf{z} = \mathbf{0}$ as well (from Eq. (4.39)).

4.10 TRANSIENT PERFORMANCE OF MULTIPLE-FREQUENCY DISTURBANCE REJECTION

In this section, we evaluate the effectiveness of the multi-frequency adaptive piezoelectric absorber for suppressing harmonic excitations with time-varying frequencies. For the purpose of comparison, the filtered-x adaptive feedforward control law (Fuller et $al.$, 1996) is again selected as a baseline. The excitation signal used in this section is a summation of two linear chirp signals, with nominal frequencies of ω_{1o} = 20 Hz and ω_{2o} = 40 Hz. The bandwidth of both chirp signals is fixed at Δ = .05, which means that the excitation frequencies increase linearly from $.95\omega_{1o}$ to $1.05\omega_{1o}$ and from $.95\omega_{2o}$ to $1.05\omega_{2o}$. The frequency rates of change are varied over a range of $\dot{\omega}/\omega_{o}$ = .005 to .05. For a given bandwidth and frequency rate of change, the required time span can be calculated. The performance index used is the root-mean-square (RMS) of the response at the point of interest. The control effort index is the RMS control power, which is calculated using the apparent power (the product of the RMS current and RMS voltage) of the source.

We begin by fixing the adaptation gain α of the filtered-x controller and optimizing the parameters of the adaptive blocking filter for the active-passive absorber. For simplicity, the closed-loop damping ratios of the two circuit branches, represented by ζ_i, are selected to be equal. Fig. 4.20 shows the RMS response of the system with the active-passive absorber versus the parameters β_{21} and ζ_{21} for a moderately fast changing excitation case. The optimal blocking filter parameters for this case are $\beta_{21} = .12$ and $\zeta_{21} = .0005$. If we examine the control effort instead of the performance we find that these parameters also provide a near-minimum power requirement. Investigating faster and slower excitation cases shows that the optimal blocking filter parameters are also quite insensitive to the frequency rate of change. It is interesting to note that the optimal β_{21} for the transient excitation considered here is significantly lower than the β_{21} that appeared to be optimal from the steady-state response shown in Fig. 4.19. However, the steady-state results did not take into account the transient dynamics of the blocking filter, which is a lightly damped resonant system. It is reasonable to expect that these transient dynamics could have a negative effect on the transient response of the system, and a larger value of β_{21} would amplify the blocking filter response at all frequencies.

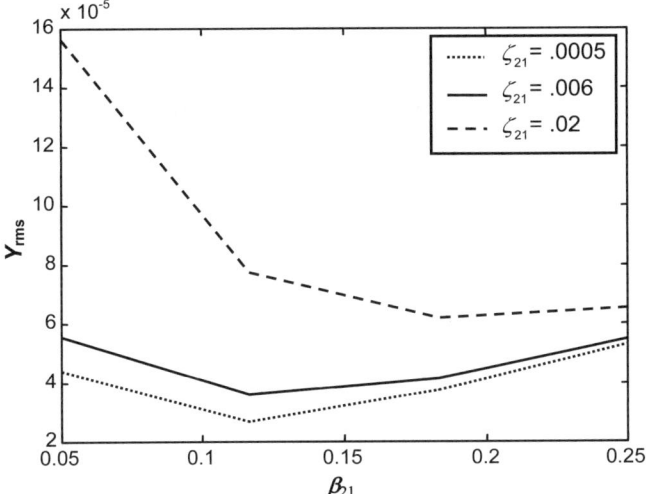

Fig. 4.20. Performance effects of β_{21} and ζ_{21}. ($\dot{\omega} / \omega_o = .0275$, $\zeta_i = .02$, $G_{ac} = 14$).

Next we examine the effects of the absorber damping ratio and the frequency rate of change on the performance of the multi-frequency piezoelectric absorber, as shown in Fig. 4.21. This figure shows that the performance of the multi-frequency absorber is slightly better for slowly changing excitations, although the performance loss due to more rapidly changing excitations is not very significant. The optimal absorber damping ratios obtained from this plot range from $\zeta_i = .017$ for the slowest case to $\zeta_i = .026$ for the fastest case. The parametric study on the effects of ζ_i also shows that the control power requirement of the active-passive multi-frequency absorber is affected very little by this parameter. For this reason, we select the optimal parameters for the multi-frequency absorber as the ones that provide the best performance.

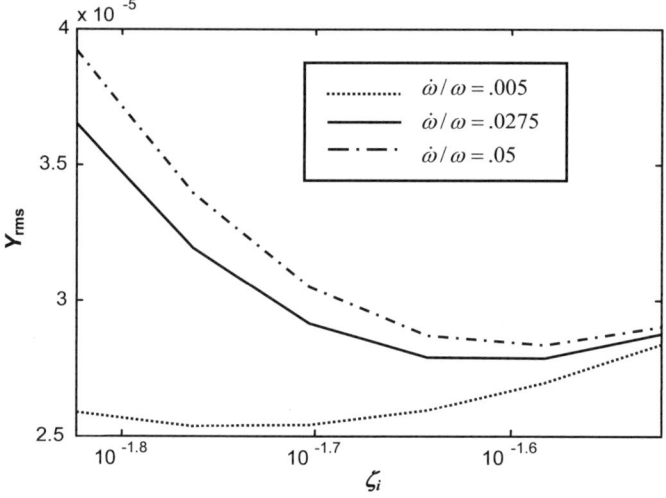

Fig. 4.21. Performance effects of ζ_i and $\dot{\omega}$. ($\zeta_{21} = .0005$, $\beta_{21} = .12$, $G_{ac} = 14$).

Finally, we examine the relative performance and control effort of the optimized multi-frequency absorber compared to the multi-frequency filtered-x baseline system. The performance index is selected to be the percentage reduction in structural RMS response of the active-passive absorber system relative to the filtered-x baseline. Similarly, the percentage reduction in RMS control power is selected as an index of the control effort requirement. Since both the performance and the control effort of the baseline system are strongly dependent on the adaptation gain α, we plot the results versus this gain, as shown in Fig. 4.22. This figure shows a pair of lines representing the percentage reduction in RMS response and RMS control power for each of the three excitation speeds considered. In each case, a range of α exists where the active-passive absorber outperforms the filtered-x baseline while requiring less control power (both percentage reductions indices are positive). The reverse situation, where the filtered-x baseline has better performance and less control power requirement (both indices are negative), does not occur. This is due to the compromise between the performance and control power requirement of the filtered-x controller. As mentioned previously, this trade-off in the optimal design of the multi-frequency absorber is almost insignificant. This is most likely due to the active-passive nature of the piezoelectric absorber, which is inherently more energy efficient than a purely active system.

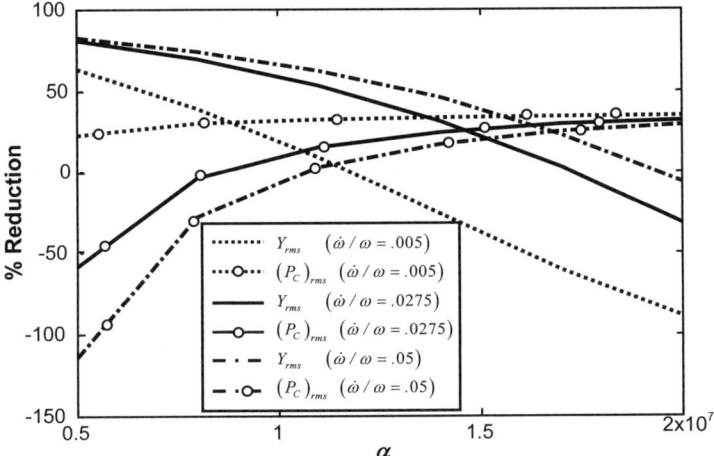

Fig. 4.22. Relative performance (Y_{rms}) and control power ((P_c)$_{rms}$) comparison. ($\zeta_i = \zeta_{opt}$, ζ_{21} =.0005, β_{21} = .12, G_{ac} = 14).

Fig. 4.22 also shows that at high filtered-x gains the relative performance of the active-passive absorber decreases significantly while the relative control power requirement decreases only slightly. From these results, it would appear that the filtered-x controller is a better choice when high performance is required. However, evaluating only the relative performance can be somewhat misleading in this case. Compared to the short-circuit system, the active-passive absorber system shown above produces a 92% reduction in the structural RMS response. At the highest gains shown in Fig. 4.22, the filtered-x controller produces a 92-96% reduction in the RMS response compared to the short-circuit system, with the largest reductions corresponding to the slowly changing excitation cases. Because both methods are so effective, the absolute performance gain produced by the larger filtered-x gains is actually quite insignificant. Thus, it can be said that the active-passive absorber design can achieve a better balance between performance and efficiency than the filtered-x method.

4.11 FURTHER EXTENSION TO THREE OR MORE EXCITATION FREQUENCIES

To this point, we have restricted the analysis of the adaptive multi-frequency piezoelectric absorber to the dual-frequency example for simplicity. In practice, the methods and analysis presented in the previous sections can be extended to higher order cases in a straightforward manner. Consider the circuit diagram shown in Fig. 4.17. For the case where n excitation frequencies are to be suppressed by the absorber, the active-passive circuit must contain n branches, each containing a passive inductance and a control input. In addition, the ith circuit branch will include i-1 active blocking filters that are tuned to the excitation frequencies below ω_i. For example, the third branch in a triple frequency absorber configuration will contain two active blocking filters of the form shown in Eq. (4.30) with gains β_{31} and β_{32} and resonant frequencies ω_1 and ω_2. The general form of the control input for the ith branch is given in Eq.(4.47).

$$V_{iC} = -L_{ia}(t) \cdot \dot{I}_i + R_{ia} I_i - (G_{ac} - 1) \mathbf{K}_C^T \mathbf{q} + V_{ib} \tag{4.47}$$

$$\text{where, } V_{ib}(s) = -\left(\sum_{j=1}^{i-1} \frac{\beta_{ij}}{C_p^S} \frac{s + 2\zeta_{ij}\omega_j}{s^2 + 2\zeta_{ij}\omega_j s + \omega_j^2} \right) I_i(s)$$

The optimal tuning derivation for higher order piezoelectric absorbers can also be performed in a manner analogous to that shown in this chapter. Once again, the optimal tuning ratios for each branch are solved successively starting with the lowest frequency branch and the known circuit dynamics are lumped together with the structure model. For branch i, the known circuit dynamics include the i-1 previous branch circuit equations as well as the i-1 active blocking filter equations contained in branch i. In principle, the Lyapunov stability analysis could also be extended to higher-order absorber designs in a similar manner.

4.12 CONCLUDING REMARKS

This chapter presents an extensive study of utilizing active-passive piezoelectric transducer circuitry for adaptive disturbance rejection under varying excitation frequencies. Such an approach is capable of much higher performance than passive or semi-active piezoelectric absorbers, and is also simple to implement.

The study first focuses on systems with a single excitation frequency source. An active inductance is used to tune the piezoelectric absorber online via feedback control. The performance and robustness of the absorber are further enhanced by using an active negative resistance to reduce the absorber damping and active coupling feedback to enhance the electro-mechanical coupling. An optimal tuning law for a MDOF system is derived using a quasi-steady-state assumption. Conditions under which the stability of the closed-loop system can be guaranteed are derived. Experimental investigations are conducted for both non-resonant and near-resonant excitations. In both cases, the vibration reduction performance of the proposed active-passive absorber is quite significant compared to the passive baseline systems, even when the excitation frequency is changing rapidly. The parametric studies illustrate the effects of the absorber parameters and excitation characteristics on the performance of the active-passive adaptive absorber. Through simulation studies, it is also shown that the proposed adaptive circuitry design can outperform the filtered-x active control method and an optimized APPN approach while requiring less control effort.

The investigation is then expanded to control systems with multiple excitation frequencies. A multi-frequency adaptive piezoelectric vibration absorber design is presented. It uses a combination of a simple passive circuit along with an adaptive active control law. The active control law enhances the passive piezoelectric absorber in four ways: it adds an adaptive tuning ability by imitating a variable inductance; it reduces the effective resistance in the absorber circuit to increase performance; it increases the effective coupling of the system to increase robustness and performance; and it effectively decouples the dynamics of the individual circuit branches. This decoupling action allows the tunings of the multi-frequency absorber to be calculated using an analytical optimal tuning law. The stability criteria of the multi-frequency piezoelectric absorber device are also derived. The proposed design is shown to be effective for simultaneously suppressing two harmonic excitations with time-varying frequencies. For the purpose of comparison, the adaptive feedforward filtered-x algorithm is used as a baseline. It is shown that the multi-frequency adaptive piezoelectric absorber can achieve better performance while requiring less control power, compared to the filtered-x algorithm. The design and analysis presented can be extended in a straight-forward manner to cases with three or more excitation frequencies.

5 Nonlinear High-Precision Robust Control with Hysteresis Compensation

As stated in the preceding chapters, piezoelectric actuators have well-know advantages that include high bandwidth, compactness, and easy integration with the host structures and with the circuitry elements. While high precision is often claimed as one of the advantages as well, the actual performance of piezoelectric actuators in this regard is clearly dependent upon the modeling accuracy and the control algorithms. Most of the studies related to piezoelectric actuators have been based upon a linear strain-field constitutive relation assumption (IEEE, 1988). The presence of nonlinearities in the response of piezoelectric materials, however, has been well documented since the early description of ferroelectrics (Devonshire, 1954). The physics involved in piezoelectric theory may be regarded as a coupling between the Maxwell's equations of electro-magnetism and the elastic stress equations of motion. The coupling takes place through the piezoelectric constitutive equations. Normally, the electrical field (against the poling direction) applied to the piezoelectric actuator should be kept below the coercive field to avoid depoling. Experiments have revealed that even in cases where the applied fields are not sufficient to completely re-orient the remnant polarization in the entire actuator, a small number of domains can still be switched (Sirohi and Chopra, 2000). Thus both the material states and the electro-mechanical coupling are changed, giving rise to the nonlinear hysteretic strain-field behavior.

The hysteresis phenomenon obviously affects the piezoelectric control performance especially for applications that require high precision, and there has been significant interest in modeling such behavior (King *et al.*, 1990). In general, the hysteresis can be characterized by using the Preisach model (Mayergoyz, 1991; Ge and Jouaneh, 1995 and 1997) or its variant, the Maxwell resistive capacitor (MRC) model (Goldfarb and Celanovic, 1997a and 1997b; Lee and Royston, 2000). The Preisach model consists of a weighted summation of an infinite number of the simplest hysteresis operator (Fig. 5.1), each representing a rectangular loop in the input-output diagram. To avoid the amplification of measurement error in the classical Preisach model that involves the differential operation, Ge and Jouaneh (1995 and 1997) proposed a new calculation scheme, which is essentially a two-dimensional interpolation MRC model is composed of a number of elasto-slide elements connected in parallel (Fig. 5.2). Each of these elements is subject to a Coulomb friction force. This model can represent the hysteresis relation between the force and the displacement. Both the Preisach model and the MRC model, albeit complicated, can be used to accurately describe the piezoelectric hysteresis. Researchers have also attempted to model the piezoelectric hysteresis behavior using polynomial approximation (Chonan *et al.*, 1996) and time delay process (Tsai and Chen, 2003).

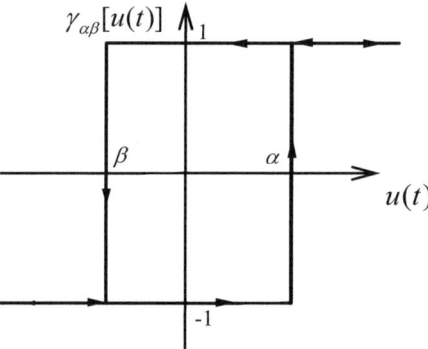

Fig. 5.1. The simplest hysteresis operator $\gamma_{\alpha\beta}$ is a rectangular loop with an up switch at α and a down switch at β.

Following the modeling studies, several control strategies have been proposed to handle the hysteresis nonlinearity involved in piezoelectric actuators, which include hysteresis cancellation using inverse model (Ge and Jouaneh, 1996; Main and Garcia, 1997; Croft and Devasia, 1998; Kung and Fung, 2004), feedback linearization (Choi *et al.*, 2002), and Smith predictor (Tsai and Chen, 2003). Tang and Wang (2000) explored the piezoelectric robust control using a sliding mode theory, where linear constitutive relation is used as the baseline and all nonlinearities are considered as uncertainties. A unique feature of their idea is to introduce actuator dynamics to the system by connecting resistance-inductance (*RL*) elements to the piezoelectric transducer, *i.e.*, to form a piezoelectric circuitry, which yields the possibility of direct compensation of piezoelectric nonlinearity. Xue and Tang (2006) further advanced such idea by explicitly incorporating the hysteresis model into the control design.

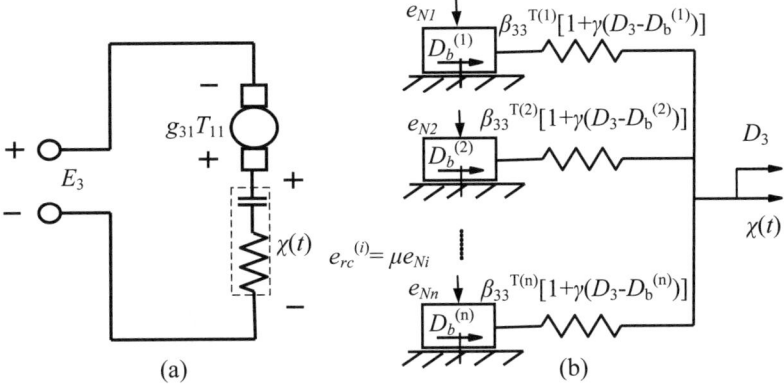

Fig. 5.2. Schematic representation of (MRC) Hysteresis Model: (a) Equivalent electric circuit, (b) Equivalent mechanical analogy.

5.1 PROBLEM STATEMENT AND OBJECTIVE

Most of the precision control studies concerning the piezoelectric hysteresis resort to the inverse modeling approach. In these approaches, however, several factors may deteriorate the control performance. For example, very complex coupling effects exist among the stress, strain, electrical field, and electrical displacement of a piezoelectric actuator. Thus the nonlinear hysteretic strain-field relation of the actuator actually also depends on the electrical displacement/charge and stress which are typically treated as internal variables in the aforementioned control designs. The estimation of these internal variables and hence the characterization of hysteresis become extremely complicated and even unreliable in practical applications when the actuator is bonded to a host structure and thus also undergoes deformation. Moreover, the forward physical process (hysteresis) is usually an integral process, and in inverse modeling/cancellation methods, differential operations are involved in the hysteresis inverse calculation. As a result, the inverse model obtained may not be reliable when the hysteresis measurement data contains noise.

In this chapter, we present a direct method for piezoelectric hysteresis compensation, which is built upon the piezoelectric circuitry concept. In this approach, a resistance-inductance (RL) shunt circuit is connected in series with the piezoelectric actuator to form an actuator network (see Fig. 2.6a or Fig. 3.1). In addition to the increased passive damping and active control authority that have been analyzed comprehensively in Chapter 2, the main advantage of this actuator network for high precision control development is that the charge and/or current in the piezoelectric actuator now become independent state variables that can be directly measured and fed back. With the introduction of the *RL* shunt circuit, the controlled system now consists of two coupled second-order dynamic subsystems, the mechanical subsystem and the electrical subsystem, whereas the nonlinearity only appears explicitly as a function of the electrical charge. This nonlinear dynamic system can be readily cast into the standard state-space form for nonlinear control design. Throughout this chapter, we will highlight the characteristics of the new approach by extensively discussing the following issues:

(a) What is the fundamental difference between the direct hysteresis compensation by means of piezoelectric circuitry and the inverse hysteresis cancellation approach?

(b) For a nonlinear dynamic system, the sliding mode theory is very appealing for controller design. Can we overcome some drawbacks (*e.g.*, chattering due to non-smooth switching) of the currently available sliding mode control methods, to achieve high-precision performance?

An important feature of the approach outlined in this chapter is an improved, integral continuous sliding mode control (ICSMC) algorithm that can avoid the control action switching. Detailed analysis and case studies in this chapter will demonstrate that this approach can lead to improved precision for both tracking control and vibration attenuation, enhanced control robustness, and smoother control action.

5.2 PIEZOELECTRIC HYSTERESIS AND SYSTEM MODELING

Piezoelectric materials are ferroelectric, and their displacement responses under an applied electrical field are intrinsically nonlinear. The physical explanation of this phenomenon provided by Chen and Montgomery (1980) shows that the effective number of dipoles aligned in the direction of the applied field changes over time as domains switch under the action of an external electric field. This essentially gives rise to the hysteresis behavior (Fig. 5.3).

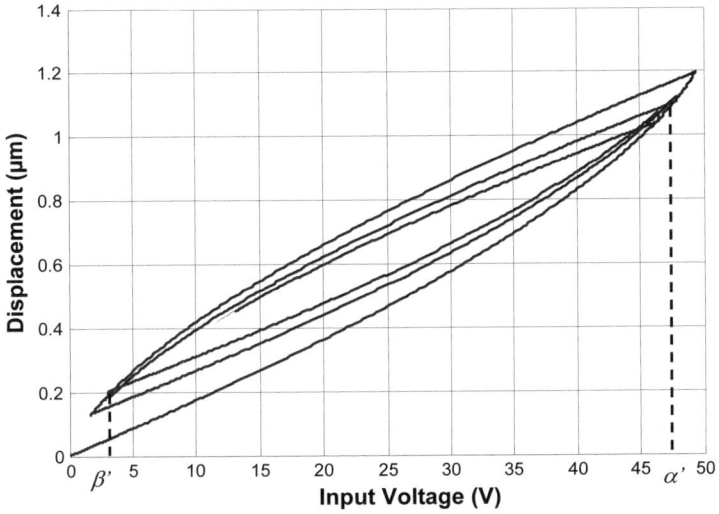

Fig. 5.3. Hysteresis loop of a piezoelectric actuator.

5.2.1 Piezoelectric Hysteresis Characterization

The Preisach model basically is the continuous analog of a finite parallel connection of relays. Generally, two aspects of nonlinearities are involved in hysteretic behavior (Mayergoyz, 1991): one is the hysteretic nonlinearity with local memory where the future output depends only on the future input; the other is the hysteretic nonlinearity with non-local memory where the future output depends not only upon the current output and future input but also on the past history of input switching values (Ge and Jouaneh, 1995). When used to describe the piezoelectric hysteresis, the classical Preisach model (CPM) can be written as (Ge and Jouaneh, 1995 and 1997),

$$\chi(t) = \iint_{\alpha \geq \beta} \mu(\alpha, \beta) \gamma_{\alpha\beta} \big[u(t) \big] \mathrm{d}\alpha \, \mathrm{d}\beta \tag{5.1}$$

where $\chi(t)$ is the displacement response of the piezoelectric actuator, $u(t)$ is the input voltage, $\mu(\alpha,\beta)$ is a weighting function, $\gamma_{\alpha\beta}$ is the hysteresis relay operator whose value is determined by the input operation, and α and β represent "up" and "down" switching values of the input (Fig. 5.1). In order to calculate the displacement response, the weighting function $\mu(\alpha,\beta)$ needs to be known, which is traditionally calculated by differentiating the Preisach function $\Gamma(\alpha',\beta')$ in the following manner (Ge and Jouaneh, 1995),

$$\mu(\alpha',\beta') = -\frac{\partial^2\Gamma(\alpha',\beta')}{\partial\alpha'\partial\beta'} \qquad (5.2)$$

where $\Gamma(\alpha',\beta')$, obtained from experimental data, represents the change in the hysteresis loop of the displacement response when the input voltage $u(t)$ changes from α' (or β') to β' (or α') (Fig. 5.1). Since the experimental data usually contains measurement noise, the differential operation in Eq. (5.2) will amplify the error. As a result, the model might not be reliable. To bypass the differential operation, Ge and Jouaneh proposed an alternative calculation scheme which is implemented through the two-dimensional interpolation using experimental data (Ge and Jouaneh, 1995 and 1997). A typical hysteretic response under a decaying sinusoidal input excitation described by the CPM model is shown in Fig. 5.3. It is worth noting that the trajectory exhibits a jump (discontinuous point) at the switching point due to the interpolation variables switching from β (or α) to α (or β).

The hysteresis behavior can also be modeled using the MRC representation (Goldfarb and Celanovic, 1997a and 1997b; Lee and Royston, 2000), which consists of a number of elasto-slide elements subject to Coulomb friction connected in parallel (Fig. 5.2b). If the element number becomes infinite, the model is referred to as Generalized Maxwell slip. This model can be extended to its electrical analogy to represent the piezoelectric hysteresis behavior (Fig. 5.2a). By introducing a parameter γ to account for a reversible nonlinear stiffness effect (Fig. 5.2b), Lee and Royston extended the MRC model (2000), which is represented as,

$$\chi(t) = \sum_{i=1}^{n} E_{rc}^{(i)} \tag{5.3a}$$

where

$$
E_{rc}^{(i)} =
\begin{cases}
\beta_{33}^{T(i)}[1 + \gamma(D_3 - D_b^{(i)})](D_3 - D_b^{(i)}); \\
\qquad \text{if } \left| \beta_{33}^{T(i)}[1 + \gamma(D_3 - D_b^{(i)})](D_3 - D_b^{(i)}) \right| < e_{rc}^{(i)} \\
e_{rc}^{(i)} \, \mathrm{sgn}[\dot{D}_3], \text{ where } \left| \beta_{33}^{T(i)}[1 + \gamma(D_3 - D_b^{(i)})](D_3 - D_b^{(i)}) \right| = e_{rc}^{(i)}; \\
\qquad \text{otherwise}
\end{cases}
\tag{5.3b}
$$

Correspondingly, β_{33}^{T}, e_N, μ, e_{rc} and D_b are, respectively, the electrical analogies to the mechanical spring stiffness, normal force, Coulomb friction coefficient, Coulomb friction force, and the displacement from the equilibrium position of a massless box (Figure 5.2b). These notations will be further discussed in the next section. Here n is the number of massless boxes.

Both CPM and MRC have been successfully utilized for the modeling of piezoelectric hysteresis. It is worth mentioning that the new control strategy developed in this research does not depend upon a specific hysteresis model, *i.e.*, either CPM (Eq. (5.1)) or MRC (Eq. (5.3)) or other hysteresis models can be inserted into the control design. Without loss of generality, in what follows we use MRC to demonstrate the control design.

5.2.2 Nonlinear Dynamic Model of Integrated System

The purpose of the research presented in this chapter is to outline a control strategy for the robust and high precision control using piezoelectric actuator with hysteresis compensation. To illustrate the methodology development, we use a cantilevered beam as an example for control design (e.g., see Fig. 2.6a or Fig. 3.1). The control objective can be tracking control or vibration suppression. For the beam problem, the following linear piezoelectric constitutive relation has been widely used (IEEE, 1988),

$$S_1 = s_{11}^D T_1 + g_{31} D_3 \qquad (5.4)$$
$$E_3 = -g_{31} T_1 + \beta_{33}^T D_3$$

where 1 designates the beam longitudinal direction, 2 the width direction, and 3 the transversal direction; S is the strain, T the stress, D the electrical displacement, and E the electrical field. Here it is usually assumed that the stress components at the beam width direction and at the transversal direction are both zero, and the in-plane components of the electrical displacement and electrical field are all zero, $i.e.,$ $T_3 = T_2 = D_1 = D_2 = E_1 = E_2 = 0$. It is worth noting that the notations used here, which are consistent with the relevant literature (Xue and Tang, 2006), may be slightly different from those used in the previous chapters.

Recent studies have revealed that even in cases that the applied fields are not sufficient to completely re-orient the remnant polarization in the entire actuator, a small number of domains can still be switched. Thus both the material states and the electro-mechanical coupling are changed, giving rise to the hysteretic strain-field behavior. A number of experimental observations show that the electrical displacement D and the electrical field E exhibits a strong hysteretic behavior (Main et $al.$, 1995; Goldfarb and Celanovic, 1997a; Lee and Royston, 2000). Under zero stress T, while the electrical displacement D versus strain S relation is reversible ($i.e.$, without hysteresis), the applied electrical field E versus S is not. It has been identified that the mechanical stress-strain relation under constant electrical displacement is reversible, but this relation under constant electrical field is hysteretic (Goldfarb and Celanovic, 1997a). In addition, the relation between the applied stress and the electrical displacement is hysteretic (Damjanovic, 1997; Taylor and Damjanovic, 1997). In order to account for all these characteristics, a nonlinear term should be introduced to the original linear constitutive relation. It is worth mentioning that the Poisson's ratio of the piezoelectric actuator is generally different from that of the beam and, consequently, the stress component T_2 is not zero. Combining all these observations, we obtain the following constitutive relation for a piezoelectric actuator (Lee and Royston, 2000),

$$T_1 = \bar{c}_{11}^D S_1 - h_{31} D_3 \qquad (5.5)$$
$$E_3 = -g_{31}(\bar{c}_{11}^D + \bar{c}_{12}^D) S_1 + 2h_{31} g_{31} D_3 + \{\beta_{33}^T D_3\}$$

where $\bar{c}_{11}^D = c_{11}^D - v_p c_{13}^D$ and $\bar{c}_{12}^D = c_{12}^D - v_p c_{23}^D$, and the Poisson's ratio of the piezoelectric material is $v_p = -S_3 / S_1$. In Eq. (5.5), {} is the hysteresis operator, and $\{\beta_{33}^T D_3\}$ denotes the hysteresis effect which can be either the MRC model, CPM model or other nonlinear models.

The mathematical model of a cantilevered beam integrated with a piezoelectric actuator is developed under the following assumptions:

1. The bonding between the beam and the piezoelectric actuator is perfect, *i.e.*, the beam and the actuator have the same displacement at the bonding location;
2. The poling direction of the piezoelectric actuator is in the positive transversal direction of the beam;
3. The piezoelectric actuator is thin and short compared to the beam.

The system equations can then be derived using Hamilton's principle and the assumed mode method. In this study, we use a single mode, the dominant mode of the beam without the circuitry, for discretization. The transversal displacement of the beam can be expressed as,

$$w(x,t) = \phi(x)q(t) \tag{5.6}$$

where ϕ is the first mode of the cantilevered beam, and q is the generalized mechanical displacement. We can then obtain the system equations (see Appendix 2.2 for similar derivations),

$$m\ddot{q} + g\dot{q} + kq + k_1 Q = F_m \tag{5.7a}$$

$$k_2 Q + k_3 q + f(Q) = V_a \tag{5.7b}$$

where m is the equivalent mass, q the generalized mechanical displacement, g the beam equivalent damping, k the equivalent stiffness, k_1 and k_3 the cross coupling coefficients, F_m the external disturbance force, Q the charge flow to the piezoelectric actuator, k_2 the inverse of capacitance of the piezoelectric actuator, h_p the actuator thickness, and V_a the voltage across the piezoelectric actuator. It is worth emphasizing that while Appendix 2.2 only concerns the linear portion of the system, the derivation of the nonlinear term $f(Q) = h_p\{\beta_{33}^T Q\}$ is straightforward based on the analysis given in the preceding section.

Observe Eqs. (5.7a, b) and note that the input voltage across the piezoelectric actuator is V_a. When the hysteresis effect $f(Q)$ is neglected, one can combine the mechanical (dynamic) equation (5.7a) and the electrical (static) equation (5.7b) together by eliminating the generalized electrical coordinate Q. The presence of piezoelectric hysteresis, however, rules out the possibility of mathematically eliminating Q. The main difficulty in developing a control algorithm for the nonlinear system described by Eqs. (5.7a, b) appears to be that the second equation is a static one that contains an extremely complicated nonlinear term that may be subject to modeling error and uncertainties.

In literature, a variety of control approaches have been proposed to deal with the piezoelectric nonlinear behavior with different complexity of hysteresis models (Chonan et al., 1996; Ge and Jouaneh, 1996; Main and Garcia, 1997; Croft and Devasia, 1998; Tsai and Chen, 2003). One approach is based upon the inverse hysteresis modeling in the feedforward controller, and a feedback is used to regulate the output error. Polynomial approximation and neural network were employed to model the inverse hysteresis (Main and Garcia, 1997; Croft and Devasia, 1998; Chang and Sun, 2001; Kung and Fung, 2004). Another approach used is the traditional feedback linearization (Choi et al., 2002). Based on a time delay model, a Smith predictor is employed to compensate the hysteresis nonlinearity (Tsai and Chen, 2001). In these approaches, several factors may deteriorate the control performance. For example, the forward physical process (hysteresis) is usually an integral process, while in inverse modeling/cancellation methods, the differential operation will be involved in the hysteresis inverse calculation. As a result, the inverse model obtained may not be reliable when the hysteresis measurement data contains noise. In addition, typical inverse design approach treats the hysteresis and structural dynamics separately, i.e., the hysteresis is decoupled from structural dynamics. Such separation, however, is difficult to achieve in the practical measurement of piezoelectric actuation.

In this chapter, we outline a different approach for dealing with the piezoelectric hysteresis by introducing dynamics to the electrical part of the controlled system. Specifically, we integrate the piezoelectric actuator with a resistance-inductance (*RL*) circuit. Consequently, the system equations become (see Appendix 2.2 for similar derivations),

$$m\ddot{q} + g\dot{q} + kq + k_1Q = F_m \tag{5.8a}$$

$$L\ddot{Q} + R\dot{Q} + k_2Q + k_3q + f(Q) = V_i \tag{5.8b}$$

where L and R are the inductance and resistance, respectively, and V_i is the control input. Observe Eqs. (5.8a, b) and compare them with (5.7a, b). Cleary, the main advantage of introducing dynamics to the piezoelectric actuator is that the charge and/or current in the piezoelectric actuator now become independent state variables that can be directly measured and fed back. Not only can this actuator network configuration improve the hysteresis characterization accuracy, the control design can also be greatly simplified. With the introduction of the circuitry elements, the controlled system now consists of two coupled second-order *dynamic* subsystems, the mechanical subsystem (5.8a) and the electrical subsystem (5.8b), whereas the nonlinearity only appears explicitly as a function of the electrical charge. As will be shown later, this nonlinear dynamic system can be readily cast into the standard state-space form, and one may then use various nonlinear control methods to handle the hysteresis problem. It should be noted that the aforementioned piezoelectric circuitry has shown enhanced passive damping and active control authority. As discussed in Chapter 2, under the linear constitution relation assumption, one may find optimal resistance and inductance values for the maximum damping and/or active authority amplification. Throughout this chapter, without loss of generality, the inductance L and resistance R are chosen in such a manner that they yield the maximum vibration damping/absorbing under passive situation.

The piezoelectric hysteresis effect $f(Q)$ in Eq. (5.8) (illustrated in Fig. 5.3) is demonstrated in Fig. 5.4, where an initial velocity is imposed on the tip of the beam with both external disturbance F_m and input voltage V_i being zero. Here we use the MRC model in the simulation, and all system parameters are listed in Tables 5.1 and 5.2.

Table 5.1. MRC parameters of a monolithic piezoelectric actuator (Lee and Roys-ton, 2000)

$\beta_{33}^{T(i)} \times 10^{-6}$	0.30868	0.23188	0.18356	0.41700	1.67796
$e_{rc}^{(i)}$	0.08996	0.13515	0.16048	0.16529	∞

Table 5.2. System parameters used in simulation

$l_b = 0.3\text{m}$	$w_b = 0.0381\text{m}$
$h_b = 0.003175\text{m}$	$\rho_b = 7.8335 \times 10^3 \text{kg/m}^3$
$E_b = 1.9818 \times 10^{11} \text{N/m}^2$	$C_b = 2.0576$
$w_p = 0.0343\text{m}$	$h_p = 0.000267\text{m}$
$x_l = 0.02\text{m}$	$x_r = 0.0724\text{m}$
$\rho_p = 7.8 \times 10^3 \text{kg/m}^3$	$E_p = 6.2 \times 10^{10} \text{N/m}^2$
$h_{31} = -1.35 \times 10^9 \text{V/m}$	$g_{31} = -9.5 \times 10^{-3} \text{Vm/N}$
$\bar{c}_{11}^D = 10.64 \times 10^{10} \text{N/m}^2$	$\bar{c}_{12}^D = -5.68 \times 10^{10} \text{N/m}^2$
$L = 76\text{H}$	$R = 3090\Omega$

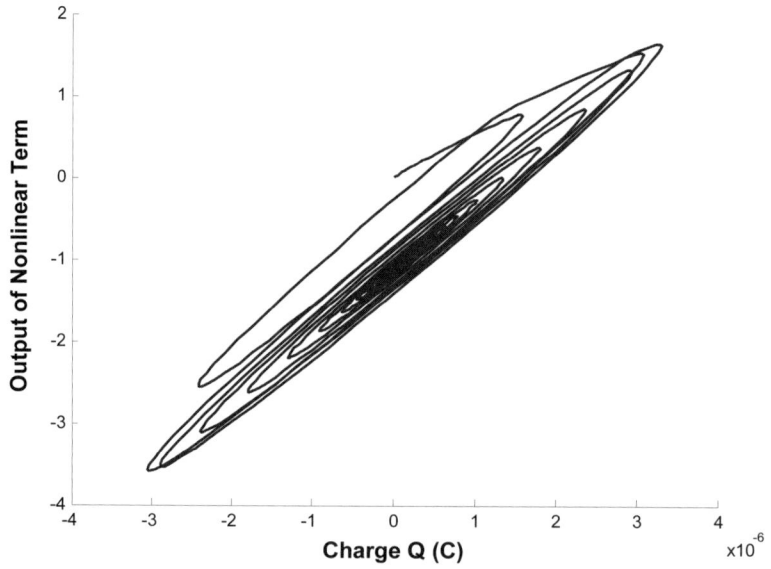

Fig. 5.4. Hysteresis behavior (MRC) of a piezoelectric actuator due to the free vibration of cantilevered beam.

5.3 INTEGRAL CONTINUOUS SLIDING MODE CONTROL (ICSMC) DESIGN

The system (5.8) can be readily cast into the standard state-space form,

$$\dot{\mathbf{x}} = \mathbf{A}\mathbf{x} - \mathbf{B}f(x_3) + \mathbf{B}u + \mathbf{D}F_m \tag{5.9}$$

where

$$\mathbf{x}^{\mathrm{T}} = \begin{bmatrix} x_1 & x_2 & x_3 & x_4 \end{bmatrix}^{\mathrm{T}} = \begin{bmatrix} q & \dot{q} & Q & \dot{Q} \end{bmatrix}^{\mathrm{T}} \tag{5.10a}$$

$$\mathbf{A} = \begin{bmatrix} 0 & 1 & 0 & 0 \\ -\dfrac{k}{m} & -\dfrac{g}{m} & -\dfrac{k_1}{m} & 0 \\ 0 & 0 & 0 & 1 \\ -\dfrac{k_3}{L} & 0 & -\dfrac{k_2}{L} & -\dfrac{R}{L} \end{bmatrix} \tag{5.10b}$$

$$\mathbf{B}^{\mathrm{T}} = [0,0,0,\dfrac{1}{L}]^{\mathrm{T}} \tag{5.10c}$$

$$\mathbf{D}^{\mathrm{T}} = [0,\dfrac{1}{m},0,0]^{\mathrm{T}} \tag{5.10d}$$

where u represents the control input V_i. It's worth mentioning that all the system parameters are subject to uncertainties. In this chapter, full state feedback is assumed. For system (5.9), various nonlinear control methods may be implemented. While the traditional feedback linearization technique is straightforward, parameter uncertainties of the system and modeling error/uncertainty on the hysteresis effect might deteriorate the control performance. In this study, we present an integral continuous sliding mode control (ICSMC) (Xue and Tang, 2006) that is improved from the conventional continuous sliding mode control (CSMC) (Zhou and Fisher, 1992) and the integral variable structure control (IVSC) (Chern and Wu, 1991). This method can directly deal with the hysteresis nonlinearity in forward cancellation, and also allows us to implement cubic state feedback which improves the system robustness as compared with IVSC.

The unique feature of the variable structure control is that the desired system dynamics with required performances (static and dynamic characteristics) can be designed using a sliding manifold (Hung et al., 1993; Utkin, 1993). The designed system dynamics is generally of lower order than the original system. Once the reduced order system with the sliding manifold is provided, a variable structure control action, which consists of the equivalent control action and the switching control action, is developed so that the switching control action constrains the system to follow the designed sliding manifold while the equivalent control action controls the performance of the reduced order dynamic system. Therefore, a typical variable structure control design consists of two modes, the reaching mode and the sliding mode (Hung et al., 1993; Utkin, 1993). For the reaching mode, the control action is to force the response of the system to reach the sliding manifold in finite time. Once it enters the sliding manifold, the response of the system will be constrained to follow the sliding manifold and approach the steady-state. Consequently, the performances of the system will depend on not only the reaching mode but also the characteristics of the sliding manifold. One of the fundamental requirements of the sliding manifold definition is that the system dynamics should be stable and robust once it enters the sliding motion, so that the system can finally reach the zero point (the steady-state of the system). While the traditional manifold design using linear combination of system states can meet certain performance requirement such as robust stability, it is difficult to satisfy the tracking performance when the system is required to follow a reference input signal (Chern and Wu, 1991; Cheng and Liu, 1999). The integral control (based on internal model principle), as an effective approach of eliminating errors, has been widely employed in tracking control (Davison and Goldenberg, 1975). The combination of integral control and variable structure control, which leads to the integral variable structure control (IVSC), was first proposed to control the electrohydraulic velocity servo-systems (Chern and Wu, 1991), and then generalized to a multi-input-multi-output (MIMO) system (Cheng and Liu, 1999). Essentially, the integral of the tracking error is used in the definition of the sliding manifold. On the other hand, the control action of traditional variable structure is discontinuous in nature due to the switching control at the reaching mode, which leads to the chattering phenomenon and in turn may trigger the high frequency un-modeled dynamics. To improve, Zhou and Fisher (1992) developed a continuous sliding mode control (CSMC). This approach retains the positive properties such as robustness and disturbance rejection capability of traditional variable structure control, and in the meantime uses a continuous control law to completely eliminate the chattering problem on the other. However, this CSMC cannot be directly applied when the system is commanded to track a reference signal.

In what follows, we outline an integral continuous sliding mode controller (ICSMC), which is a combination of CSMC and ISVC. This approach can directly deal with both the tracking problem and the hysteresis nonlinearity, which keeps the merits of CSMC such as robustness and continuous control action as long as the reference signal is continuous.

5.3.1 Determination of Control Action

We define

$$z = \int_0^t [r_d - \phi(l)x_1] \mathrm{d}t \qquad (5.11)$$

where r_d is the reference input signal, and l the beam length. Clearly, z indicates the tracking error between the command input signal and the actual beam tip displacement. The block diagram of the ICSMC to minimize z is shown in Fig. 5.5. We then define the sliding manifold as follows (Hung et al., 1993; Utkin, 1993),

$$s = \sum_{i=1}^{4} c_i x_i + c_5 \int [r_d - \phi(l)x_1] \mathrm{d}t = \mathbf{C}^\mathrm{T} x + c_5 \int [r_d - \phi(l)x_1] \mathrm{d}t \qquad (5.12)$$

where $\mathbf{C} = [c_1, c_2, c_3, c_4]^\mathrm{T}$. Essentially, the sliding manifold is a linear combination of the system states and the tracking error term.

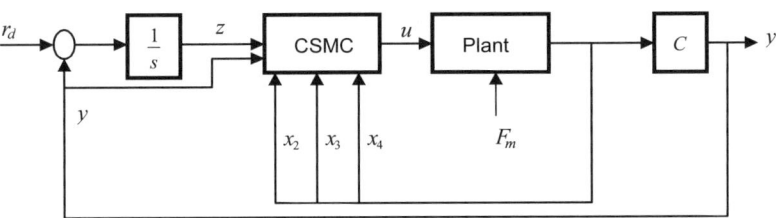

Fig. 5.5. Block diagram of ICSMC.

In order to compensate for the piezoelectric hysteresis and attenuate the external disturbance, we propose the following control action for system (5.9),

$$u = -\mathbf{K}x + \hat{f}(x_3) - u_d \tag{5.13}$$

where $\mathbf{K} = [K_1, K_2, K_3, K_4]$ denotes state gain, $\hat{f}$ is the hysteresis estimation, and u_d is the disturbance rejection term. As will be shown later, this control action actually leads to a cubic state feedback. Therefore, once the states move away from the sliding manifold, the proposed ICSMC will provide more powerful control and faster reaction due to its cubic feedback nature. It is still worth mentioning that, following along the similar approach developed by Zhou and Fisher (1992) for nonlinear systems, one may end up with a feedback control that is of first or second order, which, in general, leads to less efficient control actions.

We now apply the reaching condition,

$$s\dot{s} < 0 \tag{5.14}$$

Recalling Eqs. (5.13) and (5.9), and assuming $\mathbf{C}^T\mathbf{B} \neq 0$ with a constant sign (e.g., $\mathbf{C}^T\mathbf{B} > 0$), we obtain

$$s\dot{s} = s\{\mathbf{C}^T\dot{\mathbf{x}} + c_5[r_d - \phi(l)x_1]\} = s\mathbf{C}^T\mathbf{B}\{[\boldsymbol{\alpha} - \mathbf{K}]x + [\beta + \gamma c_5 r_d - u_d]\} \tag{5.15}$$

where

$$\boldsymbol{\alpha} = (\mathbf{C}^T\mathbf{B})^{-1}(\mathbf{C}^T\mathbf{A} - \mathbf{A}_\phi) = [\alpha_1, \alpha_2, \alpha_3, \alpha_4] \tag{5.16a}$$

$$\beta = (\mathbf{C}^T\mathbf{B})^{-1}\mathbf{C}^T\mathbf{D}F_m + \Delta f \tag{5.16b}$$

$$\gamma = (\mathbf{C}^T\mathbf{B})^{-1} \tag{5.16c}$$

$$\mathbf{A}_\phi = [c_5\phi(l), 0, 0, 0] \tag{5.16d}$$

$$\Delta f = \hat{f}(x_3) - f(x_3) \tag{5.16e}$$

Define

$$\bar{\alpha}_i = \frac{\sup \alpha_i + \inf \alpha_i}{2} \qquad i = 1, 2, 3, 4 \tag{5.17}$$

$$\bar{\beta} = \frac{\sup \beta + \inf \beta}{2} \tag{5.18}$$

$$\bar{\gamma} = \frac{\sup \gamma + \inf \gamma}{2} \tag{5.19}$$

$$\bar{\mathbf{B}} = \frac{\sup \mathbf{B} + \inf \mathbf{B}}{2} \tag{5.20}$$

Using the above equations, we have,

$$\mathbf{C}^T\mathbf{B} = \varepsilon(\mathbf{C}^T\bar{\mathbf{B}}), \qquad 0 < \varepsilon \le 1 \tag{5.21}$$

In order to satisfy the reaching condition (5.14), we may choose the following control parameters,

$$K_i = \bar{\alpha}_i + (\sup \alpha_i - \bar{\alpha}_i + \delta_i)\frac{s\mathbf{C}^T\bar{\mathbf{B}}x_i}{\lambda}, \quad \delta_i > 0, \quad i = 1, 2, 3, 4 \tag{5.22}$$

$$u_d = \bar{\beta} + \bar{\gamma}c_5 r_d + \{[\sup \beta - \bar{\beta}] + (\tilde{\gamma} - \bar{\gamma})c_5 r_d + \delta_d\}\frac{s\mathbf{C}^T\bar{\mathbf{B}}}{\lambda}, \quad \delta_d > 0 \tag{5.23}$$

where

$$\tilde{\gamma} = \begin{cases} \sup \gamma, & when \ c_5 r_d > 0 \\ \inf \gamma, & when \ c_5 r_d < 0 \end{cases} \tag{5.24}$$

and λ, δ_i ($i = 1,2,3,4$) and δ_d are design parameters. Indeed, λ is the boundary layer thickness of the sliding manifold (Zhou and Fisher, 1992).

Substituting K_i and u_d into Eq. (5.15), we obtain,

$$s\dot{s} = s\mathbf{C}^T\mathbf{B}\{\sum_{i=1}^{4}[(\alpha_i - \bar{\alpha}_i) - (\sup\alpha_i - \bar{\alpha}_i + \delta_i)\frac{s\mathbf{C}^T\bar{\mathbf{B}}x_i}{\lambda}]x_i$$

$$+[(\beta - \bar{\beta}) + (\gamma - \bar{\gamma})c_s r_d] - [(\sup\beta - \bar{\beta}) + (\tilde{\gamma} - \bar{\gamma})c_s r_d + \delta_d]\frac{s\mathbf{C}^T\bar{\mathbf{B}}}{\lambda}\}$$

(5.25)

Combining Eqs. (5.25) and (5.21) yields,

$$s\dot{s} = \varepsilon\{\sum_{i=1}^{4}[\alpha_i - \bar{\alpha}_i]s\mathbf{C}^T\bar{\mathbf{B}}x_i + [(\beta - \bar{\beta}) + (\gamma - \bar{\gamma})c_s r_d]s\mathbf{C}^T\bar{\mathbf{B}}$$

(5.26)

$$-\sum_{i=1}^{4}(\sup\alpha_i - \bar{\alpha}_i + \delta_i)(s\mathbf{C}^T\bar{\mathbf{B}}x_i)^2\frac{1}{\lambda}$$

$$-[(\sup\beta - \bar{\beta}) + (\tilde{\gamma} - \gamma)c_s r_d + \delta_d](s\mathbf{C}^T\bar{\mathbf{B}})\frac{1}{\lambda}\}$$

One can see from the above equation that the reaching condition (5.14) is guaranteed if the following conditions are satisfied,

$$\left|s\mathbf{C}^T\bar{\mathbf{B}}x_i\right| > \lambda \qquad i = 1,2,3,4$$

(5.27)

$$\left|s\mathbf{C}^T\bar{\mathbf{B}}\right| > \lambda$$

(5.28)

Observe Eqs. (5.13), (5.22) and (5.23). One may see that the proposed control action will cancel the hysteresis effect and also achieve a cubic order state feedback. It is worth mentioning that the control action of (5.13) with parameters given in Eqs. (5.22) and (5.23) is continuous as long as the reference input r_d is a continuous signal. If the reference signal has discontinuity (e.g., a step input), the control action will correspondingly be discontinuous. However, in the overall control action shown in Eq. (5.13), the effect of reference input r_d only appears as a linear term (see Eq. (5.23)). Compared to the cubic state feedback (the first term in the right hand side of Eq. (5.13)), one may envision that the control action corresponding to a reference discontinuity will have a relatively small effect on the overall control action. This is further illustrated in the simulations that follow. In comparison, the conventional IVSC (Chern and Wu, 1991; Cheng and Liu, 1999) (see Appendix 5.2) uses a linear state feedback and, as a result, the discontinuous effect is more significant. In addition, the switching control also could play its role at this discontinuous point, *i.e.*, attempting to confine the states within the boundary layer of the sliding manifold, which will intensify the discontinuity of control action. It is clear that the control action of ICSMC will be smoother than that of the conventional IVSC under the same discontinuous reference input. The detailed comparisons will be provided in the simulation section.

5.3.2 Manifold Coefficients Determination and Tracking Performance Analysis

It is often recognized that $s = 0$ when the system enters the sliding manifold (Chern and Wu, 1991; Utkin, 1993). However, generally, $s \neq 0$, because of the non-zero boundary layer thickness and the tracking error. Here we set $s = \delta(t)$ where $|\delta(t)|$ is less than the boundary layer thickness, when the system is confined within the boundary layer of sliding manifold. The system reaches the steady-state when $\delta(t)$ approaches to a small constant value, *e.g.*, δ^c. Combining Eqs. (5.9), (5.11) and (5.12), we may obtain,

$$\dot{\overline{\mathbf{x}}} = \mathbf{A}_{sm}\overline{\mathbf{x}} + \mathbf{B}_{sm}\begin{bmatrix} r_d \\ F_m \end{bmatrix} \qquad (5.29a)$$

$$y = \mathbf{C}_{sm}\overline{\mathbf{x}} \qquad (5.29b)$$

where

$$\overline{\mathbf{x}} = \begin{bmatrix} x_1 \\ x_2 \\ x_3 \\ z - \dfrac{c_4}{c_5}\delta^c \end{bmatrix}, \quad \mathbf{A}_{sm} = \begin{bmatrix} 0 & 1 & 0 & 0 \\ -\dfrac{\hat{k}}{\hat{m}} & -\dfrac{\hat{g}}{\hat{m}} & -\dfrac{\hat{k}_1}{\hat{m}} & 0 \\ -\dfrac{c_1}{c_4} & -\dfrac{c_2}{c_4} & -\dfrac{c_3}{c_4} & -\dfrac{c_5}{c_4} \\ -\phi(\hat{l}) & 0 & 0 & 0 \end{bmatrix}$$

$$\mathbf{B}_{sm} = \begin{bmatrix} 0 & 0 \\ 0 & \dfrac{1}{\hat{m}} \\ 0 & 0 \\ 1 & 0 \end{bmatrix}, \quad \mathbf{C}_{sm} = \begin{bmatrix} \phi(\hat{l}) & 0 & 0 & 0 \end{bmatrix}$$

Clearly, the system dynamics within the boundary layer is determined by c_i, $i = 1, \cdots, 5$. The determination of c_i should lead to: 1) stability of (5.29); and 2) tracking performance and external disturbance attenuation capability.

In order to satisfy the stability requirement, the eigenvalues of matrix A should all have negative real parts. The standard pole placement approach can be employed to determinate the values of c_i, $i = 1, \cdots, 5$. Imposing the Laplace transform to Eq. (5.29) under zero initial conditions yields,

$$Y(s) = \frac{\dfrac{\hat{k}_1}{\hat{m}}\phi(\hat{l})\dfrac{c_5}{c_4}R_d(s) + (s^2 + \dfrac{c_3}{c_4}s)\dfrac{\phi(\hat{l})}{\hat{m}}F_m(s)}{s^4 + (\dfrac{\hat{g}}{\hat{m}} + \dfrac{c_3}{c_4})s^3 + (\dfrac{\hat{k}}{\hat{m}} + \dfrac{\hat{g}}{\hat{m}}\dfrac{c_3}{c_4} - \dfrac{\hat{k}_1}{\hat{m}}\dfrac{c_2}{c_4})s^2 + (\dfrac{\hat{k}}{\hat{m}}\dfrac{c_3}{c_4} - \dfrac{\hat{k}_1}{\hat{m}}\dfrac{c_1}{c_4})s + \dfrac{\hat{k}_1}{\hat{m}}\phi(\hat{l})\dfrac{c_5}{c_4}}$$

$$= G_{r_d y}(s)R_d(s) + G_{F_m y}(s)F_m(s)$$

$$(5.30)$$

The characteristic equation of the sliding mode dynamics is given as

$$s^4 + (\frac{\hat{g}}{\hat{m}} + \frac{c_3}{c_4})s^3 + (\frac{\hat{k}}{\hat{m}} + \frac{\hat{g}}{\hat{m}}\frac{c_3}{c_4} - \frac{\hat{k}_1}{\hat{m}}\frac{c_2}{c_4})s^2 \tag{5.31}$$

$$+ (\frac{\hat{k}}{\hat{m}}\frac{c_3}{c_4} - \frac{\hat{k}_1}{\hat{m}}\frac{c_1}{c_4})s + \frac{\hat{k}_1}{\hat{m}}\phi(\hat{l})\frac{c_5}{c_4} = 0$$

Let the desired characteristic equation under the desired eigenvalues ω_i ($i = 1, \cdots, 4$) be

$$\prod_{i=1}^{4}(s - \omega_i) = s^4 + \sigma_1 s^3 + \sigma_2 s^2 + \sigma_3 s + \sigma_4 = 0 \tag{5.32}$$

Comparing Eqs. (5.31) and (5.32), we obtain the following sliding manifold coefficients,

$$\frac{c_1}{c_4} = -\frac{\bar{k}}{\bar{k}_1}(\frac{\bar{g}}{\bar{m}} - \sigma_1) - \frac{\bar{m}}{\bar{k}_1}\sigma_3, \qquad \frac{c_1}{c_4} = -\frac{\bar{m}}{\bar{k}_1}\sigma_2 + \frac{\bar{k}}{\bar{k}_1} - \frac{\bar{g}}{\bar{k}_1}(\frac{\bar{g}}{\bar{m}} - \sigma_1) \tag{5.33}$$

$$\frac{c_3}{c_4} = -\frac{\bar{g}}{\bar{m}} + \sigma_1, \qquad \frac{c_5}{c_4} = \frac{\bar{m}}{\bar{k}_1\phi(l)}\sigma_4$$

where the over-bar indicates the nominal value of the corresponding parameter. Without loss of generality, we may set $c_4 = 1$, and the rest of c_i can be solved from Eq. (5.33).

When all the desired eigenvalues are placed in the left-hand side of the complex plane, the stability of (5.29) is guaranteed. In order to examine the tracking performance, we may use the final value theorem,

$$\lim_{t \to \infty} y(t) = \lim_{s \to 0} sY(s) = \lim_{s \to 0} s[G_{r_d y}(s)R_d(s) + G_{F_m y}(s)F_m(s)] \tag{5.34}$$

Assume that both the reference r_d and the external disturbance F_m are step signals. In virtue of Eq. (5.30), we have,

$$\lim_{t \to \infty} y(t) = \lim_{s \to 0}[sG_{r_d y}(s)\frac{r_d}{s} + sG_{F_m y}(s)\frac{F_m}{s}] = r_d + 0 \cdot F_m \qquad (5.35)$$

Clearly, we can achieve zero steady-state tracking error and completely reject the external disturbance for step signals.

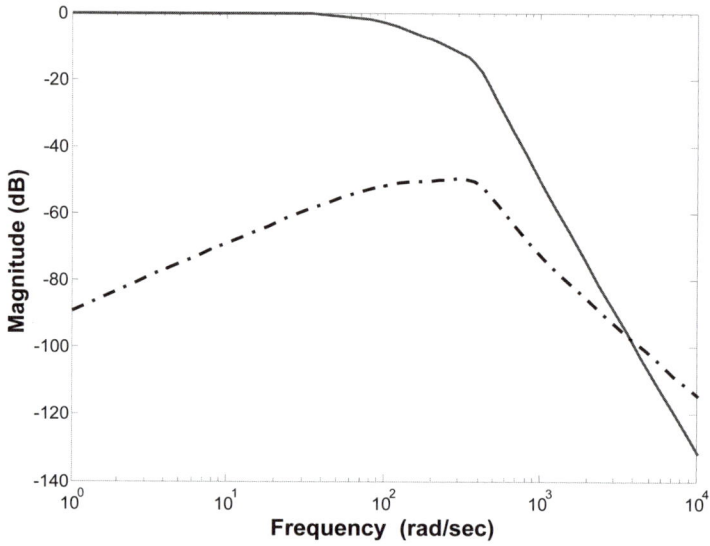

Fig. 5.6. Bode diagram of sliding manifold.
——— : signal transfer function; — · — ·: disturbance transfer function.

The above analysis provides a theoretical explanation of system tracking and disturbance attenuation capability for some specific input and external disturbance (both being step signals). In practical situation, both input signal and external disturbance could be more complicated. It has been mentioned that once the system is confined in the sliding manifold, the performance will depend on the manifold design. From Eq. (5.30), one can see that the transfer function from inputs (for both reference signal and disturbance) to output is strictly proper. If the poles of Eq. (5.30) are placed in the open left hand side of complex plane, the magnitude of the transfer function after cutoff frequency will roll off. Typical transfer functions of the sliding manifold are illustrated in Fig. 5.6. In other words, the transfer function from the reference input signal to the output has low-pass filter characteristic, while the transfer function from the disturbance to output plays an attenuation role. These characteristics of sliding manifold are expected, because generally the tracked reference signal is in low frequency band while the disturbance signal is in high frequency band.

5.4 SIMULATION RESULTS AND DISCUSSIONS

In this section, we carry out analyses to illustrate the control design and demonstrate the system performance. We first compare the proposed ICSMC with a regular optimal control case, to verify the hysteresis cancellation performance of the former. We then compare the ICSMC with conventional CSMC and IVSC under a variety of operating conditions to confirm the performance improvement in terms of tracking accuracy, robustness and control action smoothness. The system parameters used in the analyses are listed in Tables 5.1 and 5.2.

5.4.1 Comparison with respect to LQR on Hysteresis Compensation

If one uses a linear constitution relation for the piezoelectric actuator (such as the one shown in Eq. (5.4)), after introducing the circuitry elements one will end up with a linear system that is similar to the one described by Eq. (5.8) except that the nonlinear hysteresis effect $f(Q)$ does not appear. In that sense, one could treat the hysteresis effect in Eq. (5.8) as disturbance/noise, and then resort to linear control algorithms such as LQR (linear quadratic regulator) for control development. It is worth mentioning that the piezoelectric hysteresis could be explicitly treated as disturbance/noise, only after we introduce the circuitry. Meanwhile, using linear control algorithms is indeed a very natural choice in most of the applications so far, when one neglects the nonlinearity in the piezoelectric constitutive relation. In order to illustrate the hysteresis effect and demonstrate the system performance improvement due to ICSMC, in this first case study we compare the proposed ICSMC with a conventional LQR where the hysteresis nonlinearity is not considered in control design and only treated as a disturbance. Here the reference signal is assumed to be a sinusoid signal. In order to obtain a fair comparison, we adjust the weightings in the LQR design so that the peak control voltage requirements for both controllers are the same under zero external excitations.

The simulation results are shown in Fig. 5.7. Compared to the reference signal, the beam displacement exhibits a significant delay under LQR control, which is obviously caused by the hysteresis that is treated as a disturbance in the LQR design. ICSMC, on the other hand, leads to a very fast controlled response and much reduced delay, due to the hysteresis compensation explicitly incorporated into the control development as well as the cubic feedback. Most importantly, ICSMC significantly reduces the tracking error. As shown in Fig. 5.7, ICSMC has a 10% peak tracking error as compared to the 25% tracking error under LQR. This example clearly shows the negative effect of piezoelectric hysteresis under linear system assumption, and demonstrates the necessity of hysteresis compensation for high-precision control.

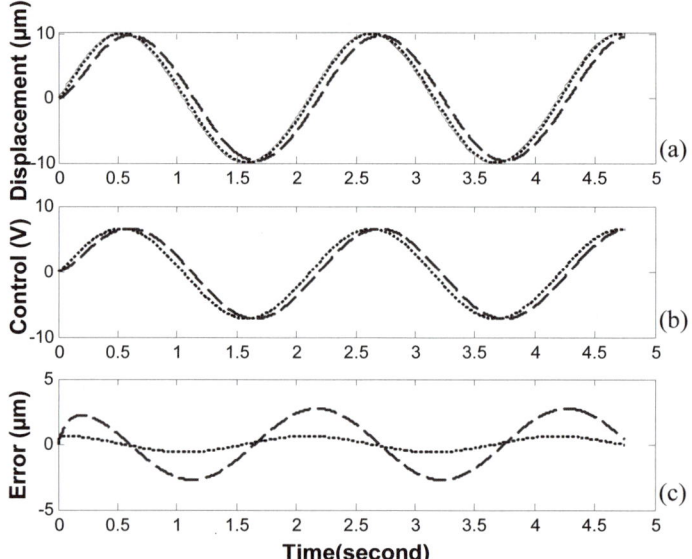

Fig. 5.7. Tracking control comparisons between LQG and ICSMC: (a) control performance; (b) control voltage; (c) tracking error.
——— : reference signal, — — — : LQG, ·············· : ICSMC.

5.4.2 Comparisons with respect to CSMC and IVSC on Robustness and Control Action Smoothness

One important feature of the proposed ICSMC is that it has improved tracking performance and control action smoothness as compared to other nonlinear controls such as CSMC and IVSC. In what follows we demonstrate the performance improvement in this regard. We assume the system described by Eq. (5.8) has the following bounds of uncertainties due to modeling error,

$$|\Delta m| \le m \times 10\% \qquad |\Delta g| \le g \times 10\% \qquad |\Delta k| \le k \times 10\% \qquad |\Delta k_1| \le k_1 \times 10\%$$
$$|\Delta L| \le L \times 10\% \qquad |\Delta R| \le R \times 10\% \qquad |\Delta k_2| \le k_2 \times 10\% \qquad |\Delta k_3| \le k_3 \times 10\%$$

In all the following case studies, the piezoelectric hysteresis is modeled and included in control development and in the analyses. We also assume that the bounds for the parameters related to the hysteresis modeling (listed in Table 5.1) are 20%.

Table 5.3. Parameters of CSMC

Parameter	Value	Parameter	Value
$\bar{\alpha}_1$	-6.5813×10^7	$\bar{\alpha}_2$	3.6869×10^4
$\bar{\alpha}_3$	2.9569×10^8	$\bar{\alpha}_4$	2.6307×10^5
$\sup \alpha_1$	-4.8459×10^7	$\sup \alpha_2$	4.0644×10^4
$\sup \alpha_3$	3.9934×10^8	$\sup \alpha_4$	2.8999×10^5
δ_1	0.1	δ_2	0.1
δ_3	0.1	δ_4	0.1
δ_d	0.1	$\bar{\beta}$	1.4764
$\sup \beta$	2.9528	$\bar{\gamma}$	-6.8472×10^7
$\sup \gamma$	-4.8459×10^{-4}	$\inf \gamma$	-8.8485×10^7
$\bar{\tau}$	-4.1234×10^5	$\sup \tau$	-3.7110×10^5
$\inf \tau$	-4.5357×10^5	c_1	5418.9
c_2	1.4623	c_3	3497.8
c_4	1	λ	8.1×10^{-4}

Table 5.4. Parameters of IVSC

Parameter	Value	Parameter	Value
c_1	23.174	c_2	1.2
c_3	3581.8	c_4	1
c_5	-3959.7	ξ_0	20
ξ_1	100	η	42688

The detailed derivation of CSMC approach is outlined in Appendix 5.1. The control parameters of CSMC (listed in Table 5.3) are chosen in a manner such that the system can follow the command step signal when external disturbance is absent. Here the reference input is a step signal with magnitude rising from 0 to 10 μm at 0.1 second. We then apply an external disturbance that is also a step signal with magnitude rising from 0 to 100 μN at 2.5 second. The control result under CSMC is shown in Fig. 5.8. Clearly, the beam tip displacement can indeed follow the command signal from 0 to 2.5 second when the external disturbance is absent. Once the external step disturbance is imposed onto the system, however, CSMC fails to eliminate the tracking error. The control voltage input is also illustrated in Fig. 5.8. Corresponding to the reference signal step-up, the control input has a very large jump at 0.1 second. In practical situation, the control input voltage will saturate before it reaches such large magnitude. Under the assumption that the saturation voltage is $\pm 80\text{V}$, the control result is shown in Fig. 5.9. Obviously the chattering phenomenon is present due to the saturation of control input voltage, and the system performs very badly. This second case study clearly demonstrates the limitation of CSMC in terms of the command signal tracking under disturbance.

Tab. 5.5. Parameters of ICSMC

Parameter	Value	Parameter	Value
$\bar{\alpha}_1$	-4.8345×10^6	$\bar{\alpha}_2$	1.5651×10^3
$\bar{\alpha}_3$	2.4198×10^8	$\bar{\alpha}_4$	2.6946×10^5
$\sup \alpha_1$	-3.5498×10^6	$\sup \alpha_2$	1.7937×10^3
$\sup \alpha_3$	3.2713×10^8	$\sup \alpha_4$	2.9702×10^5
$\bar{\gamma}$	76.092	$\sup \gamma$	83.702
$\inf \gamma$	68.483	c_1	23.174
c_2	1.2	c_3	3581.8
c_4	1	c_5	-3959.7
λ	10^{-5}	δ_1	0.1
δ_2	0.1	δ_3	0.1
δ_4	0.1	δ_d	0.1
$\bar{\beta}$	1.2119	$\sup \beta$	2.4237

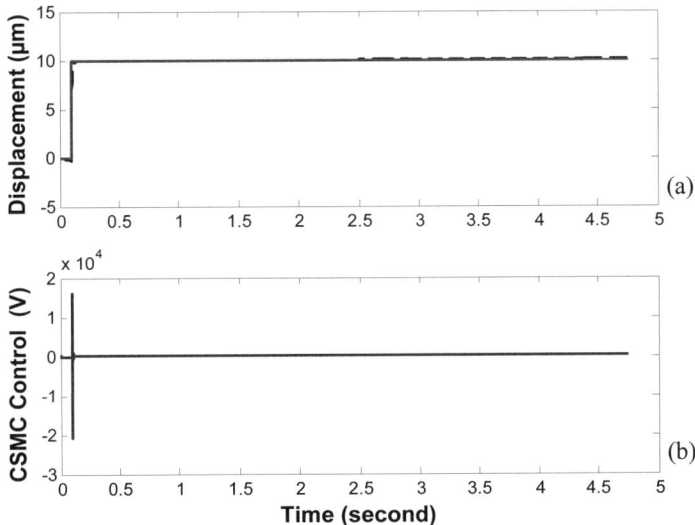

Fig. 5.8. CSMC result under step reference input and step external disturbance without considering the saturation of control input voltage: (a) control performance; ———— : reference signal, — — — : CSMC; (b) control voltage.

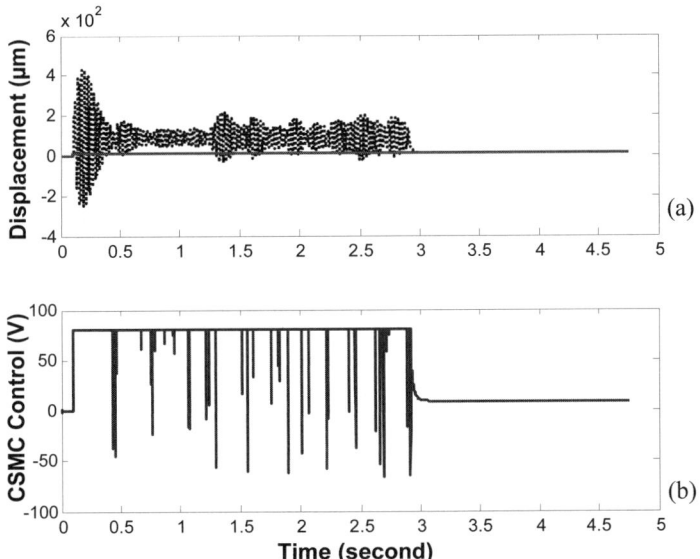

Fig. 5.9. CSMC result under step reference input and step external disturbance with control input voltage saturation: (a) control performance; ———— : reference signal, — — — : CSMC; (b) control voltage.

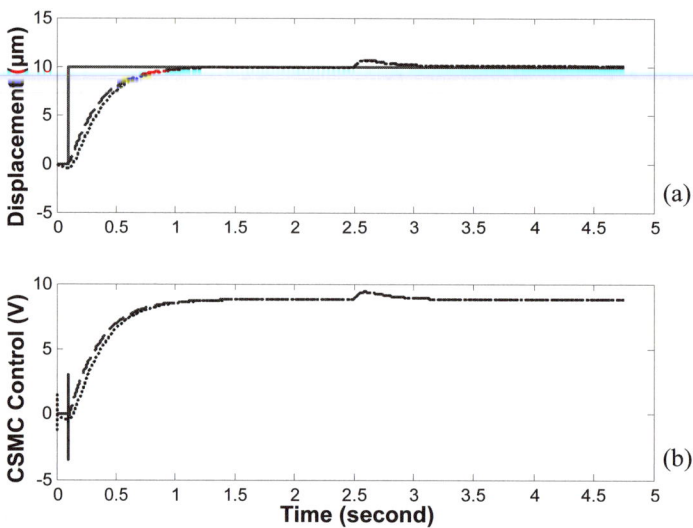

Fig. 5.10. IVSC and ICSMC comparison: (a) control performance; (b) control voltage. ——————: reference signal, — — —: IVSC, ⋯⋯⋯⋯⋯: ICSMC.

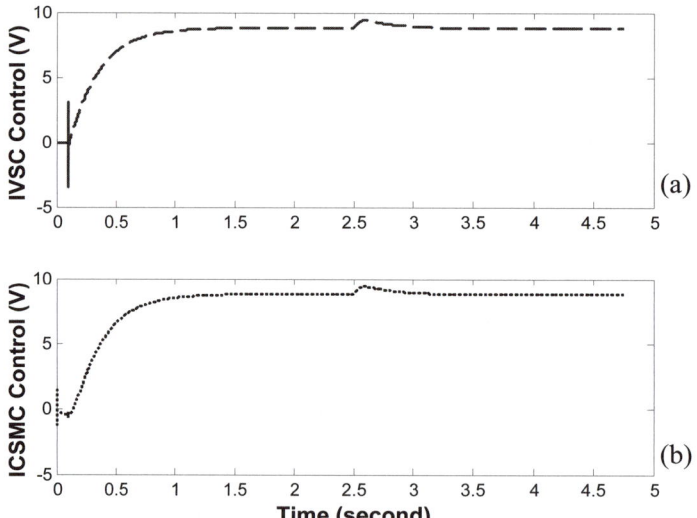

Fig. 5.11. Control voltage comparison: (a) IVSC; (b) ICSMC.

It is generally recognized that IVSC is suitable for robust tracking control (Chern and Wu, 1991; Cheng and Liu, 1999). In the third case study, we compare the proposed ICSMC with IVSC (see Appendix 5.2 for derivations) under various operating conditions. In order to obtain a fair comparison, we use the same coefficients of sliding manifolds for both controllers. The other control parameters are selected in such a manner that the control voltage input and tip displacement of ICSMC are very close to those of IVSC under the condition that both reference input and external disturbance are step signals. The control parameters are listed in Tables 5.4 and 5.5, respectively. The reference signal and disturbance situation are the same as assumed in the previous case study.

From Fig. 5.10, we can see that both IVSC and ICSMC can follow the step command input and reject the external step disturbance. Further inspection of the control voltage inputs (shown in Fig. 5.11) shows that IVSC exhibits control voltage sudden change at 0.1 second when the command signal has a sudden jump. It is worth mentioning that in the IVSC design, we have already incorporated a modified proper continuous function to alleviate the chattering phenomenon (Chern and Wu, 1991). The current simulation is under single mode discretization, and thus the sudden and drastic change of control voltage does not lead to significant tracking performance deterioration. Nevertheless, it can be envisioned that in practical implementation where the beam has infinitely many degrees of freedom, such voltage chattering would trigger the high frequency unmodeled dynamics. The control action of ICSMC, on the other hand, is much smoother than that of IVSC as expected. At 0.1 second, the control voltage of ICSMC still exhibits a small discontinuity due to the reference jump. However, the magnitude is very small as compared to that of IVSC. After a step external disturbance is imposed onto the system at 2.5second, both IVSC and ICSMC can eliminate the tracking error and their control voltage inputs are similar.

Fig. 5.12 illustrates a square wave tracking control comparison between IVSC and ICSMC. A Gaussian random noise is added to the system as an external disturbance. Both IVSC and ICSMC can follow the square wave. Once again, the control input of IVSC exhibits large jumps at changing edges of square wave. Close inspection of the beam tip displacement shows that ICSMC has better control robustness as compared to IVSC under the same random disturbance. In order to more clearly demonstrate the random disturbance attenuation, we present another case study in Fig. 5.13, where the reference input is set to be zero and the system is under Gaussian random disturbance. This is a typical case of vibration suppres-

sion. It is obvious from Fig. 5.13 that ICSMC has much improved disturbance attenuation capability than IVSC, which is due to the cubic state
feedback that can respond faster and more effective at early stage when the
beam tip is disturbed away from the zero position. Fig. 5.14 illustrates another beam vibration attenuation comparison. In this case, an initial velocity is imposed on beam tip, and external disturbance is still Gaussian random noise. Both IVSC and ICSMC show good vibration attenuation
capability, but the result of ICSMC is better than that of IVSC. ICSMC
has smaller over-shoot and requires less time to reach steady-state.

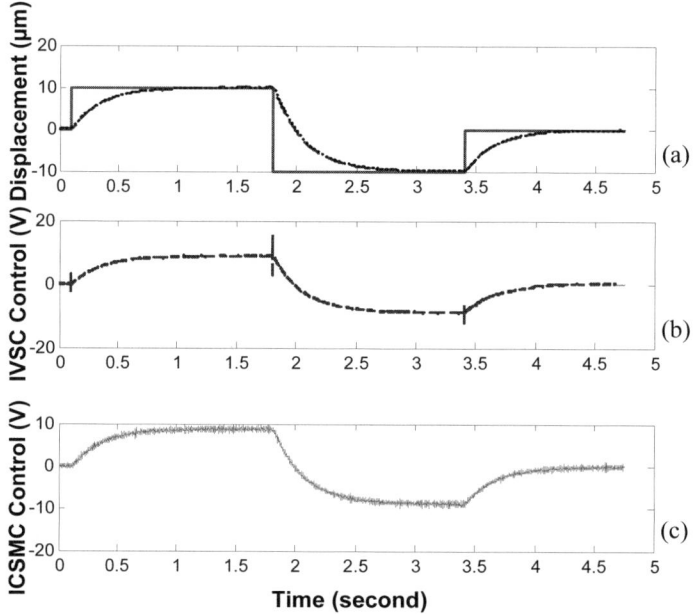

Fig. 5.12. Square wave tracking result comparison between IVSC and ICSMC:
(a) control performance; ———— : reference signal,   — — — : IVSC,   ·············· :
ICSMC; (b) IVSC control voltage; (c) ICSMC control voltage.

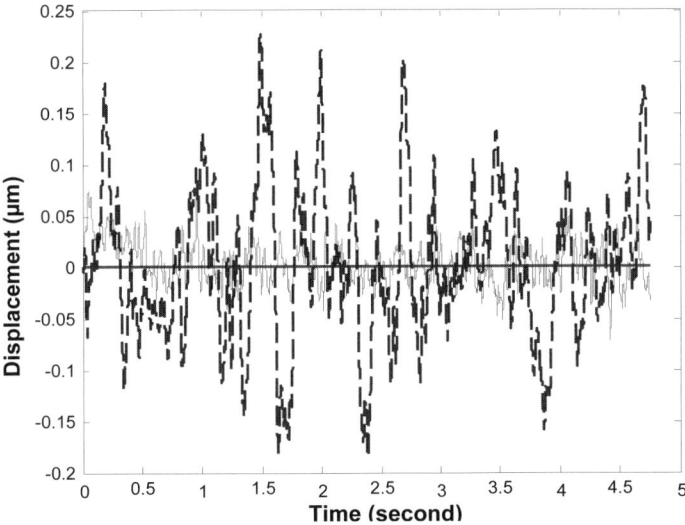

Fig. 5.13. Comparison of random disturbance attenuation between IVSC and ICSMC. ———— : reference signal, — — — : IVSC, ·············· : ICSMC.

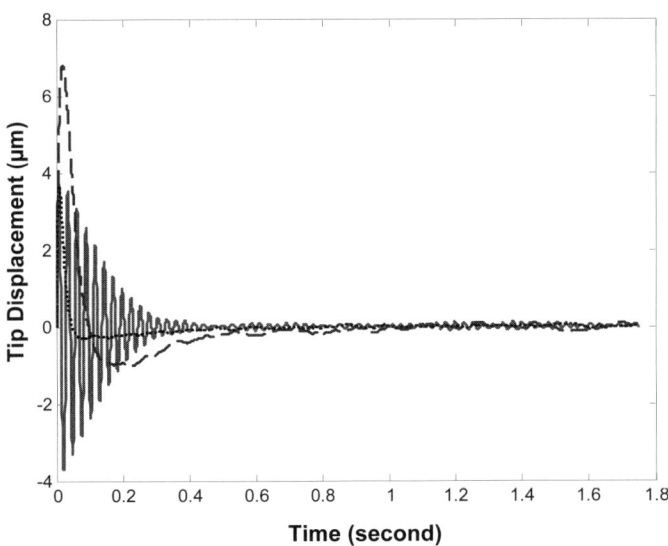

Fig. 5.14. Comparison of free vibration attenuation between IVSC and ICSMC. ———— : uncontrolled, — — — : IVSC, ·············· : ICSMC.

5.5 CONCLUDING REMARKS

In this chapter, a robust control methodology for piezoelectric actuators with hysteresis compensation is presented. A resistance-inductance (RL) circuitry is introduced to the piezoelectric actuator, which leads to two coupled dynamic sub-systems that can be cast into standard state-space format for control development. With this piezoelectric circuitry configuration, the charge and/or current in the piezoelectric actuator now become independent state variables that can be measured and fed back, which enables a direct hysteresis compensation.

An integral continuous sliding mode control (ICSMC) scheme, which combines the advantages of conventional continuous sliding mode control (CSMC) and integral variable structure control (IVSC), is developed and analyzed. It is shown that this ICSMC algorithm leads to much improved control precision as compared to the linear optimal control (LQR) when piezoelectric hysteresis is present. This comparison clearly demonstrates the necessity of hysteresis compensation. Analyses are also provided to illustrate the improved tracking control and vibration attenuation accuracy, enhanced control robustness under external disturbance, and better control action smoothness of the proposed ICSMC as compared to the conventional CSMC and IVSC.

APPENDIX 5.1 CONTINUOUS SLIDING MODE CONTROL (CSMC) DESIGN

The derivation is actually similar to that of ICSMC. Let

$$\bar{\mathbf{z}} = [\phi(l)x_1, x_2, x_3, x_4]^{\mathrm{T}} \tag{A5.1}$$

Then Eq. (5.9) can be re-written as,

$$\dot{\bar{\mathbf{z}}} = \mathbf{A}_z \bar{\mathbf{z}} - \mathbf{B}f(x_3) + \mathbf{B}u + \mathbf{D}F_{\mathrm{m}} \tag{A5.2}$$

where

$$
\mathbf{A}_z =
\begin{bmatrix}
0 & \phi(l) & 0 & 0 \\
-\dfrac{k}{m}\dfrac{1}{\phi(l)} & -\dfrac{g}{m} & -\dfrac{k_1}{m} & 0 \\
0 & 0 & 0 & 1 \\
-\dfrac{k_3}{L}\dfrac{1}{\phi(l)} & 0 & -\dfrac{k_2}{L} & -\dfrac{R}{L}
\end{bmatrix}
\tag{A5.3}
$$

We define the sliding manifold as,

$$
s = \mathbf{C}^\mathrm{T}\overline{\mathbf{z}} - c_1 r_\mathrm{d}
\tag{A5.4}
$$

Let the control input be

$$
u = -\mathbf{K}\tilde{\mathbf{z}} + \hat{f}(x_3) - u_\mathrm{d}
\tag{A5.5}
$$

where

$$
\tilde{\mathbf{z}} = [\overline{z}_1 - r_\mathrm{d}, \overline{z}_2, \overline{z}_3, \overline{z}_4]^\mathrm{T}
\tag{A5.6}
$$

Applying the reaching condition Eq. (5.14), and using Eqs. (A5.2) and (A5.3), we have

$$
s\dot{s} = s\mathbf{C}^\mathrm{T}[\mathbf{A}_z\overline{\mathbf{z}} - \mathbf{B}f(x_3) + \mathbf{B}u + \mathbf{D}F_\mathrm{m}] - c_1 \dot{r}_\mathrm{d}
\tag{A5.7}
$$

Substituting (A5.5) into (A5.7) yields

$$
s\dot{s} = s\mathbf{C}^\mathrm{T}\mathbf{B}\{[\boldsymbol{\alpha} - \mathbf{K}]\tilde{\mathbf{z}} + [\beta + \gamma r_\mathrm{d} + \tau\dot{r}_\mathrm{d} - u_\mathrm{d}]\}
\tag{A5.8}
$$

where

$$
\boldsymbol{\alpha} = (\mathbf{C}^\mathrm{T}\mathbf{B})^{-1}\mathbf{C}^\mathrm{T}\mathbf{A}_z = [\alpha_1, \alpha_2, \alpha_3, \alpha_4]
\tag{A5.9}
$$

$$\beta = (\mathbf{C}^{\mathrm{T}}\mathbf{B})^{-1}\mathbf{C}^{\mathrm{T}}\mathbf{D}F_{\mathrm{m}} + \Delta f \qquad (A5.10)$$

$$\gamma = -(\mathbf{C}^{\mathrm{T}}\mathbf{B})^{-1}\mathbf{C}^{\mathrm{T}}\mathbf{E} \qquad (A5.11)$$

$$\tau = -(\mathbf{C}^{\mathrm{T}}\mathbf{B})^{-1}c_1 \qquad (A5.12)$$

$$\mathbf{E} = [0, \frac{k}{m}\frac{1}{\phi(l)}, 0, \frac{k_3}{L}\frac{1}{\phi(l)}]^{\mathrm{T}} \qquad (A5.13)$$

$$\Delta f = \hat{f}(x_3) - f(x_3) \qquad (A5.14)$$

Here we define

$$\bar{\tau} = \frac{\sup \tau + \inf \tau}{2} \qquad (A5.15)$$

In order to satisfy the reaching condition Eq. (5.14), we may choose the following control parameters

$$K_i = \bar{\alpha}_i + (\sup \alpha_i - \bar{\alpha}_i + \delta_i)\frac{s\mathbf{C}^{\mathrm{T}}\bar{\mathbf{B}}\tilde{z}_i}{\lambda} \qquad \delta_i > 0 \qquad i = 1,2,3,4 \qquad (A5.16)$$

$$u_{\mathrm{d}} = \bar{\beta} + \bar{\gamma}c_5 r_{\mathrm{d}} + \{[\sup \beta - \bar{\beta}] + (\tilde{\gamma} - \bar{\gamma})r_{\mathrm{d}} + (\tilde{\tau} - \bar{\tau})\dot{r}_{\mathrm{d}} + \delta_{\mathrm{d}}\}\frac{s\mathbf{C}^{\mathrm{T}}\bar{\mathbf{B}}}{\lambda} \qquad (A5.17)$$

$$\delta_{\mathrm{d}} > 0$$

where

$$\tilde{\gamma} = \begin{cases} \sup \gamma, & \text{when } r_{\mathrm{d}} > 0 \\ \inf \gamma, & \text{when } r_{\mathrm{d}} < 0 \end{cases} \qquad (A5.18)$$

$$\tilde{\tau} = \begin{cases} \sup \tau, & \text{when } \dot{r}_d > 0 \\ \inf \tau, & \text{when } \dot{r}_d < 0 \end{cases} \tag{A5.19}$$

λ is again the boundary layer thickness of sliding manifold.

Substituting Eqs. (A5.16) and (A5.17) into Eq. (A5.8), we obtain

$$s\dot{s} = \varepsilon\{\sum_{i=1}^{4}[\alpha_i - \overline{\alpha}_i]s\mathbf{C}^T\overline{\mathbf{B}}\tilde{z}_i + [(\beta - \overline{\beta}) + (\gamma - \overline{\gamma})r_d + (\tau - \overline{\tau})\dot{r}_d]s\mathbf{C}^T\overline{\mathbf{B}} \tag{A5.20}$$

$$-\sum_{i=1}^{4}(\sup \alpha_i - \overline{\alpha}_i + \delta_i)(s\mathbf{C}^T\overline{\mathbf{B}}\tilde{z}_i)^2\frac{1}{\lambda}$$

$$-[(\sup \beta - \overline{\beta}) + (\tilde{\gamma} - \gamma)r_d + (\tilde{\tau} - \overline{\tau})\dot{r}_d + \delta_d](s\mathbf{C}^T\overline{\mathbf{B}})\frac{1}{\lambda}\}$$

Observe Eqs. (A5.16) and (A5.17). Clearly, if the following conditions are satisfied, the reaching condition Eq. (5.14) can be guaranteed.

$$\left|s\mathbf{C}^T\overline{\mathbf{B}}x_i\right| > \lambda \qquad i = 1,2,3,4 \tag{A5.21}$$

$$\left|s\mathbf{C}^T\overline{\mathbf{B}}\right| > \lambda \tag{A5.22}$$

The sliding manifold (A5.4) can be derived using a similar pole placement procedure outlined in Section 5.3.2.

APPENDIX 5.2 INTEGRAL VARIABLE STRUCTURE CONTROL (IVSC) DESIGN

The classical design of IVSC involves the reaching mode and the sliding mode. In virtue of Eqs. (5.12) and (5.9), we have,

$$\dot{s} = \mathbf{C}^{\mathrm{T}}[\mathbf{A}\mathbf{x} - \mathbf{B}f(x_3) + \mathbf{B}u + \mathbf{D}F_{\mathrm{m}}] + c_5[r_{\mathrm{d}} - \phi(l)x_1] \qquad \text{(A5.23)}$$
$$= [\mathbf{C}^{\mathrm{T}}\mathbf{A} - \mathbf{A}_\phi]x - \mathbf{C}^{\mathrm{T}}\mathbf{B}f(x_3) + \mathbf{C}^{\mathrm{T}}\mathbf{B}u + \mathbf{X}^{\mathrm{T}}\mathbf{D}F_{\mathrm{m}} + c_5 r_{\mathrm{d}}$$

In what follows we let the control input be decomposed into the equivalent control input u_{eq} and the switching control input u_{s},

$$u = u_{\mathrm{eq}} + u_{\mathrm{s}} \qquad \text{(A5.24)}$$

The equivalent control input is defined as the solution of $\dot{s} = 0$ under conditions that both the external disturbance and the uncertainties are absent,

$$u_{\mathrm{eq}} = -(\mathbf{C}^{\mathrm{T}}\overline{\mathbf{B}})^{-1}[(\mathbf{C}^{\mathrm{T}}\overline{\mathbf{A}} - \mathbf{A}_\phi)\mathbf{x} + c_5 r_{\mathrm{d}}] + \hat{f}(x_3) \qquad \text{(A5.25)}$$

The switching part is used primarily for satisfying the reaching condition and also for overcoming the external disturbance and the uncertainties. When the trajectory of the system is in the vicinity of the sliding manifold, the switching part of the controller constraints the system to follow the sliding manifold. In theory, the sign discontinuous function can be used as the switching control action (Hung et al., 1993; Utkin, 1993). However, the control action triggered by the sign function is in the high frequency band and can lead to the chattering phenomenon and excite the high frequency un-modeled dynamics of the system (Chern and Wu, 1991). In order to alleviate the chattering phenomenon, sign function is generally replaced by a proper continuous function, which is further improved by using a modified proper continuous function to account for different operating conditions (Chern and Wu, 1991),

$$u_{\mathrm{s}} = -\eta \cdot M_\xi(s) \qquad \text{(A5.26)}$$

where

$$M_\xi(s) = \frac{s}{|s| + \xi_0 + \xi_1|\phi(l)x_1 - r_d|} \tag{A5.27}$$

Here ξ_0 and ξ_1 are positive constants.

It can be easily verified that when the constant switching gain η is chosen as following, the reaching condition Eq. (5.14) can be guaranteed,

$$\eta = \sum_{i=1}^{4}(\sup a_i)|x_i| + \sup\Delta f + (\sup\Delta\mathbf{D})|F_m| + (\sup\Delta\mathbf{B})|r_d| \tag{A5.28}$$

where

$$\Delta\mathbf{A} = (\mathbf{C}^T\bar{\mathbf{B}})^{-1}(\mathbf{C}^T\bar{\mathbf{A}} - \mathbf{A}_\phi) - [\mathbf{C}^T\mathbf{B}]^{-1}[\mathbf{C}^T\mathbf{A} - \mathbf{A}_\phi] = [a_1, a_2, a_3, a_4] \tag{A5.29}$$

$$\Delta\mathbf{B} = \{(\mathbf{C}^T\bar{\mathbf{B}})^{-1} - [\mathbf{C}^T\mathbf{B}]^{-1}\}c_5 \tag{A5.30}$$

$$\Delta\mathbf{D} = [\mathbf{C}^T\mathbf{B}]^{-1}\mathbf{C}^T\mathbf{D} \tag{A5.31}$$

The sliding manifold can be derived using the pole placement procedure outlined in Section 5.3.2.

6 Active Vibration Confinement Enhancement

The design of vibration control schemes for structural systems can be facilitated in a variety of ways through energy dissipation and/or energy relocation. Many structures are spatially distributed, yielding the possibility of achieving vibration control through utilizing the vibration confinement technique. The underlying principle of vibration confinement can be considered as altering the structural vibration modes in such a manner that the corresponding modal components have much smaller amplitude in concerned regions than in other regions of the structure. As a result, the vibration energy will be confined or relocated to areas that are less important or where damping can be applied more easily. As pointed out by many researchers (Song and Jayasuriya, 1993; Choura, 1995; Shelly and Clark, 2000a), such confinement techniques could suppress vibration in the regions of concern more effectively than many traditional methods, and could reduce control power requirement when active means is involved.

It has been found that the presence of irregularities in a spatially periodic structure can inhibit the propagations of vibration energy within the structure, causing normal mode localization. Vakakis (1994), Vakakis *et al.* (1999) and Keane (1995) utilized this idea to realize passive vibration control. Essentially, they took advantage of the high sensitivity of vibration modes in periodic structures with respect to structural property change. (The vibration localization phenomenon will be further discussed in the next chapter.) For general non-periodic structures, Allaei and Tarnowski (1997) explored vibration confinement by using additional stiffening elements to alter the structure mode distribution. The performance of such idea, however, is limited, mainly because passive alteration of vibration modes for general non-periodic structures requires significant modification of the structure in most cases and thus oftentimes might not be practical. Researchers have also been resorting to active feedback controls to realize closed-loop vibration confinement. Choura and Yigit (1995) and Choura (1995) proposed solving feedback gains for vibration confinement by using an inverse eigenvalue problem method. Using a similar strategy, Shelley and Clark (1996) demonstrated that, for a special class of discrete systems where coupling in the equation of motion only occurs between neighboring elements, only two actuators are needed to shape all the ei-genvectors.

The flexibility offered by state feedback in multi-input systems beyond closed-loop eigenvalue assignment was first recognized by Moore (1976). It was shown that in those systems, not only the eigenvalues could be as-signed; one also has the freedom to adjust the eigenvectors, which is now generally referred to as eigenstructure assignment. Andry *et al.* (1983) dis-cussed some key issues in eigenstructure assignment in linear control sys-tems. In general, the assignment of eigenvector is possible only with mul-tiple inputs, and the assigned eigenvector must fall into an admissible space. They paid special attention to feedback gain solutions yielding the achievable eigenvectors that are the closest to the desired ones. Kwon and Youn (1987) extended the previous results by removing the assumption that the closed-loop eigenvalues are distinct and are different from any ei-genvalues of the open-loop system. Kim and Kim (1999) considered the case of partial eigenstructure assignment. Datta *et al.* (2000) developed an algorithm for the partial eigenstructure assignment of quadratic matrix pencil (mass, stiffness, and damping matrices in physical space). Song and Jayasuriya (1993) considered using an eigenstructure assignment algorithm for multi-input-multi-output systems to realize active vibration confine-ment. Their method requires that the number of actuators equals to the number of the structural degrees of freedom. Shelley and Clark (2000b and 2000c) used a singular value decomposition-based method for assign-ing shaped eigenvectors, where the closed-loop eigenvectors are pre-selected and the feedback gains are calculated to give the closest achiev-able approximation to the desired eigenvectors. Corr and Clark (1999) suggested a hybrid active-passive scheme to realize vibration confinement, where passive mechanical absorbers were used to reduce active power re-quirements and cost. Tang and Wang (2004) advanced the vibration con-finement technique by applying the active control input through a piezo-electric transducer circuitry. With the introduction of circuitry elements, the design space for eigenstructure assignment can be greatly enlarged. They also developed a new eigenvector assignment algorithm using Rayleigh Principle to suppress vibration more directly in regions of inter-est. To maximize system performance, a simultaneous optimization/ opti-mal eigenvector assignment approach to concurrently determine the cir-cuitry elements and active control parameters was formulated. Following the idea proposed by Tang and Wang (2004), Wu and Wang (2007) prac-ticed the design of vibration isolation systems using the energy confine-ment approach.

6.1 PROBLEM STATEMENT AND OBJECTIVE

Theoretically, the eigenstructure assignment is a natural technique for designing active vibration confinement. The practical drawbacks of such approach can be inspected from a mathematical standpoint, which are summarized as follows.

(a) In such an approach, the closed-loop eigenvectors have to fall into certain admissible space that is decided by the closed-loop eigenvalues and the state matrices. Although the dimension of this admissible space can be increased with the increase of number of actuators/sensors, the design options for realizing vibration confinement is often limited. Moreover, the vibration energy for the closed-loop system, even if it is confined, is still within the mechanical structure. In fact, the part of the structure where the energy is confined to could have severe vibration.

(b) In most eigenstructure assignment methods, one needs to predetermine the desired closed-loop eigenvectors a priori. This will create two problems. First, because the closed-loop eigenvectors have to fall into certain admissible space, the desired eigenvectors may be very different from the achievable eigenvectors. Second, in many cases, one is only concerned with the vibration suppression of certain degrees of freedom of the structure, and the vibration level of the remaining part of the structure is of little interest. Therefore, a precise match between the desired and the achievable eigenvectors is not necessary.

The objective of this chapter is to present a methodology for vibration confinement by integrating piezoelectric transducer circuitry, which can effectively address the aforementioned issues and advance the state-of-the-art. Through electro-mechanical tailoring of the piezoelectric circuitry that is integrated with the structure, one can confine the vibration energy to the circuitry part of the system and to other unimportant structural regions. By using the circuitry elements as design variables, one can re-design the electrical system to achieve vibration confinement without changing the mechanical structure. This can greatly enlarge the design space in eigenstructure assignment with active inputs. By confining the system energy to the electrical circuit part of the integrated system, one will be able to quiet a much larger portion of the mechanical structure more effectively. To fully utilize the piezoelectric circuitry idea, an eigenvector assignment algorithm is developed. After obtaining the base of the achievable eigenvector subspace, the Rayleigh Principle is used to solve for the optimal achievable eigenvectors that have the minimal vector 2-norm in the concerned regions. In other words, we are designing a closed-loop system that directly minimizes vibration energy in the concerned regions. Following the methodology development, two case studies, one demonstrating the methodology development of active-passive vibration control and one illustrating the implementation of the method for vibration isolation, are outlined in this chapter.

6.2 BASIC IDEA AND EIGENSTRUCTURE ASSIGNMENT ALGORITHM

6.2.1 System Model and Vibration Confinement Idea

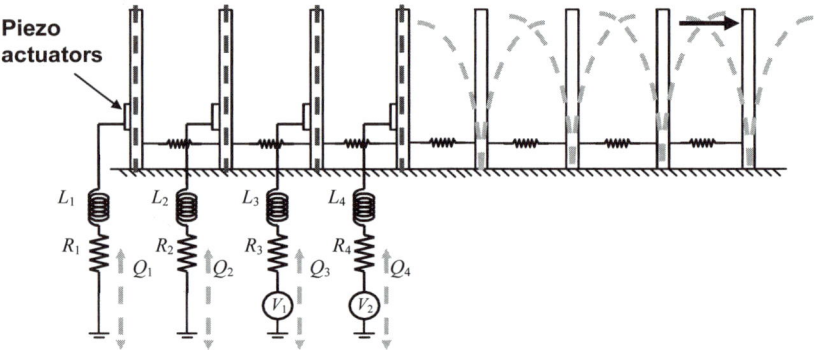

Fig. 6.1. Illustrative system consisting of mechanical structure, multi-branch piezoelectric circuitry, and active control voltage inputs. The disturbance acts on the 8-th beam, and the vibration of the first four beams needs to be suppressed.

An illustrative system consisting of the host mechanical structure and an active-passive piezoelectric circuitry network is shown in Fig. 6.1. It should be noted that there will be many possible piezoelectric circuitry configurations, and one may select the network topology and circuitry dimension based on specific application. Without loss of generality, we first use the example system shown in Fig. 6.1 to demonstrate the merits of the proposed new ideas. In general, the discretized equation of motion for an integrated system can be expressed as (Tang and Wang, 2004)

$$\begin{bmatrix} \mathbf{M} & \mathbf{0} \\ \mathbf{0} & \mathbf{L} \end{bmatrix} \begin{Bmatrix} \ddot{\mathbf{q}} \\ \ddot{\mathbf{Q}} \end{Bmatrix} + \begin{bmatrix} \mathbf{C_d} & \mathbf{0} \\ \mathbf{0} & \mathbf{R} \end{bmatrix} \begin{Bmatrix} \dot{\mathbf{q}} \\ \dot{\mathbf{Q}} \end{Bmatrix} + \begin{bmatrix} \mathbf{K_0} & \mathbf{K_1} \\ \mathbf{K_1^T} & \mathbf{K_2} \end{bmatrix} \begin{Bmatrix} \mathbf{q} \\ \mathbf{Q} \end{Bmatrix} = \begin{Bmatrix} \mathbf{F_d} \\ \mathbf{0} \end{Bmatrix} + \begin{Bmatrix} \mathbf{0} \\ \mathbf{B_0 V} \end{Bmatrix} \qquad (6.1)$$

where, $\mathbf{M}$, $\mathbf{C_d}$, and $\mathbf{K_0}$ are $n \times n$ mass, damping, and stiffness matrices of the structure, $\mathbf{q}$ is the n- dimensional structural displacement vector, $\mathbf{Q}$ is the m-dimensional electrical charge flow vector, $\mathbf{L}$, $\mathbf{R}$, and $\mathbf{K_2}$ are $m \times m$ inductance, resistance, and inverse capacitance matrices of the piezoelectric circuitry, $\mathbf{K_1}$ is an $n \times m$ matrix reflecting the electro-mechanical coupling effect of the piezoelectric transducer, $\mathbf{F_d}$ is the external disturbance vector acting on the structure, $\mathbf{B_0}$ is the $m \times l$ control input matrix, and $\mathbf{V}$ is the l-dimensional active control input (voltage) vector.

Equation (6.1) can be easily cast into the following state-space form,

$$\dot{\mathbf{x}} = \mathbf{A}\mathbf{x} + \mathbf{B}\mathbf{u} + \mathbf{E}\mathbf{f} \qquad (6.2\text{a-c})$$
$$\mathbf{y} = \mathbf{C}\mathbf{x}$$
$$\mathbf{u} = \mathbf{K}\mathbf{y}$$

where

$$\mathbf{x} = \begin{Bmatrix} \mathbf{q} \\ \mathbf{Q} \\ \dot{\mathbf{q}} \\ \dot{\mathbf{Q}} \end{Bmatrix}, \quad \mathbf{A} = \begin{bmatrix} \mathbf{0}_{(n+m)\times(n+m)} & \mathbf{I}_{(n+m)} \\ \begin{bmatrix} \mathbf{M}^{-1}\mathbf{K_0} & \mathbf{M}^{-1}\mathbf{K_1} \\ \mathbf{L}^{-1}\mathbf{K_1^T} & \mathbf{L}^{-1}\mathbf{K_2} \end{bmatrix} & \begin{bmatrix} \mathbf{M}^{-1}\mathbf{C_d} & \mathbf{0} \\ \mathbf{0} & \mathbf{L}^{-1}\mathbf{R} \end{bmatrix} \end{bmatrix} \qquad (6.3\text{a-f})$$

$$\mathbf{B} = \begin{Bmatrix} \mathbf{0}_{(n+m)\times l} \\ \mathbf{0}_{n\times l} \\ \mathbf{L}^{-1}\mathbf{B_0} \end{Bmatrix}, \quad \mathbf{E} = \begin{Bmatrix} \mathbf{0}_{(m+n)\times n} \\ \mathbf{M}^{-1} \\ \mathbf{0}_{m\times n} \end{Bmatrix}$$

$$\mathbf{f} = \mathbf{F_d}, \quad \mathbf{u} = \mathbf{V}$$

For simplicity, in the following derivations we let $N = 2(n+m)$. The dimensions of matrices $\mathbf{A}$, $\mathbf{B}$, and $\mathbf{E}$ are, respectively, $N \times N$, $N \times l$, and $N \times N$. The solution to Eq. (6.2) can be obtained as (Zhang et al., 1990),

$$\mathbf{x}(t) = \mathbf{\Phi} e^{\mathbf{\Lambda} t}\mathbf{\Psi}^{\mathrm{T}}\mathbf{x}(0) + \mathbf{\Phi}\int_0^t e^{\mathbf{\Lambda}(t-\tau)}\mathbf{\Psi}^{\mathrm{T}}\mathbf{E}\mathbf{f}(\tau)d\tau \tag{6.4}$$

where $\mathbf{\Lambda} = \underset{N}{\mathrm{diag}}(\lambda_i)$ is the diagonal matrix of the closed-loop eigenvalues, and $\mathbf{\Phi} = [\boldsymbol{\varphi}_1, \cdots, \boldsymbol{\varphi}_N]$ and $\mathbf{\Psi} = [\boldsymbol{\psi}_1, \cdots, \boldsymbol{\psi}_N]$ are the corresponding right and left closed-loop eigenvector sets in the state space, *i.e.*,

$$(\mathbf{A} + \mathbf{BKC} - \lambda_i \mathbf{I}_N)\boldsymbol{\varphi}_i = \mathbf{0} \tag{6.5a,b}$$

$$(\mathbf{A} + \mathbf{BKC} - \lambda_i \mathbf{I}_N)^{\mathrm{T}}\boldsymbol{\psi}_i = \mathbf{0}$$

Clearly, if one can assign the closed-loop eigenvectors such that the components of $\mathbf{\Phi}$ corresponding to the concerned displacement coordinates are as small as possible, the response amplitudes at these regions will be reduced, which is the basic idea of vibration confinement.

6.2.2 Eigenstructure Assignment Algorithm

The new eigenvector assignment algorithm to be presented in this section is an extension of the traditional right eigenvector assignment via SVD (singular value decomposition) technique that was proposed by Cunningham (1980) and later utilized by Corr and Clark (1999) and Shelley and Clark (2000b and 2000c). Here we assume that the closed-loop eigenvalues λ_i all differ from the open-loop eigenvalues, which means $\mathbf{A} - \lambda_i \mathbf{I}_N$ is non-singular. An important result for the right eigenvector assignment is given as follows (Moore, 1976; Andry *et al.*, 1983). Let $\{\lambda_i\}$ be a self-conjugate set of distinct complex numbers (closed-loop eigenvalues). There exists a real matrix $\mathbf{K}$ (feedback gain) satisfying

$$(\mathbf{A} + \mathbf{BKC})\boldsymbol{\varphi}_i = \lambda_i \boldsymbol{\varphi}_i, \qquad i = 1, \cdots, N \tag{6.6}$$

if and only if
  (a) $\boldsymbol{\varphi}_1, \cdots, \boldsymbol{\varphi}_N$ are linearly independent in C^N;
  (b) $\boldsymbol{\varphi}_i = \boldsymbol{\varphi}_j^*$ when $\lambda_i = \lambda_j^*$;
  (c) $\boldsymbol{\varphi}_i \in \mathrm{span}(N_{\lambda_i})$.

Hereafter * denotes the complex conjugate.

We define

$$\mathbf{S}_{\lambda_i} = [\mathbf{A} - \lambda_i \mathbf{I}_N \,|\, \mathbf{B}] \qquad (6.7)$$

and a compatibly partitioned matrix,

$$\mathbf{T}_{\lambda_i} = \begin{bmatrix} \mathbf{N}_{\lambda_i} \\ \mathbf{M}_{\lambda_i} \end{bmatrix} \qquad (6.8)$$

where the columns of $\mathbf{T}_{\lambda_i}$ form a basis of the null space of $\mathbf{S}_{\lambda_i}$. We may then re-write Eq. (6.6) in the following form,

$$[\mathbf{A} - \lambda_i \mathbf{I}_N \,|\, \mathbf{B}] \begin{Bmatrix} \boldsymbol{\varphi}_i \\ \mathbf{KC}\boldsymbol{\varphi}_i \end{Bmatrix} = 0 \qquad (6.9)$$

Recall that the columns of $\mathbf{T}_{\lambda_i}$ form a basis of the null space of $\mathbf{S}_{\lambda_i}$. Clearly, the (assigned) right eigenvector must fall into the subspace spanned by the column vectors of $\mathbf{N}_{\lambda_i}$, and, moreover, the vector $\begin{Bmatrix} \boldsymbol{\varphi}_i \\ \mathbf{KC}\boldsymbol{\varphi}_i \end{Bmatrix}$ must fall into the subspace spanned by the column vectors of $\mathbf{T}_{\lambda_i}$. The former condition imposes the restriction on the achievable right eigenvector, and the latter condition can be used, after the desired eigenvector is selected from the achievable eigenvector set, to determine the control gain. Let

$$\mathbf{S}_{\lambda_i} = [\mathbf{A} - \lambda_i \mathbf{I}_N \,|\, \mathbf{B}] = \mathbf{U}_i [\mathbf{D}_i \,|\, \mathbf{0}_{N \times l}] \mathbf{V}_i^* \qquad (6.10)$$

Here, $\mathbf{U}_i$ and $\mathbf{V}_i$ are the left and right singular vector matrices (Klema and Laub, 1980). Since $\mathbf{A} - \lambda_i \mathbf{I}_N$ is non-singular, $\mathbf{D}_i$ is a positive definite diagonal matrix containing all the singular values of $[\mathbf{A} - \lambda_i \mathbf{I}_N \,|\, \mathbf{B}]$. The left and right singular vector matrices are both unitary, *i.e.*,

$$\mathbf{U}_i^* \mathbf{U}_i = \mathbf{I}_N \qquad\qquad (6.11a,b)$$

$$\mathbf{V}_i^* \mathbf{V}_i = \mathbf{I}_{N+l}$$

We then partition $\mathbf{V}_i$ in the following form,

$$\mathbf{V}_i = \begin{bmatrix} \mathbf{V}_{11}^{(i)} & \mathbf{V}_{12}^{(i)} \\ \mathbf{V}_{21}^{(i)} & \mathbf{V}_{22}^{(i)} \end{bmatrix} \qquad\qquad (6.12)$$

where $\mathbf{V}_{11}^{(i)}$, $\mathbf{V}_{22}^{(i)}$, $\mathbf{V}_{12}^{(i)}$, and $\mathbf{V}_{21}^{(i)}$ are $N \times N$, $l \times l$, $N \times l$, and $l \times N$ sub-matrices, respectively. Post-multiplying Eq. (6.10) by $\begin{bmatrix} \mathbf{V}_{12}^{(i)} \\ \mathbf{V}_{22}^{(i)} \end{bmatrix}$, in virtue of Eq. (6.11b) one obtains

$$[\mathbf{A} - \lambda_i \mathbf{I}_N \,|\, \mathbf{B}] \begin{bmatrix} \mathbf{V}_{12}^{(i)} \\ \mathbf{V}_{22}^{(i)} \end{bmatrix} = \mathbf{0} \qquad\qquad (6.13)$$

Clearly, the linearly independent column vectors of matrix $\begin{bmatrix} \mathbf{V}_{12}^{(i)} \\ \mathbf{V}_{22}^{(i)} \end{bmatrix}$ span the null space of $\mathbf{S}_{\lambda_i}$. In other words, any achievable eigenvector must be a linear combination of the column vectors of $\mathbf{V}_{12}^{(i)}$.

At this point, one can easily see the benefit of using piezoelectric transducer circuitry to realize vibration confinement. As discussed earlier, one of the main limitations of the current active vibration confinement methods is due to the restricted choice of the closed-loop eigenvectors. Now with the introduction of the circuitry elements, the state matrices $\mathbf{A}$ and $\mathbf{B}$ can be tailored and reformed. The dimension of the system can be significantly increased with the addition of circuitry loops in the electrical network (see Eq. (6.3)). As a result, much more design freedom can be expected in choosing the closed-loop eigenvectors from the achievable eigenvector space. While adding mechanical components could have similar effects, integrating/adjusting electrical circuits would have practical advantages, as it is much easier to implement than modifying/adding mechanical components and has the potential to develop sophisticated circuit design for specific applications.

In the traditional eigenvector assignment approach, one needs to first determine the desired eigenvector, denoted as $\boldsymbol{\varphi}_i^d$. In general, however, a desired right eigenvector $\boldsymbol{\varphi}_i^d$ will not reside in the prescribed subspace, and hence cannot be accurately achieved. One way to resolve this matter is to find the closest choice that is the projection of $\boldsymbol{\varphi}_i^d$ onto the subspace spanned by the column vectors of $\mathbf{V}_{12}^{(i)}$. Let the "closest" achievable right eigenvector be

$$\boldsymbol{\varphi}_i^a = \mathbf{V}_{12}^{(i)} \mathbf{r}_i \tag{6.14}$$

One may determine $\mathbf{r}_i$ by minimizing (Meirovitch, 1990)

$$J_{ri} = \left\| \boldsymbol{\varphi}_i^d - \mathbf{V}_{12}^{(i)} \mathbf{r}_i \right\|^2 \tag{6.15}$$

It can be easily shown, by letting $\mathrm{d}J_{ri}/\mathrm{d}r_i = 0$, that

$$\tilde{\mathbf{r}}_i = (\mathbf{V}_{12}^{(i)*} \mathbf{V}_{12}^{(i)})^{-1} \mathbf{V}_{12}^{(i)*} \boldsymbol{\varphi}_i^d \tag{6.16}$$

and hence the closest achievable right eigenvector is

$$\boldsymbol{\varphi}_i^a = \mathbf{V}_{12}^{(i)} (\mathbf{V}_{12}^{(i)*} \mathbf{V}_{12}^{(i)})^{-1} \mathbf{V}_{12}^{(i)*} \boldsymbol{\varphi}_i^d \tag{6.17}$$

Comparing Eq. (6.9) with Eq. (6.13) and recalling Eq. (6.14), one can obtain

$$\mathbf{K} \mathbf{C} \boldsymbol{\varphi}_i^a = \mathbf{V}_{22}^{(i)} \mathbf{r}_i \tag{6.18}$$

Define $\mathbf{w}_i = \mathbf{V}_{22}^{(i)}\mathbf{r}_i$. We repeat the above procedure for all the desired right eigenvectors. Finally, we will have

$$\mathbf{KC\Phi}^a = \mathbf{W} \qquad (6.19)$$

where, $\mathbf{\Phi}^a = [\mathbf{\phi}_1^a, \cdots, \mathbf{\phi}_N^a]$, and $\mathbf{W} = [\mathbf{w}_1, \cdots, \mathbf{w}_N]$. The feedback gain matrix is then obtained as

$$\mathbf{K} = \mathbf{W}(\mathbf{C\Phi}^a)^*[\mathbf{C\Phi}^a(\mathbf{C\Phi}^a)^*]^{-1} \qquad (6.20)$$

The pre-determination of the closed-loop eigenvectors, however, is an *ad hoc* procedure. In general, it is very difficult to estimate the closed-loop eigenvectors *a priori*, especially for complex structures. Such difficulty often leads to a large discrepancy between the desired eigenvectors and the achievable eigenvectors. The closest achievable eigenvector approach described above is a common practice used in many vibration confinement problems. Although it guarantees that the final achievable eigenvectors are the closest to the desired ones in the vector 2-norm sense, it does not guarantee that the eigenvector components in the concerned regions (coordinates) have the smallest norms. On the other hand, in many cases, one is only concerned with the vibration suppression of certain degrees of freedom of the structure, and the vibration level of the remaining part of the structure is of little interest. Therefore, a precise match between the desired eigenvectors and the achievable eigenvectors is not necessary. In fact, the pre-determination of those 'unimportant' eigenvector components could lead to unsatisfactory performance because the controller will try to match the desired and achievable eigenvector components at these unimportant degrees of freedom, thus trading off the closeness at the important degrees of freedom.

Here we outline an approach to directly finding the optimal achievable eigenvectors from the achievable eigenvector subspace by minimizing a relative energy ratio. Let an achievable eigenvector be given as Eq. (6.14). Here we minimize the ratio between the square of norm of the eigenvector components at the concerned coordinates and that of the entire eigenvector (hereafter referred to as modal energy ratio), *i.e.*, to find the minimum of the following expression,

$$\min \frac{\boldsymbol{\varphi}_i^{ac*} \boldsymbol{\varphi}_i^{ac}}{\boldsymbol{\varphi}_i^{a*} \boldsymbol{\varphi}_i^{a}} = \min \frac{\mathbf{r}_i^{*} \mathbf{V}_{12}^{(i)*} \mathbf{b}^T \mathbf{b} \mathbf{V}_{12}^{(i)} \mathbf{r}_i}{\mathbf{r}_i^{*} \mathbf{V}_{12}^{(i)*} \mathbf{V}_{12}^{(i)} \mathbf{r}_i} \qquad (6.21)$$

where

$$\boldsymbol{\varphi}_i^{ac} = \mathbf{b} \boldsymbol{\varphi}_i^{a} \qquad (6.22)$$

$\boldsymbol{\varphi}_i^{ac}$ is a vector formed by the eigenvector components at the concerned coordinates, and $\mathbf{b}$ is a Boolean matrix relating $\boldsymbol{\varphi}_i^{ac}$ to $\boldsymbol{\varphi}_i^{a}$. Clearly, the denominator in Eq. (6.21), $\boldsymbol{\varphi}_i^{a*} \boldsymbol{\varphi}_i^{a}$, can be regarded as a total energy measure of the mode, and the numerator, $\boldsymbol{\varphi}_i^{ac*} \boldsymbol{\varphi}_i^{ac}$, characterizes the modal energy level at the concerned coordinates. We define

$$\boldsymbol{\alpha} = \mathbf{V}_{12}^{(i)*} \mathbf{b}^T \mathbf{b} \mathbf{V}_{12}^{(i)} \qquad (6.23a,b)$$

$$\boldsymbol{\beta} = \mathbf{V}_{12}^{(i)*} \mathbf{V}_{12}^{(i)}$$

A lower bound to the minimum energy ratio can be found by minimizing (6.21) over all non-trivial $\mathbf{r}_i$, or equivalently, by finding the smallest eigenvalue and the corresponding eigenvector of the following eigenvalue problem,

$$\boldsymbol{\alpha} \mathbf{r}_i = \gamma \boldsymbol{\beta} \mathbf{r}_i \qquad (6.24)$$

Indeed, from the Rayleigh Principle (Meirovitch, 1990), we obtain

$$\min \frac{\boldsymbol{\varphi}_i^{ac*} \boldsymbol{\varphi}_i^{ac}}{\boldsymbol{\varphi}_i^{a*} \boldsymbol{\varphi}_i^{a}} = \frac{\tilde{\mathbf{r}}_i^{*} \boldsymbol{\alpha} \tilde{\mathbf{r}}_i}{\tilde{\mathbf{r}}_i^{*} \boldsymbol{\beta} \tilde{\mathbf{r}}_i} = \gamma_1 \qquad (6.25)$$

where γ_1 is the smallest eigenvalue given by (6.24), and $\tilde{\mathbf{r}}_i$ is the corresponding eigenvector. With this procedure, one no longer needs to predetermine the desired eigenvectors. Instead, the obtained optimal achievable eigenvectors will have the minimal vector 2-norm at the concerned regions, leading to vibration confinement.

The algorithm for the optimal eigenvector assignment can thus be summarized as follows:

1) We perform the SVD for $\mathbf{S}_{\lambda_i} = [\mathbf{A} - \lambda_i \mathbf{I}_N \mid \mathbf{B}]$ for all λ_i, $i = 1, \cdots, N$, and obtain $\begin{bmatrix} \mathbf{V}_{12}^{(i)} \\ \mathbf{V}_{22}^{(i)} \end{bmatrix}$;

2) For each $\mathbf{V}_{12}^{(i)}$, we solve for the eigenvalue problem (6.24). After obtaining the smallest eigenvalue γ_1 and the corresponding eigenvector $\tilde{\mathbf{r}}_i$, the optimal achievable eigenvector is given as Eq. (6.14).

3) We finally calculate the feedback gain by using Eq. (6.20).

6.3 SIMULTANEOUS OPTIMIZATION / OPTIMAL EIGENVECTOR ASSIGNMENT FOR ACTIVE-PASSIVE VIBRATION CONTROL

This section presents a case study that utilizes the vibration confinement idea outlined in Section 6.2 for active-passive vibration control. The generic example structure shown in Fig. 6.1 is used to illustrate the concept.

6.3.1 Algorithm Development for Active-Passive Vibration Control Using Vibration Confinement

Section 6.2 provides an algorithm for the optimal selection of eigenvectors that have minimized vector 2-norm at the concerned region of the structure. As one can see from the derivation, the introduction of the circuitry elements can increase the dimension of the state matrices and thus enlarge the design space in the eigenstructure assignment. The specific selection of those passive circuitry elements warrants further discussion here.

As in many active-passive hybrid control designs, one obvious choice for the passive circuitry elements is the one that can provide optimal passive damping. In other words, the inductance and resistance values of the circuits can be selected as the optimal passive values that are determined by numerically minimizing the frequency responses of the system (see Chapters 2 and 3). After the passive parameters are determined, one can sequentially use the optimal eigenvector assignment algorithm (see Section

6.2) to solve for the active gains and the design for the control system is then completed. The merit of this sequential approach is that the system will have the optimal passive design in case the active part failed. In other words, it is a fail-safe design. On the other hand, with such an approach, the energy absorbing/storage feature of the piezoelectric circuits is not explicitly taken into account in the design of active control, and the passive parameters so determined could be in conflict with the eventual goal of vibration confinement. That is, the circuitry elements that provide the optimal passive damping might not be the best choice for active vibration confinement. It should be noted in Eq. (6.25) that, for a given set of passive parameters, after we select the closed-loop eigenvalues we could directly obtain the minimized energy ratio for all the modes. This actually implies that we can develop a streamlined approach to determine the passive and active parameters simultaneously. We seek the minimum of an overall objective function for a concurrent active-passive hybrid control design, such as a weighted sum of modal energy ratios as shown below

$$J_{min} = \min_{\text{passive parameters}} \sum_i \theta_i \min_{\text{active control gain}} \frac{\varphi_i^{ac*} \varphi_i^{ac}}{\varphi_i^{a*} \varphi_i^{a}} \tag{6.26}$$

where θ_i is the weighting on the i-th mode. In other words, there are two loops in the hybrid control design. The inner loop consists of the optimal eigenvector assignment algorithm that gives the minimum modal energy ratio at concerned region under a set of given passive circuitry parameters. The outer optimization loop consists of a search for the passive circuitry parameters to further minimize the overall objective function that is the summation of individual modal energy ratios. It is worth mentioning that, in the search of optimal passive parameters, the open-loop eigenvalues of the passive system will vary. Therefore, one will need to adjust the closed-loop eigenvalues accordingly. Such adjustment can be easily incorporated in the design process, by setting up a uniform criterion for the shift of eigenvalues (this will be illustrated in the next section). With this procedure, the active and passive design variables (active gains and passive circuitry parameters) are determined simultaneously through iteration between the inner and outer loop optimizations (Fig. 6.2). An interesting feature of this formulation is that the inner and outer loops share virtually the same objective function, which means that the passive and active mechanisms directly complement each other to reach the same goal. Clearly, such a simultaneous optimization /optimal eigenvector assignment method integrates together the advantages of both the passive and active approaches.

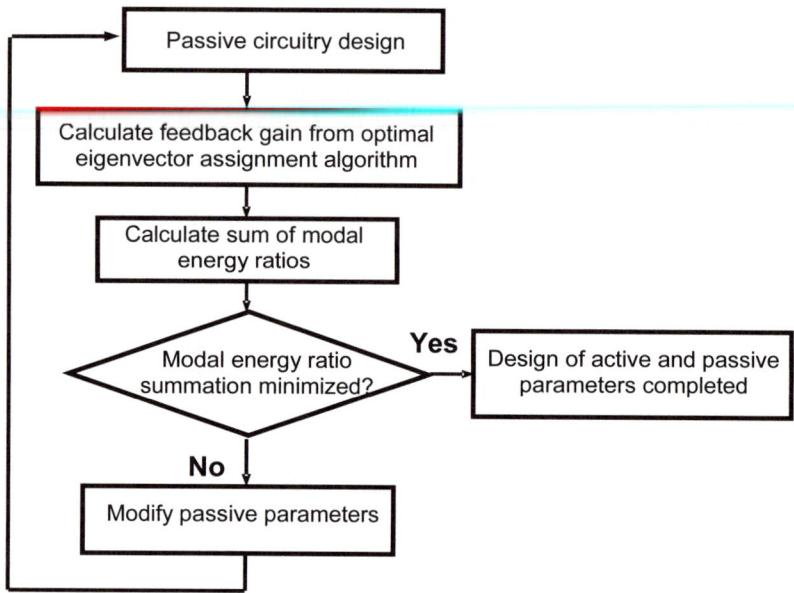

Fig. 6.2. Simultaneous optimization / optimal eigenvector assignment procedure for active-passive hybrid system design.

Table 6.1. Parameters used in active-passive vibration control simulation

$l_b = 0.2915 \, \text{m}$	$x_l = 0.012 \, \text{m}$	$x_r = 0.0844 \, \text{m}$
$w_b = 0.0508 \, \text{m}$	$w_p = 0.0343 \, \text{m}$	$t_b = 0.003175 \, \text{m}$
$t_p = 0.000267 \, \text{m}$	$\rho_b = 2663 \, \text{kg/m}^3$	$\rho_p = 7800 \, \text{kg/m}^3$
$E_b = 6.0 \times 10^{10} \, \text{N/m}^2$	$E_p = 6.6 \times 10^{10} \, \text{N/m}^2$	$h_{31} = 9.3022 \times 10^8 \, \text{N/C}$
$\beta_{33} = 9.5022 \times 10^7 \, \text{V}$	$k_c = 2k_b$	

6.3.2 Case Analysis

For the purpose of illustration and without loss of generality, we use the coupled beam structure shown in Fig. 6.1 for the case study demonstrating vibration control. This illustrative structure consists of eight identical cantilevered beams coupled with identical springs. We apply piezoelectric circuitry networks to achieve vibration control. The first four beams are augmented with resonant circuits, and the third and fourth circuits are further integrated with control voltage inputs to form two active-passive hybrid piezoelectric actuator networks. The locations and sizes of all the piezoelectric patches are the same. The system equation in the physical space is given as Eq. (6.1) or (6.2). We use a one-mode assumption to characterize each beam's motion. All the relevant derivations can be found in the paper by Tang and Wang (2004), and the parameters used in simulation are listed in Table 6.1. The original structure has eight degrees of freedom, corresponding to the mechanical displacements of the beams. The circuits provide additional four degrees of freedom, which are the charge in the circuits. For this specific system, we assume that we have full state feedback, *i.e.*, $\mathbf{C} = \mathbf{I}$. In practical applications, various sensors can be used to obtain the structural displacement and velocity signals, and the current and charge signals in the circuits can be obtained by measuring the voltage across the inductor and resistor elements. The control objective here is to suppress the vibration of the first four beams.

Table 6.2. Optimal passive circuitry element values

$L_1 = 225.15\,\mathrm{H}$	$L_2 = 239.49\,\mathrm{H}$	$L_3 = 74.50\,\mathrm{H}$
$L_4 = 140.80\,\mathrm{H}$	$R_1 = 7189.98\,\Omega$	$R_2 = 6674.59\,\Omega$
$R_3 = 6941.74\,\Omega$	$R_4 = 6461.15\,\Omega$	

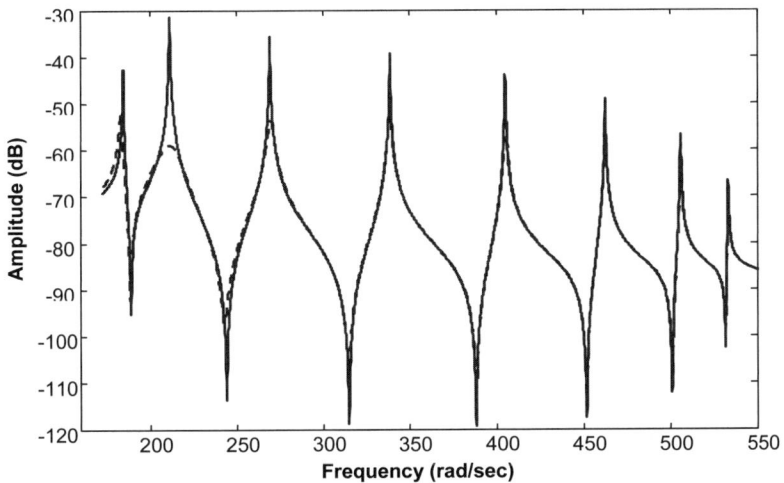

Fig. 6.3. Frequency responses of the first beam.
——— : original structure without circuits; − − − − − : mechanical structure integrated with piezoelectric circuitry network.

We first verify the optimal eigenvector assignment algorithm and compare the result with those obtained by using the traditional eigenstructure assignment approach. In this first case study, the inductance and resistance values of the circuits are selected as the optimal passive values that are determined by numerically minimizing the frequency responses of the system (Hollkamp, 1994), which are listed in Table 6.2. This provides a fail-safe situation, and leads to a sequential approach for optimal eigenvector assignment. Fig. 6.3 shows the frequency response comparison of the first beam, with and without the passive circuits. Clearly, the passive circuits provide significant amount of damping to the system. We then determine the closed-loop eigenvalues for the system. Different from other vibration control methods, in the eigenvector assignment approach we focus our interest primarily on the modal/response distribution, and treat the locations of closed-loop eigenvalues as secondary requirement. The closed-loop eigenvalues are selected by forming an LQR problem to minimize the following objective function (Lewis and Syrmos, 1995)

$$J = \int (\mathbf{x}^T \mathbf{Q} \mathbf{x} + \mathbf{u}^T \mathbf{R} \mathbf{u}) dt \qquad (6.27)$$

where the performance and control effort weighting matrices are set as,

$$\mathbf{Q} = 10^2 \mathbf{I}_{24}, \qquad \mathbf{R} = \mathbf{I}_2 \qquad (6.28)$$

Note that in the above selection the weighting on control performance is very small. In order to highlight the effect of the eigenvector assignment and vibration confinement on the system control performance, the closed-loop eigenvalues we select are only slightly different to the open-loop eigenvalues, i.e., there is only small shift in eigenvalues and thus little active damping is introduced to the system to damp out the structure vibration. The open-loop and closed-loop eigenvalues for this first case study are listed in Tables 6.3 and 6.4. After the closed-loop eigenvalues are determined, we can use the algorithm presented in Section 6.2 to solve for the feedback gain that gives the optimal eigenvector assignment.

In a traditional eigenstructure assignment approach, one needs to pre-determine the desired closed-loop eigenvectors. A method was provided by Corr and Clark (1999) and Shelley and Clark (2000b), where the desired eigenvectors were obtained by multiplying the open-loop eigenvectors by a shaping function, such as the probability distribution function,

Table 6.3. Open-loop eigenvalues for optimal passive system (sequential design)

-0.7922-183.5294i	-9.6595-205.0257i
-13.3032-210.0006i	-6.5576-214.0217i
-21.4647-268.1023i	-1.6421-269.5988i
-0.8613-338.2961i	-45.5705-367.4588i
-0.4407-405.3772i	-0.2073-462.4464i
-0.2374-505.7782i	-0.2287-533.0331i

Table 6.4. Closed-loop eigenvalues under sequential design

-0.7959-183.5294i	-9.6595-205.0253i
-13.3032-210.0005i	-6.5584-214.0219i
-21.4630-268.1021i	-1.6678-269.5990i
-0.8830-338.2958i	-45.5704-367.4590i
-0.4563-405.3773i	-0.2091-462.4464i
-0.2428-505.7783i	-0.2333-533.0331i

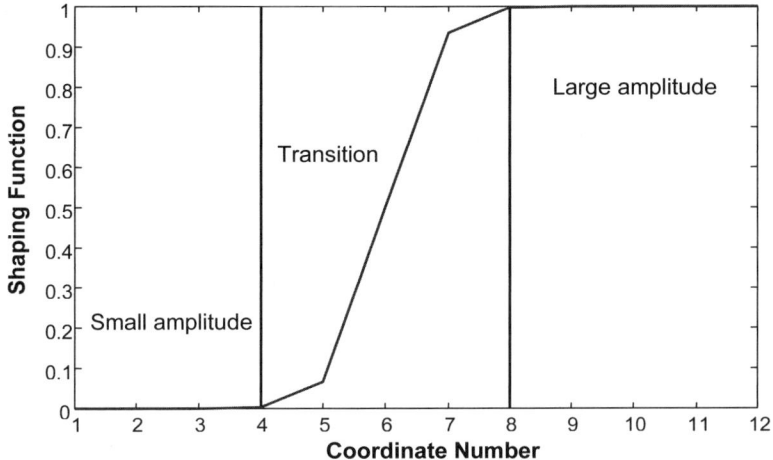

Fig. 6.4. Shaping function used for traditional eigenstructure assignment, $\mu = 6$, $3\sigma = 2$.

$$D(x) = 0.5 + 0.5\,\mathrm{erf}(\frac{x - \mu}{\sqrt{2}\sigma})$$

(6.29)

The above function is illustrated in Fig. 6.4. The first eight coordinates correspond to the beam displacement, and the last four coordinates correspond to the electrical displacement in the circuits. It's worth noting that the amplitudes of the electrical displacement coordinates are of the "unconcerned" or "unconfined" type. Clearly, a closed-loop eigenvector will have three regions, small amplitude region, transition region, and large amplitude region. The mid-point and width of the transition region are determined by μ and σ, respectively. After determining the desired closed-loop eigenvectors, we can use Eqs. (6.14), (6.16) and (6.20) to solve for the closest achievable eigenvectors and feedback gain under the traditional eigenstructure assignment approach.

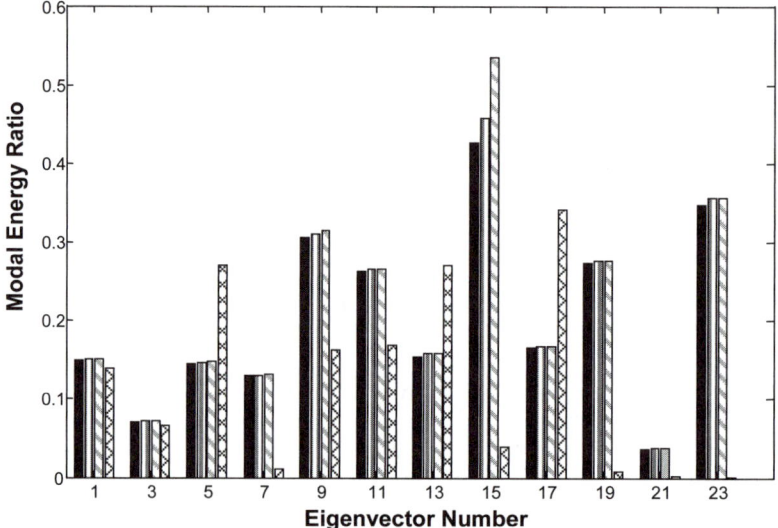

Fig. 6.5. Modal energy ratio $\dfrac{\varphi_i^{ac*}\varphi_i^{ac}}{\varphi_i^{a*}\varphi_i^{a}}$ corresponding to different eigenstructure assignment approaches.

■ : optimal eigenvector assignment, sequential approach;

▯ : traditional eigenstructure assignment, shaping function $\mu = 6$, $3\sigma = 2$;

▧ : traditional eigenstructure assignment, shaping function $\mu = 8$, $3\sigma = 4$;

▨ : optimal eigenvector assignment, concurrent approach.

The comparison of the proposed optimal eigenvector assignment method and the traditional eigenvector assignment method is then performed. Plotted in Fig. 6.5 are the modal energy ratios obtained with the different methods, and smaller value indicates better confinement for that specific mode. Only the results for the odd number eigenvectors are plotted, because even number eigenvectors are the corresponding conjugates.

For a fair comparison, we present two sets of results for the traditional ei-genvector assignment method, which correspond to two different options of transition region, $\mu = 6$, $3\sigma = 2$, and $\mu = 8$, $3\sigma = 4$. As shown in Fig. 6.5, under the same passive circuitry elements selection, the optimal ei-genvector assignment method (sequential approach) gives smaller modal energy ratio on all modes. This is not a surprise because, for the same set of passive parameters, the optimal eigenvector assignment method yields the eigenvector with the smallest modal energy ratio among all achievable eigenvectors. It should be pointed out that, although the difference of mo-dal energy ratios between the proposed optimal eigenvector assignment approach and the traditional approach seems to be small, the vibration con-trol performance could differ much more significantly.

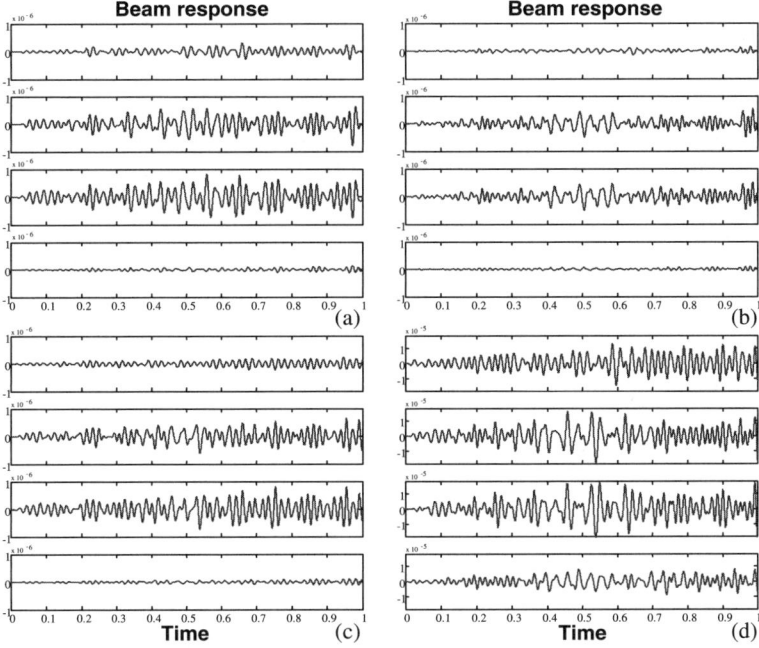

Fig. 6.6. Beam time-responses. (a) first beam; (b) second beam; (c) third beam; (d) fourth beam. Top sub-plot: optimal eigenvector assignment, sequential ap-proach; Second sub-plot: traditional assignment method, shaping parameters $\mu = 6$, $3\sigma = 2$; Third sub-plot: traditional assignment method, shaping parame-ters $\mu = 8$, $3\sigma = 4$; Bottom sub-plot: optimal eigenvector assignment, concur-rent approach.

Assume that there is zero-mean white-noise disturbance acting on the last beam. Fig. 6.6 shows the time responses of the first four beams under active controls. One can clearly see that the optimal eigenvector assignment approach outperforms the traditional approach. In fact, the time-average energy levels (kinetic and potential energy) of the first four beams under optimal eigenvector assignment approach are 1.9×10^{-6} J, 4.1×10^{-7} J, 1.9×10^{-6} J, and 5.5×10^{-3} J. As a comparison, those under the traditional eigenstructure assignment approach are, for $\mu = 6$ and $3\sigma = 2$, 1.1×10^{-5} J, 5.7×10^{-6} J, 1.1×10^{-5} J, and 5.7×10^{-3} J. The results for $\mu = 8$ and $3\sigma = 4$ are quite similar, which are 1.7×10^{-5} J, 6.4×10^{-6} J, 1.8×10^{-5} J, and 5.9×10^{-3} J.

While in the above case study we have verified that the optimal eigenvector assignment method can outperform traditional eigenstructure assignment method, such design is based upon circuitry elements that provide optimal passive damping. On the other hand, since the goal of vibration confinement is to confine energy to unimportant regions, a circuit with optimal passive damping might not be the best choice for active confinement control. This is because although such damping would help to damp out the vibration, it could also hinder the propagation of energy from the important region (region needs low vibration) to the unimportant region, thus reducing the energy confinement performance. In order to develop a hybrid control system where the passive and active parameters can truly complement each other, here in the next case study we implement the simultaneous optimization / optimal eigenvector assignment method outlined in the preceding section.

Without loss of generality, in the numerical simulation we use the optimal passive circuit elements as the initial values in the concurrent design. Since the open-loop eigenvalues will inevitably vary during the process of searching for passive parameters, a universal selection of closed-loop eigenvalues for the hybrid system (where the passive parameters keep varying during the optimization) might not serve the purpose of vibration confinement well. In fact, little knowledge could be foreseen about the result of passive parameters and the corresponding open-loop eigenvalues, and therefore a pre-selection of closed-loop eigenvalues outside the optimization loop may very well contradict the vibration confinement objective. That is, either the closed-loop control would introduce too much damping to the system and try to damp out the vibration of the entire structure which defeats the original idea of vibration confinement, or the closed-loop control would spend effort on moving the open-loop eigenvalues back towards to their original set values and which may, instead of suppressing vibration, add energy to the vibrating structure.

To resolve this issue, we insert an LQR subroutine inside the passive parameter optimization to decide the corresponding closed-loop eigenvalues. For the present case study, under a given set of passive parameters, we use the weightings shown in Eq. (6.28) to solve for the closed-loop eigenvalues. This will (a) guarantee that the closed-loop eigenvalues are shifted only slightly from the open-loop eigenvalues and therefore the objective of vibration confinement is maintained throughout the design process; and (b) provide a uniform criterion for the selection of closed-loop eigenvalues under various selections of passive parameters. The next step is to use the proposed optimal eigenvector assignment algorithm to solve for active gains. We then calculate the overall objective function that is a weighted sum of the individual modal energy ratios, shown in Eq. (6.26). In this specific case we let all the weightings be unity, *i.e.*, we have no preference on any mode. The simultaneous optimization / optimal eigenvector assignment design completes if the minimum of the overall objective function (Eq. (6.26)) is achieved. Otherwise, we choose another set of passive parameters and perform LQR to get the corresponding closed-loop eigenvalues, and repeat the above process. In the present case study, the optimization is realized by a Sequential Quadratic Programming subroutine provided in MATLAB6.1.

Table 6.5. Passive circuitry element values under concurrent design

$L_1 = 4.21\,\mathrm{H}$	$L_2 = 1.79\,\mathrm{H}$	$L_3 = 108.88\,\mathrm{H}$
$L_4 = 0.67\,\mathrm{H}$	$R_1 = 9271.17\,\Omega$	$R_2 = 374.95\,\Omega$
$R_3 = 41.56\,\Omega$	$R_4 = 242.52\,\Omega$	

Table 6.6. Open-loop eigenvalues for passive system under concurrent design

-0.2131-184.2426i	-0.2676-210.5265i
-0.2511-268.5642i	-0.1929-305.3982i
-0.2372-339.1938i	-0.2272-404.9641i
-0.2166-462.0094i	-0.2037-505.4269i
-0.1975-532.7365i	-1100.71-1102.70i
-104.701-2386.80i	-180.674-3897.79i

Table 6.7. Closed-loop eigenvalues under concurrent design

-0.2218-184.2426i	-0.2766-210.5265i
-0.2669-268.5640i	-0.3249-305.3969i
-0.3283-339.1953i	0.2358-404.9641i
-0.2218-462.0094i	-0.2061-505.4269i
-0.2063-532.7365i	-1100.71-1102.10i
-104.701-2386.80i	-180.827-3897.81i

The results for the concurrent design of the hybrid system are shown in Tables 6.5-6.7 and Figs. 6.5-6.7. Shown in Table 6.5 are the passive circuitry parameters determined by the simultaneous optimization / optimal eigenvector assignment approach. These passive parameters are quite different from those that provide the optimal passive damping. In fact, such passive circuit selection provides little damping to the system. The open-loop and closed-loop eigenvalues under the concurrent design are listed in Tables 6.6 and 6.7. The corresponding modal energy ratios are plotted in Fig. 6.5. Except for three eigenvectors (5th, 13th and 17th eigenvectors), the modal energy ratios of all the other eigenvectors are significantly smaller than those obtained by the sequential approach. It should be noted that the overall objective function here is the sum of individual modal energy ratios and in this case study we let the weightings on all eigenvectors be unity. Although for three pairs of modes the modal energy ratios do become higher, the modal energy ratio sum for the concurrent design is 1.4797, which represents a significant reduction from the sequential approach result 2.4691. The time responses of the first four beams are plotted in Fig. 6.6. The time-average energy levels for the first four beams under concurrent control design are, respectively, 3.4×10^{-7} J, 2.1×10^{-7} J, 4.0×10^{-7} J, and 1.8×10^{-3} J, which clearly indicate significant reduction from the results obtained by the sequential design. It should be noted that the concurrent design also yields reduction in control power requirements. The control power inputs for the two actuators in the sequential approach are 0.13W and 1.84W, while those in the concurrent approach are 0.02W and 0.45W.

The integration of circuitry elements to the mechanical structure plays an important role in the successful implementation of eigenstructure assignment technique. In Fig. 6.7, we compare the energy levels of the beam structures and circuits with optimal passive situation, sequential design, and concurrent design, under the action of the aforementioned zero-mean white-noise excitation. The energy levels are plotted in logarithm scale. Recall that the first four and last four coordinates correspond to, respectively, the beam displacements of concern and the electrical displacements in the circuits. For the optimal passive system, the energy distribution is quite even among the beams and the circuits. After we apply the sequential and concurrent optimal eigenvector assignment controls, the energy levels of the first four beams are reduced significantly, and, while the energy levels of the other four beams remain about the same, the energy levels of the last two circuits are greatly increased. This implies that, for this specific structure, it's much easier to move the energy from the first four beams to the circuits than to the last four beams. It clearly shows the benefit of introducing the circuitry elements into the eigenstructure assignment technique. On the other hand, with the help of active control, the circuitry elements now store much larger portion of the system energy than under the purely passive situation.

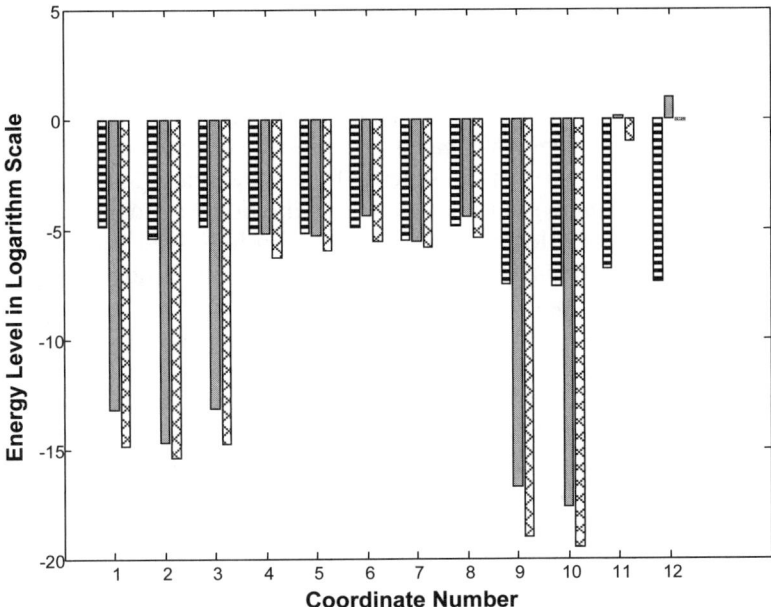

Fig. 6.7. Energy level comparison (logarithm scale).

⊟ : Optimal passive system;

▮ : Optimal eigenvector assignment, sequential approach;

⊠ : Optimal eigenvector assignment, concurrent approach.

6.4 PERIODIC ISOLATOR DESIGN ENHANCEMENT USING VIBRATION CONFINEMENT

The preceding section provides a detailed demonstration of the algorithm development on a generic vibration control/confinement example. This section presents another case study on an isolator design, which represents a possible application of vibration confinement with piezoelectric transducer circuitry.

6.4.1 Periodic Isolator with Piezoelectric Circuitry

In general, the objective of isolator design is to minimize the transfer of undesired vibration to the concerned regions of the structure over a specified frequency range of interest. The concept of periodic structure has been often utilized in traditional passive isolator design. Such a structure consists of repeating substructure cells that are designed to be identical to each other. Through building in material discontinuity, these cells are synthesized to create frequency stopbands that are capable of filtering vibratory wave transmission within the specific frequency range (Mead, 1996). The advantage of using periodic structures as vibration isolators is that the design is simple and the location of stopbands can be solved analytically. However, the conventional passive periodic isolators may not always be effective because the stopband frequency bandwidth and location may not always match with the disturbance frequencies due to design limitations and system uncertainties. It is usually not easy to tailor the mechanical periodic structure to exactly allocate the complete frequency range of excitation within the stopband. On the other hand, it has been shown in previous investigations that active isolators could achieve much higher performance than passive isolators (Sciulli and Inman, 1996; Miller *et al.*, 1995; Panza *et al.*, 1997). However, a purely active device normally requires high external power and may not be fail-safe when the active action fails. Therefore, one may envision that an alternative method to advance the state-of-the-art of vibration isolation could be to use a well designed passive periodic isolator as the baseline and then utilize active-passive actions to enhance the stopband bandwidth and performance. In this case, the active power requirement would be lower and the system would have better fail-safe property than a purely active isolator, by taking advantage of the inherent characteristics of spatial periodic structure in which the stopbands will be produced by the discontinuity of material property. Furthermore, the bandwidth and performance of the passive periodic isolator would be greatly enhanced by active control. In what follows we demonstrate that the active-passive vibration confinement idea outlined in Section 6.2 could be specifically useful in synthesizing isolation systems by confining vibration energy in the circuitry and in areas away from the attenuated end (the end that requires small vibration) of the isolator.

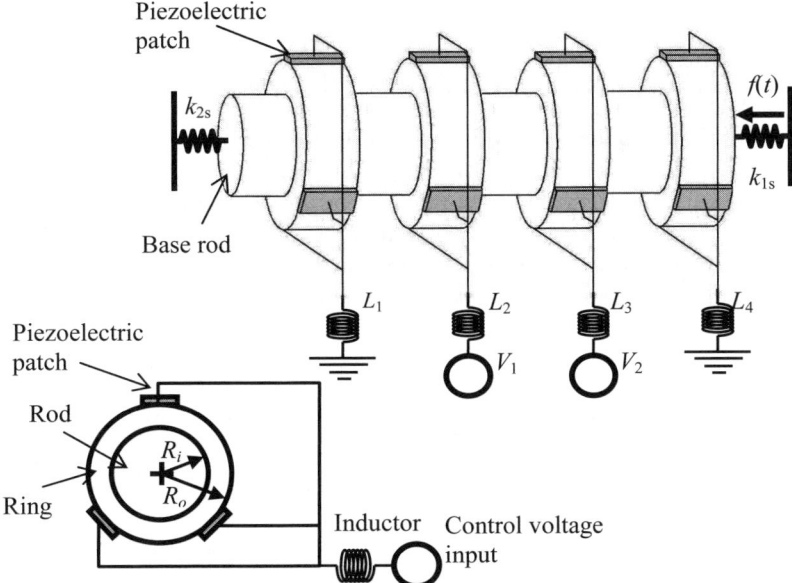

Fig. 6.8. System arrangement consisting of base rod, encircled rings, piezoelectric patches, circuits, and active control voltage inputs. External disturbance exerted at the right end of the rod.

The schematic of the isolation system is shown in Fig. 6.8. In this case study, the isolator, without losing generality, is modeled as a rod structure. The rod has circular cross section and is encircled by four identical rings at locations shown in the figure. Each of the rings is bonded by three piezoelectric patches on the ring's outer surfaces. In this investigation, only longitudinal vibration is considered and the piezoelectric patches on the same ring surface are connected to the same circuit. Each set of piezoelectric patches connects to a circuit that includes one inductor. Two voltage sources are also connected to the second and third circuits for active feedback control. In order to avoid confusion of attenuating contribution of vibration between energy confinement and dissipation in the circuit, we assume that there is no resistance in the circuits. The external excitation force is exerted on the right end of the rod (excitation end), and hence a low vibration amplitude at the left end of the rod (attenuated end) is de-

sired. To closely simulate the isolation strut applications, in which the
rods are considered having quasi-fixed-free boundary condition in most
cases, two springs, one with much higher stiffness than the other, are
mounted on the two ends of the rod. The encircled rings on the rod and the
bonded piezoelectric patches on the ring surfaces will create a periodic
structure with stopbands due to the discontinuity of structural property.
The goal of the feedback control and circuitry design is to ensure that the
vibratory energy is confined in the circuits and in regions far away from
the left end (attenuated end) of the rod. In this study, the structure is di-
vided into sixteen elements and discretized by finite element method. The
equations of motion for the integrated system can again be expressed as
Eq. (6.1), where, because of the absence of the resistance elements in this
isolator design, $\mathbf{R} = \mathbf{0}$. The concerned coordinates are chosen to be the
first eight elements in the attenuated end (left end in Fig. 6.8) of the rod.
All system parameters are listed in Table 6.8.

Table 6.8. Parameters used in isolator analysis

E_r=2.3×10^9 Pa	E_p=6.60×10^{10} Pa	ρ_r=1100 Kg/m^3
ρ_p=7800 Kg/m^3	ζ=0.001	β_{33}=9.5022×10^7 V
h_{31}=9.3022×10^8 N/C	k_{1s}=8.6708×10^5 N/m	k_{2s}=1.4451×10^8 N/m
L=0.4500 m	L_r=0.0563 m	L_p=0.0563 m
R_i=0.0200 m	R_o=0.0300 m	w_p=0.0104 m
t_p=2.67×10^{-4} m		

In the presented eigenvector assignment algorithm, we have assumed
that all the system states are available. In certain cases it might be difficult
to directly measure all the system states. Therefore the state estimator is
implemented in this study. By separation property (Chen, 1984), the design of state feedback and the design of the state estimator can be carried
out independently and the eigenvalues of the entire system are the union of
those of state feedback and those of the estimator. Combining the equations of the state feedback system and the state estimator the closed-loop
system thus becomes

$$\left\{ \begin{matrix} \dot{x} \\ \dot{\bar{x}} \end{matrix} \right\} = \left[\begin{matrix} \mathbf{A} & \mathbf{BK} \\ \mathbf{HC} & \mathbf{A - HC + BK} \end{matrix} \right] \left\{ \begin{matrix} x \\ \bar{x} \end{matrix} \right\} + \left[\begin{matrix} \mathbf{E} \\ \mathbf{E} \end{matrix} \right] \mathbf{f} \qquad (6.30)$$

$$y = \begin{bmatrix} \mathbf{C} & \mathbf{0} \end{bmatrix} \left\{ \begin{matrix} x \\ \bar{x} \end{matrix} \right\}$$

where $\bar{\mathbf{x}}$ is the estimated state vector and $\mathbf{H}$ is the output gain matrix.

6.4.2 Case Analysis

In this study, we focus on the isolator structure's longitudinal action and ignore the transverse dynamics. The rod and rings are tailored such that the first stopband of the spatial periodic structure without circuit and active control is between 4381 Hz and 8591 Hz. The desired closed-loop eigenvalues have to be decided *a priori*. In this research, we choose the desired closed-loop eigenvalues slightly different from the open-loop eigenvalues by multiplying a scalar factor 1.02. Conceptually with this approach, one will not move the closed-loop eigenvalues too far away from the open-loop eigenvalues. Therefore, we can fairly evaluate the performance improvement mainly due to vibration confinement and eigenvector assignment. The system output $\mathbf{y}$ in Eq. (6.2) and Eq. (6.30) is selected to be both the displacements and velocities at the attenuated (the very left boundary of the rod) and excitation ends (the very right boundary of the rod). The output gain matrix $\mathbf{H}$ is selected such that the eigenvalues of matrix $(\mathbf{A-HC})$ in Eq. (6.30) are 3.2 times the real part and 1.05 times the imaginary part of the desired closed-loop state feedback eigenvalues. In such a case, we can obtain faster responses in the state estimator than in the state feedback system.

To understand the effect of the inductor circuits on the vibration confinement performance, the case without the inductors (purely active control) is first considered. For this, Eq. (6.1) becomes

$$\mathbf{M\ddot{q}} + \mathbf{C_d\dot{q}} + (\mathbf{K_0} - \mathbf{K_1K_2^{-1}K_1^T})\mathbf{q} = \mathbf{F_d} - \mathbf{K_1K_2^{-1}B_0u} \qquad (6.31)$$

We define $\mathbf{K}_{eq} = \mathbf{K_0} - \mathbf{K_1K_2^{-1}K_1^T}$ and $\mathbf{B}_{eq} = -\mathbf{K_1K_2^{-1}B_0}$, and then the state space model in Eq. (6.3) becomes

$$\mathbf{x} = \begin{Bmatrix} \mathbf{q} \\ \mathbf{\dot{q}} \end{Bmatrix}, \quad \mathbf{A} = \begin{bmatrix} 0 & \mathbf{I}_n \\ -\mathbf{M^{-1}K}_{eq} & -\mathbf{M^{-1}C_d} \end{bmatrix} \qquad (6.32)$$

$$\mathbf{B} = \begin{bmatrix} 0 \\ \mathbf{M^{-1}B}_{eq} \end{bmatrix}, \quad \mathbf{E} = \begin{bmatrix} 0 \\ \mathbf{M^{-1}} \end{bmatrix}$$

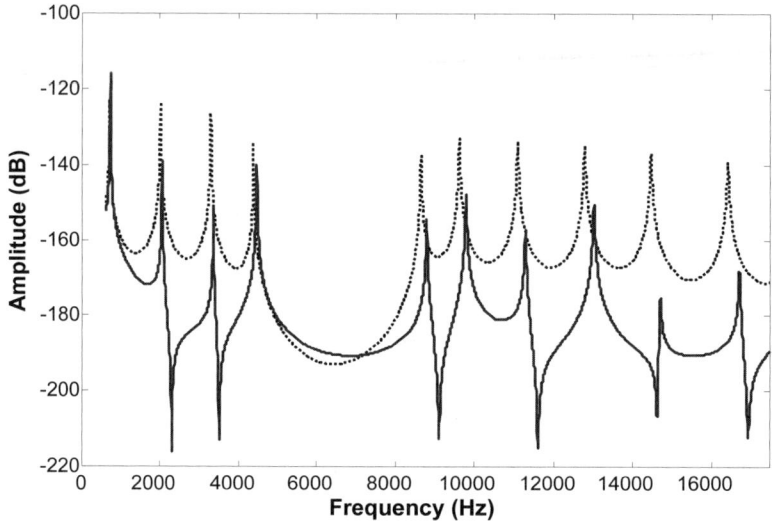

Fig. 6.9. Frequency response at the attenuated end. ················ : original system without control, ——————— : purely active control without inductors.

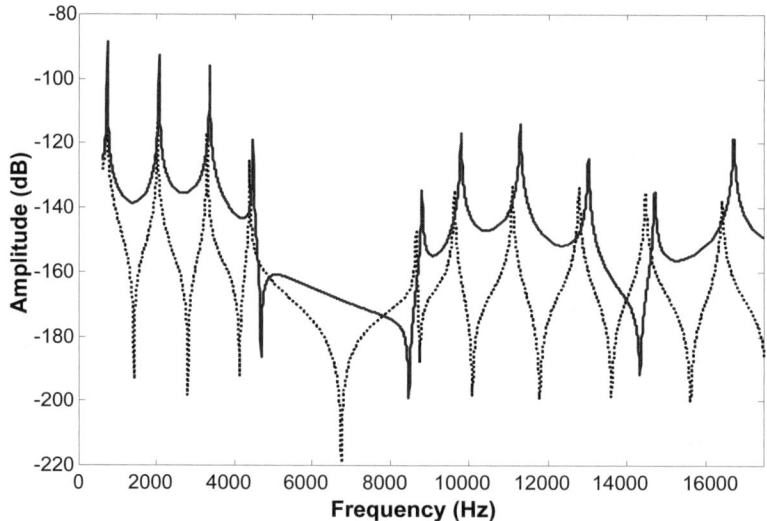

Fig. 6.10. Frequency response at the excitation end. ················ : original system without control, ——————— : purely active control without inductors.

Figs. 6.9 and Figure 6.10 show the frequency responses of displacements to external force at both ends of the isolator. The first stopband of the original periodic structure is between 4381 Hz and 8591 Hz. At the attenuated end of the isolator, although the improvement of vibration suppression is shown with active control, the vibration reductions at some of the resonance frequencies around the first stopband are not very obvious. Moreover, the amplitude of frequency response of the excitation end of the isolator has increased enormously with the active control. Although this phenomenon proves that the vibratory energy can be confined in the unconcerned areas and less vibratory energy has transmitted to the concerned areas, the significantly increased vibration at the excitation end may be still undesirable and will be detrimental to the structural health. Since there are no inductors connecting to the circuits in this purely active action case, no inertia elements exist in the electrical part. In this situation, the achievable closed-loop eigenvectors are synthesized by p linear combinations of $2n$-dimensional vectors.

We then study the results of introducing inductance parameters in the active feedback control. As mentioned earlier, the introduction of circuitry elements will increase the design space of achievable eigenvectors. However, with the active feedback control algorithm for eigenvector assignment, the inductance values need to be optimized concurrently. In this study, an optimization procedure is utilized to find the optimal set of inductance parameters. While this optimization procedure is conceptually similar to the one presented in Section 6.3, the details are different in order to address the specific issues of periodic vibration isolators. Assume the first stopband of the original system is located from ω_{ss} to ω_{se}, and the external excitation bandwidth is broader than the stopband. The optimization procedure will find the set of inductance values such that the summation of modal energy ratio of concerned modal energy to the total modal energy corresponding to the frequencies between $(\omega_{ss} - \omega_{ext})$ and $(\omega_{se} + \omega_{ext})$ is minimized. We choose ω_{ext} such that the entire excitation frequency range is covered. In this example, ω_{ext} is selected to be 800 Hz and thus $\omega_{ss} - \omega_{ext}$ = 3581 Hz and $\omega_{se} + \omega_{ext}$ = 9391 Hz. The final minimized objective function can be expressed as follows,

$$J = \min_{L_i}(\sum_{j=k}^{l} \min \frac{\mu_j^* \Lambda_j \mu_j}{\mu_j^* \Omega_j \mu_j}), \quad i = 1,2,..,m \tag{6.33}$$

$$\omega_{ss} - \omega_{ext} \leq |\operatorname{Img}(\lambda_j)| \leq \omega_{ss} + \omega_{ext}, \quad j = k, k+1, k+2,..,l$$

For a given set of inductance values (passive design parameters), one can determine μ_j from Eqs. (6.24) and (6.25). Then the optimal inductance values are derived by minimizing the expression of Eq. (6.33). The optimization (minimization) process is realized by a Sequential Quadratic Programming method (using the MATLAB subroutine software) in which the medium-scale optimization (line search) is performed to search for the optimal solution. Once the optimal inductance values are determined, the corresponding active gains can be derived from Eq. (6.20). The optimal inductance parameters are listed in Table 6.9. If the active action failed, switches can be applied so that all circuits can be open, where a fail-safe passive isolator can still be achieved due to the periodicity of the mechanical isolator. Without inductors, the minimal modal energy ratios $\left|\alpha_j\right|_{\min}$ of the 4th and 5th modes (within 3581-9391 Hz) in Eq. (6.25) are 0.2079 and 0.0450, respectively. By adding the inductance and through performing the optimization process, the minimal modal energy ratios of the two modes can be significantly reduced to 0.0050 and 0.0029, respectively.

Table 6.9. Optimal inductance values

L_1= 34.307292 mH	L_2= 30.595578 mH
L_3= 27.312091 mH	L_4= 41.222682 mH

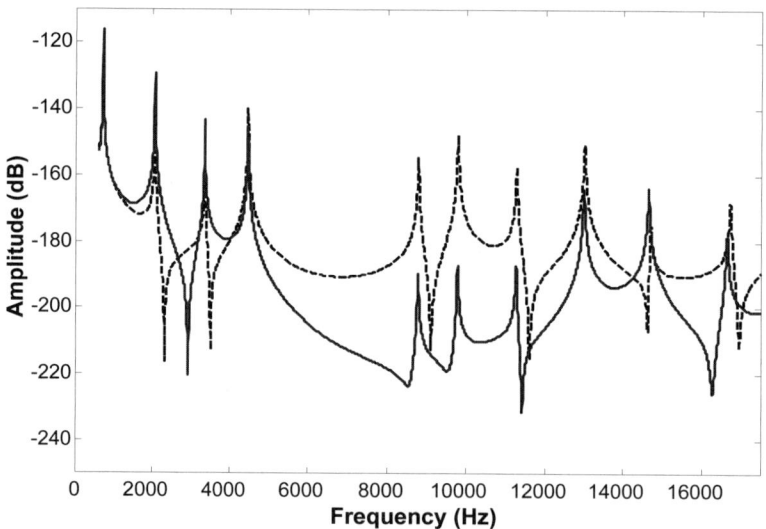

Fig. 6.11. Frequency response at the attenuated end. ················ : purely active control without inductors, ———— : active feedback control with inductors.

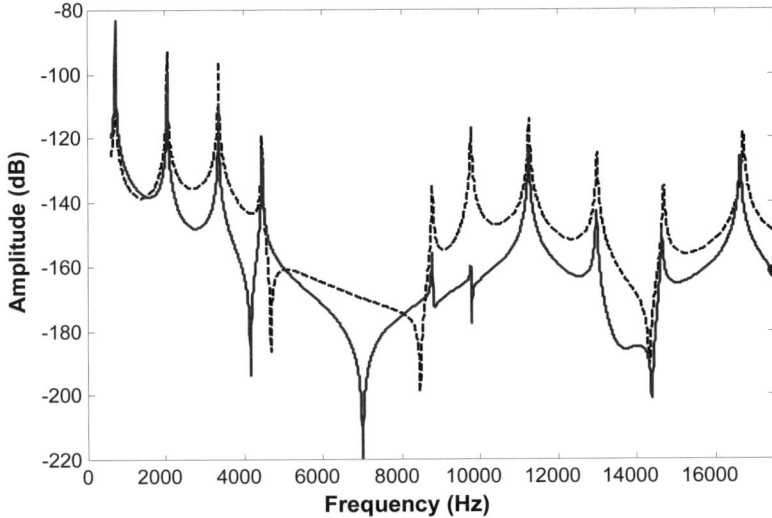

Fig. 6.12. Frequency response at the excitation end. ⋯⋯⋯⋯: purely active control without inductors, ———— : active feedback control with inductors.

Figs 6.11 and 6.12 show the frequency responses of displacements to external force at both ends of the isolator. Compared to the purely active control case, Fig. 6.11 shows more vibration suppression at the attenuated end within the designed bandwidth and throughout the broad frequency range. On the other hand, Fig. 6.12 shows that the vibration amplitude of the excitation end in this case is much smaller than that in the purely active control case. The reason for this could be the fact that the electrical degrees of freedom are absorbing some of the energy and thus the vibration amplitude of the excitation end can be reduced.

Fig. 6.13 shows the comparison of frequency responses near the first stopband by purely active control without inductance circuits and active control with inductance circuits. It is obvious that with the introduction of inductors, the vibration isolation performance can be greatly improved at the resonant frequencies within the designed range (3581 Hz to 9391 Hz). Figs. 6.14 to 6.17 show the frequency responses of the charge in the inductance circuits. Since inductors can store energy, they play a role in redistributing the system energy. The introduction of the inductance elements expands the design space of the achievable eigenvectors. The achievable space for the closed-loop eigenvectors are increased to p linear combinations of $2(n+m)$-dimensional vectors. Compared to the purely active control, the design space of the achievable closed-loop eigenvectors is significantly enlarged by introducing inductance circuits.

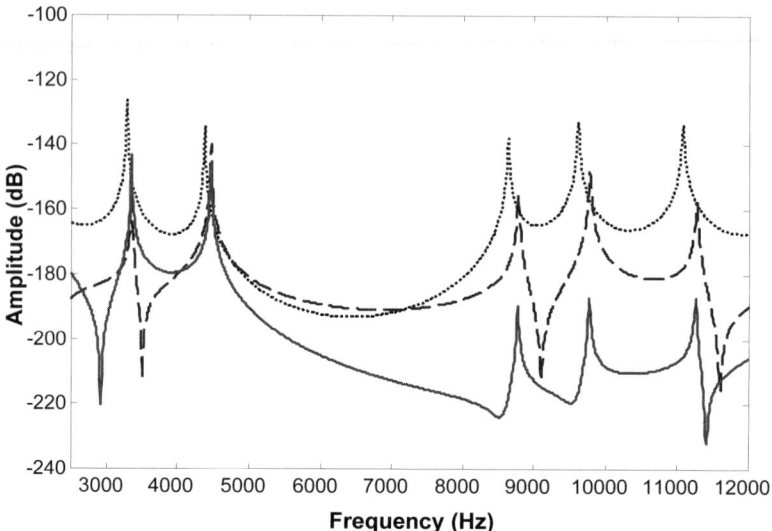

Fig. 6.13. The comparison of frequency responses near the first stopband.
···············: original system without control, ─ ─ ─ ─: purely active control without inductors, ─────── : active feedback control with inductors.

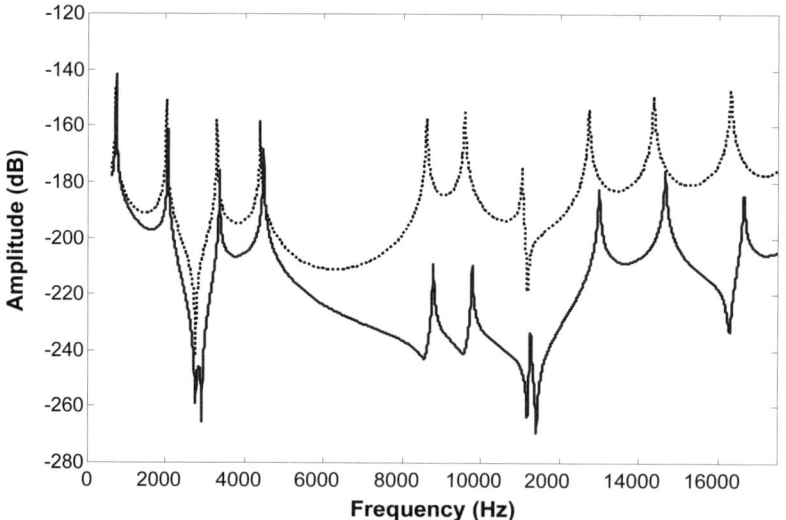

Fig. 6.14. Frequency responses of charge in inductor L_1. ···············: passive circuit, ─────── : active feedback control with inductors.

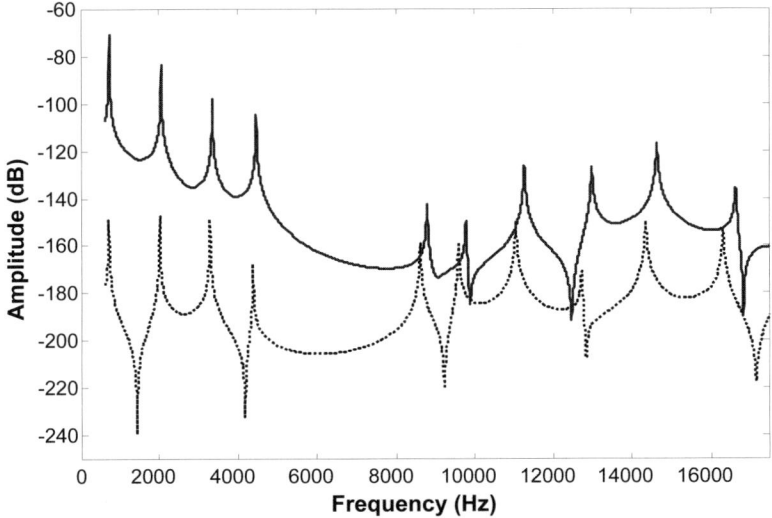

Fig. 6.15. Frequency responses of charge in inductor L_2. ·············· : passive circuit, ————— : active feedback control with inductors.

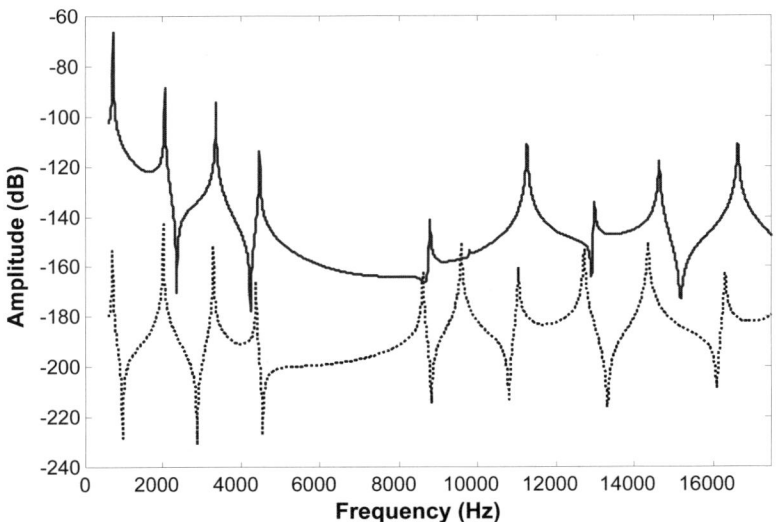

Fig. 6.16. Frequency responses of charge in inductor L_3. ·············· : passive circuit, ————— : active feedback control with inductors.

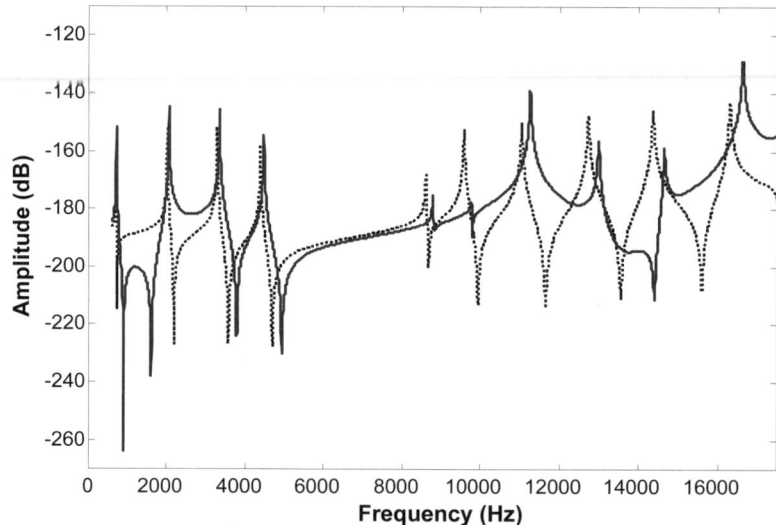

Fig. 6.17. Frequency responses of charge in inductor L_4. ················ : passive circuit, ———— : active feedback control with inductors.

Figs. 6.15 and 6.16 show that the frequency responses of electric charge increase greatly with active control, which means much more energy is transferred to the second and third inductors. That is one reason for the improvement of the isolation performance in this case as compared to the purely active control case. We define the modal energy level E_m to depict the total magnitude of all mode shapes, i.e.

$$E_m = \sum_{j=1}^{N} \left\{ \sigma_{j1}^2, \sigma_{j2}^2, ..., \sigma_{jN}^2 \right\}^T ,$$ (6.34)

$$\phi_j = \left\{ \sigma_{j1}, \sigma_{j2}, ..., \sigma_{jN} \right\}^T , \quad j = 1, 2, ..., N$$

Fig. 6.18 shows the modal energy level distribution of all states corresponding to the displacements and velocities. The states 1 through 17 are related to structural displacements, the states 22 through 38 are related to structural velocities, the states 18 through 21 are related to electric charge and the states 39 through 42 are related to the rate of charge. The 1st and 22nd states correspond to the displacement and velocity of the very left boundary (attenuated end) of the rod, respectively, and the 17th and 38th states correspond to the displacement and velocity of the very right boundary (excitation end) of the rod, respectively. The concerned regions consist of the first eight elements on the left side of the rod, which are represented by displacement states 1 through 8 and velocity states 22 through 29. It shows that the modal energy, including both potential energy (related to displacements) and kinetic energy (related to velocities), has been re-distributed after the purely active control is applied and thus the modal energy levels decrease in the concerned states. However, the energy levels in some of the unconcerned states are increased significantly, which will increase the vibration amplitude in these regions of the mechanical structure undesirably. Fig. 6.18 also shows that by introducing inductance circuits in the system, some system energy has been confined in the inductors (states 18 through 21 and 39 through 42) so that the modal energy levels in the concerned regions can be further suppressed and at the same time, the modal energy levels will be decreased in other parts of mechanical structure as well. Conclusively, the vibration isolation and suppression performance in the concerned regions can be enhanced by active vibration confinement with inductance circuitry.

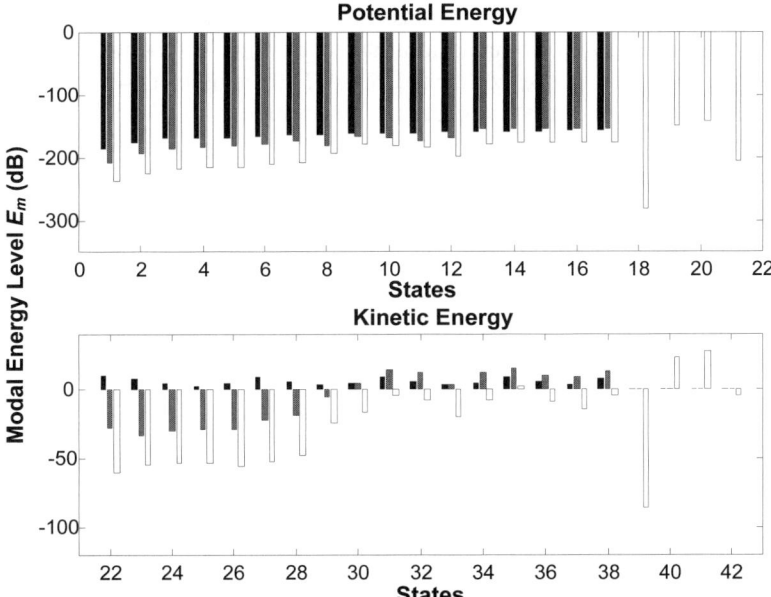

Fig. 6.18. Modal energy distribution comparison. ■ : Original system without control, ▨ : purely active control, ☐ : active feedback control with inductors.

6.5 CONCLUDING REMARKS

A methodology for vibration confinement is presented in this chapter, where the active control input is applied through a piezoelectric circuitry. With the introduction of circuitry elements, the dimension of the state matrices is increased, which can enlarge the design space for vibration confinement. An improved eigenstructure assignment algorithm is developed. This algorithm allows us to directly derive the optimal achievable eigenvectors with minimized eigenvector components at concerned region by using the Rayleigh Principle, and hence can take full advantage of the vibration confinement idea.

This methodology is applied to two case studies. First, a generic structure is used as an example to illustrate the fundamental idea. A simultaneous optimization / optimal eigenvector assignment design approach is proposed, which enables us to decide the system passive and active parameters concurrently. It is demonstrated that such an approach can further improve the vibration confinement performance. It is also shown that, compared with the traditional eigenstructure assignment technique, the proposed optimal eigenvector assignment algorithm has better vibration control performance in the concerned region. In another case study, active-passive vibration isolation design is performed on a periodic isolator structure, representing a possible application of the proposed idea. The vibration confinement control action is shown to reduce the modal transmissibility and increase the system isolation capability around the stopband of the traditional passive isolator. The introduction of inductors in the circuits enlarges the design space of achievable closed-loop eigenvectors, which enhances the vibration isolation performance.

7 Delocalization and Suppression of Vibration in Periodic Structures

Periodic structures are systems containing identical substructures that form a spatial periodicity. Such structures are very common in engineering systems; examples include turbo-machinery bladed disks, antenna, and space truss structures. In an ideal situation where the substructures are indeed identical, the vibration energy is uniformly distributed among the substructures, and thus the mode shapes are extended. However, in realistic situations, there are always small differences (hereafter referred to as mistunings) among the substructures, such as differences in geometry and material properties caused by manufacturing tolerance and in-service degradation. These mistunings could change the dynamic behavior of the periodic structures drastically (which now become nearly periodic or mistuned structures) as compared to that of the perfectly periodic structures. A phenomenon known as vibration localization could occur under certain circumstances (Hodges, 1982, Mester and Benaroya, 1995, Pierre et al., 1996) especially when the couplings between substructures are weak. In a localized vibration situation, the vibration energy is confined to a small region of the structure, resulting in increased stresses and amplitudes locally. The vibration localization could cause severe damage to the structure.

To date, most of the research on vibration localization has concentrated on exploring the cause (Pierre and Dowell, 1987, Wei and Pierre, 1988, Cornwell and Bendiksen, 1989, Pierre and Cha, 1989) and predicting the localized response (Hodges and Woodhouse, 1983, 1989, Pierre, 1990, Cha and Pierre, 1991, Bouzit and Pierre, 1992). The dynamic properties of periodic structures have often been studied from a wave propagation perspective (Mead, 1975a, 1975b) utilizing the transfer matrix approach, and the Lyapunov exponents were frequently used to characterize the localization (Kissel, 1991, Castanier and Pierre 1995, Xie and Ariaratnam, 1996a, 1996b). On the other hand, fewer studies have been performed on reducing vibration localization. Castanier and Pierre (2002) investigated the feasibility of using intentional mistuning to reduce the localization effect in turbo-machinery rotors. Since vibration localization is often related to

K.W. Wang and J. Tang, *Adaptive Structural Systems with Piezoelectric Transducer Circuitry*, doi: 10.1007/978-0-387-78751-0_7,
© Springer Science + Business Media, LLC 2008

weak coupling, Gordon and Hollkamp (2000) studied the potential of utilizing piezoelectric transducers to increase blade-to-blade coupling of bladed-disks, and found that the improvement was marginal. Cox and Agnes attempted a similar idea (1999), and then added one inductive element (Agnes, 1999) to the shorted piezoelectric transducers. Although it was observed that this addition could enhance the substructure coupling, the illustrative example only showed improvement in some of the localized modes. Tang and Wang (2003) developed a systematic approach for vibration delocalization via piezoelectric transducer circuitry (Fig. 7.1). In this approach, identical inductive piezoelectric circuits (*LC* circuits) are attached to all the substructures to absorb vibration energy and store that portion of energy in the electrical form. While in most cases directly increasing the mechanical coupling between substructures is difficult to achieve due to various design limitations, one can easily introduce strong electrical coupling, such as connecting these *LC* circuits with capacitors as shown in Figs. 7.1b or 7.1c, to create a new wave propagation channel. With this coupled piezoelectric transducer circuitry design, the otherwise localized vibration energy can now be transferred into electrical form and propagate in the integrated system through the newly created electro-mechanical wave channel. Yu *et al.* (2006) further utilized the negative capacitance, which, as demonstrated in Chapter 2, can enhance the electro-mechanical coupling effect of the piezoelectric transducer, to improve the delocalization performance of the piezoelectric circuitry.

The piezoelectric transducer circuitry concept, meanwhile, has also been adopted for the suppression of vibration in periodic structures. Tang and Wang (1999) extended the piezoelectric absorber circuit idea to periodic structures, where active feedback was used to achieve multi-harmonic excitation suppression. Yu and Wang (2007a) replaced the active feedback with coupling capacitance, and studied the vibration suppression performance under mistuning.

7.1 PROBLEM STATEMENT AND OBJECTIVE

This chapter presents a comprehensive study on how to employ the piezo-electric transducer circuitry to realize vibration delocalization and vibration suppression in mistuned periodic structures. The fundamental idea is to apply resonant piezoelectric circuits on individual substructures to absorb/store/dissipate the system energy converted from the vibration energy, and then utilize electrical coupling between these circuits to regulate the energy propagation or suppression. This circuitry architecture is different from that presented in the preceding chapters, as we are now dealing with networking multiple piezoelectric circuits on systems with multiple sub-structures. We focus our discussion on the following issues:

a) In terms of vibration delocalization, how the wave/energy is propagated in a nearly periodic structure with or without the piezoelectric transducer circuitry network, and how the circuitry tuning affects such propagation and the vibration pattern?

b) In terms of vibration suppression, how to take advantage of the periodicity to design piezoelectric damped absorbers that can optimally suppress the vibration from excitations with different spatial distributions, and how the performance is affected by mistuning?

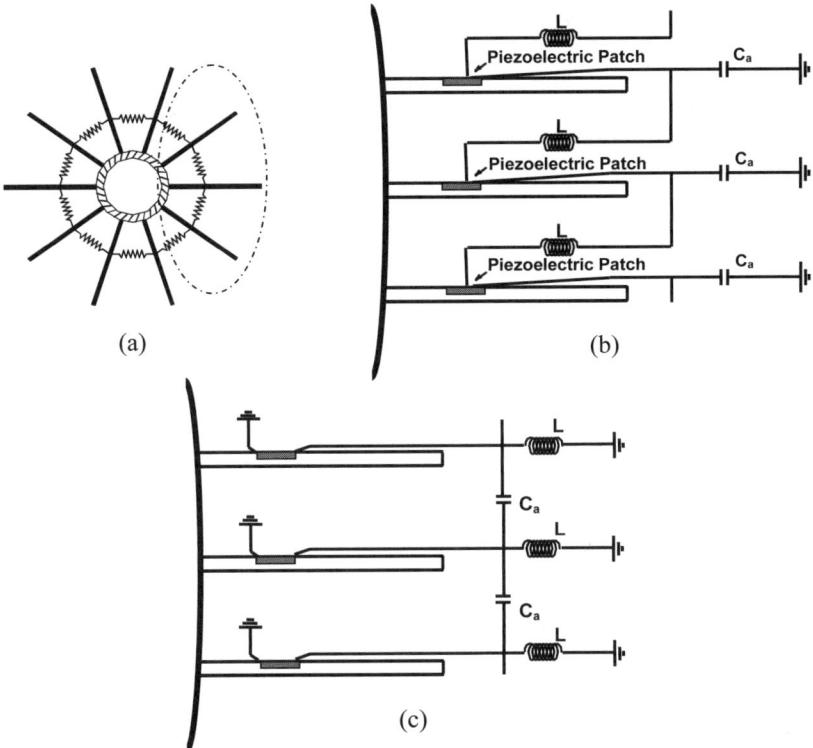

Fig. 7.1. (a) Schematic of period structure. (b) and (c) Periodic structure integrated with coupled piezoelectric circuit network.

In the studies presented in this chapter, the wave transfer matrix approach, the Lyapunov exponents, and the Monte Carlo simulations are utilized in the analysis and synthesis process. Experimental investigations are also conducted to evaluate the proposed approach.

7.2 SYSTEM MODEL

For the purpose of illustration and without loss of generality, the periodic structure that we consider in this chapter is assumed to consist of N identical cantilever beams coupled by N springs (Fig. 7.1a). Here we use coupling springs to emulate the coupling effects in generic periodic structures, such as the stiffness coupling due to the disk in turbine bladed disks. This periodic structure is integrated with a piezoelectric transducer circuitry. A variety of such circuitries can be envisioned, two of which are illustrated in Figs. 7.1b and 7.1c. Note that in both cases the inductive piezoelectric shunts applied on individual beams are all identical. Let ϕ be the first local beam mode without the piezoelectric circuitry. The transversal displacement of the jth beam is approximated as

$$w_j(x,t) \approx \phi(x)q_j(t) \tag{7.1}$$

The equations of motion for the periodic structure integrated with piezoelectric transducers can be derived using Hamilton's principle, which are

$$m\ddot{q}_j + c\dot{q}_j + kq_j + k_1 Q_j + k_c(q_j - q_{j-1}) + k_c(q_j - q_{j+1}) = f_j \tag{7.2a}$$

$$k_2 Q_j + k_1 q_j = V_j \tag{7.2b}$$

where m, c, k, k_c, k_1 and k_2 are, respectively, mass, damping, substructure stiffness, substructure coupling stiffness, cross coupling coefficient related to the electro-mechanical coupling, and inverse capacitance of the piezoelectric transducer, which are given by,

$$m = \int_0^{l_b} \rho_b A_b \phi^2 \, dx + \int_0^{l_b} \rho_p A_p \phi^2 \Delta H \, dx \ , \ c = \int_0^{l_b} c_b \phi^2 \, dx$$

$$k = \int_0^{l_b} E_b I_b \phi''^2 \, dx + \int_0^{l_b} E_p I_p \phi''^2 \Delta H \, dx \ , \ k_c = k_s \phi^2(x_s)$$

$$k_1 = \frac{1}{w_p l_p} \int_0^{l_b} F_p h_{31} \phi'' \Delta H \, dx, \quad k_2 = \frac{A_p \beta_{33}}{w_p^2 l_p}$$

Here, q_j and f_j are the generalized mechanical displacement and external force of the jth beam. Q_j and V_j are the charge flowing to and the voltage across the piezoelectric patch attached to the jth beam, respectively. $\Delta H = H(x - x_l) - H(x - x_r)$, where $H(x)$ is the Heaviside step function. x_l and x_r are distances from the piezoelectric patch edges to the root of the cantilever beam, and the length of the patch is thus $l_p = x_r - x_l$. Other relevant notations used here are: ρ_b and ρ_p – density of beam and piezoelectric patch; A_b and A_p – cross section area of beam and piezoelectric patch; E_b and E_p – elastic modulus of beam and piezoelectric patch; I_b and I_p – moment of inertial of beam and piezoelectric patch; l_b – length of beam; w_p – width of piezoelectric patch; F_p – moment of area of piezoelectric patch; c_b – beam damping constant; k_s and x_s – stiffness of coupling spring and location of coupling spring; h_{31} – piezoelectric constant; and β_{33} – dielectric constant of piezoelectric patch.

Applying circuit analysis to Figs. 7.1b and 7.1c, for the two piezoelectric circuitry configurations, we can obtain the equations for the jth circuit branch as, respectively (Tang and Wang, 2003; Yu *et al.*, 2006),

$$V_j = -L\ddot{Q}_j - k_a(Q_j - Q_{j-1}) + k_a(Q_{j-1} - Q_j) \tag{7.3}$$

$$V_j = -L\ddot{Q}_j - \frac{L}{k_a} k_2 (2\ddot{Q}_j - \ddot{Q}_{j-1} - \ddot{Q}_{j+1}) - \frac{L}{k_a} k_1 (2\ddot{q}_j - \ddot{q}_{j-1} - \ddot{q}_{j+1}) \tag{7.4}$$

where L is the circuitry inductance. $k_a = 1/C_a$ denotes the inverse of the coupling capacitance. Substituting Eq. (7.3) or (7.4) into Eq. (7.2b), we can derive the equations of motion of the electro-mechanically integrated system under different circuitry configurations as

$$m\ddot{q}_j + c\dot{q}_j + kq_j + k_1 Q_j + k_c(q_j - q_{j-1}) + k_c(q_j - q_{j+1}) = f_j \tag{7.5a,b}$$

$$L\ddot{Q}_j + k_2 Q_j + k_1 q_j + k_a(Q_j - Q_{j-1}) + k_a(Q_j - Q_{j+1}) = 0$$

for the network shown in Fig. 7.1b, or for the network in Fig. 7.1c,

$$m\ddot{q}_j + c\dot{q}_j + kq_j + k_1 Q_j + k_c(q_j - q_{j-1}) + k_c(q_j - q_{j+1}) = f_j \qquad (7.6\text{a,b})$$

$$L\ddot{Q}_j + \frac{L}{k_a}k_2(2\ddot{Q}_j - \ddot{Q}_{j-1} - \ddot{Q}_{j+1}) + \frac{L}{k_a}k_1(2\ddot{q}_j - \ddot{q}_{j-1} - \ddot{q}_{j+1}) + k_2 Q_j + k_1 q_j = 0$$

For the original periodic structure without the piezoelectric circuitry, the equation of motion is

$$m\ddot{q}_j + c\dot{q}_j + kq_j + k_c(q_j - q_{j-1}) + k_c(q_j - q_{j+1}) = f_j \qquad (7.7)$$

In the above derivation, we assume that the system is perfectly periodic. In reality, however, there will be imperfections. As the common practice in localization study, in this chapter we assume that the structural imperfection (or called mistuning) only exists in the substructure mechanical stiffness. The stiffness of the jth beam with mistuning then is

$$\tilde{k}_j = k + \Delta k_j \qquad (7.8)$$

where k is the nominal stiffness of the perfectly periodic structure, and Δk_j is the zero-mean random mistuning. In the following, we seek harmonic solutions considering free vibration and negligible damping. After nondimensionalization, we have the equations of the system with and without the piezoelectric circuitry as, respectively,

$$-\Omega^2 x_j + (1 + \Delta s_j)x_j + R_c^2(x_j - x_{j-1}) + R_c^2(x_j - x_{j+1}) + \delta\xi y_j = 0 \qquad (7.9\text{a,b})$$

$$-\Omega^2 y_j + \delta^2 y_j + R_a^2\delta^2(y_j - y_{j-1}) + R_a^2\delta^2(y_j - y_{j+1}) + \delta\xi x_j = 0$$

or

$$-\Omega^2 x_j + (1 + \Delta s_j)x_j + R_c^2(x_j - x_{j-1}) + R_c^2(x_j - x_{j+1}) + \delta\xi y_j = 0 \qquad (7.10\text{a,b})$$

$$-\Omega^2[y_j + R_a^2(2y_j - y_{j-1} - y_{j+1}) + \frac{\xi R_a^2}{\delta}(2x_j - x_{j-1} - x_{j+1}) + \delta\xi x_j + \delta^2 y_j = 0$$

and

$$-\Omega^2 x_j + (1 + \Delta s_j)x_j + R_c^2(x_j - x_{j-1}) + R_c^2(x_j - x_{j+1}) = 0 \qquad (7.11)$$

where

$$\omega_m = \sqrt{k/m}, \quad \omega_c = \sqrt{k_2/L}, \quad \delta = \omega_c/\omega_m$$
$$\Omega = \omega/\omega_m, \quad x_j = \sqrt{m}q_j, \quad y_j = \sqrt{L}Q_j$$
$$\xi = k_1/\sqrt{kk_2}, \quad R_c = \sqrt{k_c/k}, \quad R_a = \sqrt{k_a/k_2}, \quad \Delta s_j = \Delta k_j/k$$

Here ω and Ω are the dimensional and nondimensionalized harmonic frequencies, ω_m and ω_c are the natural frequencies of uncoupled substructure and circuit, respectively; δ is the frequency tuning ratio which characterizes the circuitry inductance; ξ, which has been discussed extensively in Chapter 2, is the generalized electro-mechanical coupling coefficient reflecting the energy transfer capability of the piezoelectric transducer; R_c is the mechanical coupling ratio between the substructures, R_a is the electrical coupling ratio which is related to the coupling capacitance, and Δs_j is the mistuning ratio which is a zero-mean random number with standard deviation σ_m.

7.3 LOCALIZATION ANALYSIS

In this section we present the analysis of vibration and wave/energy propagation in nearly periodic structure with and without the piezoelectric circuitry network. For the purpose of illustration and without loss of generality, here we use the circuitry configuration shown in Fig. 7.1c.

As indicated earlier, the localization phenomenon in mistuned periodic structures has been analyzed extensively in the past. The probabilistic nature of the localization phenomenon was recognized, and stochastic methods were applied to study the spatial exponential decay rate of the vibration amplitude (Pierre *et al.*, 1996). Lyapunov exponents of the global stochastic wave transfer matrix have been employed to approximate the spatial exponential decay rate of the amplitude, and have been recognized as a good measure of the localization level (Pierre *et al.*, 1996, Kissel, 1991, Castanier and Pierre 1995, Xie and Ariaratnam, 1996a, 1996b). The wave transfer matrix method and the numerical computation of the Lyapunov exponents for multi-coupled periodic structures have been discussed by Pierre *et al* (1996). To formulate the wave transfer matrix expression for the bi-coupled system presented in this paper, a displacement state vector is defined for each bay (a bay consists of two adjacent substructures). Then the transfer matrix that relates the dynamics of two adjacent bays can be derived from the equations of motion. The dimension of the state vector is always twice of the number of inter-bay coupling coordinates m, and the dimension of the square transfer matrix is $2m$ by $2m$, which is independent of the number of substructures N. The details of the wave transfer matrix approach for the delocalization study can be found in (Tang and Wang, 2003). Only the formulation for the system with integrated piezoelectric network is presented below. This integrated system is bi-coupled, thus the inter-bay coupling is $m=2$, and the dimensions of the displacement state vector and transfer matrix are 4×1 and 4×4, respectively. The 4×1 displacement state vector for the jth bay, consisting of both mechanical and electrical generalized displacements from the jth and the $j+1$th substructures, is defined as

$$\mathbf{u}_j = \begin{bmatrix} x_{j+1} & y_{j+1} & x_j & y_j \end{bmatrix}^{\mathrm{T}} \tag{7.12}$$

The system equations (7.10a,b) can be cast into the transfer matrix expression using the displacement state vector

$$\mathbf{u}_j = \mathbf{T}_j \mathbf{u}_{j-1} \tag{7.13}$$

where $\mathbf{T}_j$ is the 4×4 transfer matrix given as follows

$$
\mathbf{T}_j =
\begin{bmatrix}
\dfrac{1+2R_c^{\,2}+\Delta s_j-\Omega^2}{R_c^{\,2}} & \dfrac{\delta\xi}{R_c^{\,2}} & -1 & 0 \\[2ex]
-\left(\dfrac{\delta\xi}{\Omega^2 R_a^{\,2}}+\dfrac{\xi}{\delta}(\dfrac{1+\Delta s_j-\Omega^2}{R_c^{\,2}})\right) & \left(2+\dfrac{1}{R_a^{\,2}}-\dfrac{\xi^2}{R_c^{\,2}}-\dfrac{\delta^2}{\Omega^2 R_a^{\,2}}\right) & 0 & -1 \\[2ex]
1 & 0 & 0 & 0 \\[1ex]
0 & 1 & 0 & 0
\end{bmatrix}
$$

$$(7.14)$$

Obviously, the transfer matrix is random for mistuned structures because of the random mistuning Δs_j. For tuned structures, $\Delta s_j = 0$, and $\mathbf{T}_j$ is identical for all j's. A Lyapunov exponent is defined as

$$\gamma(\mathbf{u}_0) = \lim_{N\to\infty}\frac{1}{N}\log\|\mathbf{u}_N\| \tag{7.15}$$

where $\mathbf{u}_0$ is the initial displacement state vector and $\mathbf{u}_N$ is the displacement state vector of the Nth substructure. It can be shown that the Lyapunov exponent can be calculated from the product of the transfer matrices:

$$\gamma_k = \lim_{N\to\infty}\frac{1}{N}\log\left\{\sigma_k(\prod_{j=N}^{1}\mathbf{T}_j)\right\} \tag{7.16}$$

where $\sigma_k(\cdot)$ denotes the singular value decomposition operator.

In this study, the Lyapunov exponents of the bi-coupled periodic structure integrated with piezoelectric network are calculated using the Wolf's algorithm (Wolf et al., 1985). The number of Lyapunov exponents equals to twice of the number of coupling coordinates. For this bi-coupled system, there are four Lyapunov exponents in total. In fact, these Lyapunov exponents appear in pairs, with same magnitude but opposite signs, $\pm|\gamma|$. For a periodic structure, both these positive and negative Lyapunov exponents share the same physical meaning. Therefore, calculating the positive Lyapunov exponents is sufficient to identify the entire spectrum. In the following discussion, Lyapunov exponents by default refer to the positive ones unless otherwise noted.

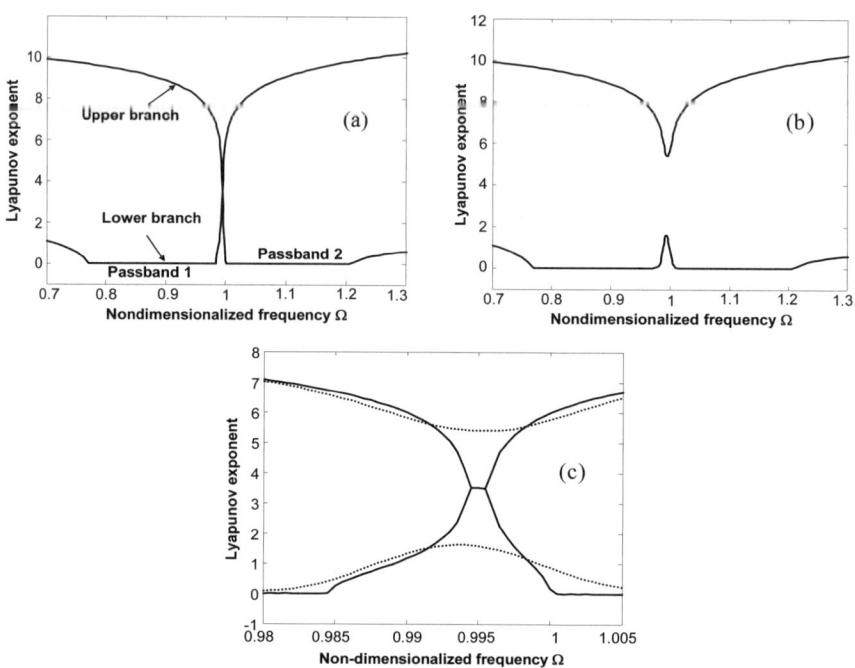

Fig. 7.2. Lyapunov exponents for the system with piezoelectric network (ζ=0.1, R_a=0.6, δ =1.2, R_c= 0.005, σ_m= 0.01). (a) tuned system; (b) mistuned system; (c) zoom-in of tuned (——) and mistuned (⋯⋯⋯) system around Ω =1.

For this investigation on vibration localization, the structural damping and circuitry resistance are neglected so that the vibration localization effect can be clearly highlighted. The original periodic structure without piezoelectric network is a mono-coupled system, so there is only one Lyapunov exponent and mode localization can be directly evaluated by the Lyapunov exponent, as demonstrated in (Tang and Wang, 2003). For the periodic structure with the piezoelectric network treatment, the introduction of the electrical coupling enabled by the capacitors between the local shunt circuits creates another wave channel. Therefore, the system becomes bi-coupled; and there are two Lyapunov exponents at each frequency. Fig. 7.2 shows the Lyapunov exponents for the system with parameters ζ=0.1, R_a=0.6, δ =1.2, R_c= 0.005, and σ_m= 0.01. Here Fig. 7.2(a) corresponds to the tuned case, Fig. 7.2(b) corresponds to the mistuned case and Fig. 7.2(c) provides a zoom-in view around Ω=1 for both the tuned and mistuned cases. The two Lyapunov exponents are referred to as upper and lower branches. For this system, as indicated in Fig. 7.2a, there are two separate passbands (referred to as passband 1 and passband 2) for the tuned case. The natural frequencies are found to be inside the passbands. It is well known that mode shapes of the tuned system are extended without spatial attenuation throughout the substructures, or in other words, having zero decay rates. This characteristic of tuned system is captured by the zero Lyapunov exponents in the passbands. When the system has mistuning, there are no longer any passbands; and the lower Lyapunov exponent branch becomes non-zero (see the lower dotted line in Fig. 7.2(c)), indicating mode localization caused by mistuning. When localization occurs, the localized modal amplitudes will have a spatial exponential decay, the rate of which can be approximated by the lower of the two Lyapunov exponents in the original separate passbands.

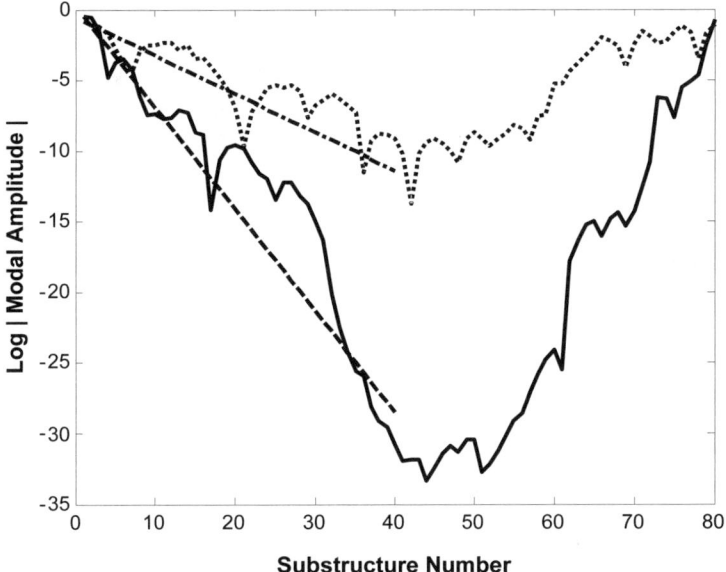

Fig. 7.3. Correlation of modal amplitude exponential decay rates to the lower Lyapunov exponents.
——— : modal amplitude of the 93rd mode ($\Omega=1.0006$); – – – : straight line with slope (-0.7202) corresponding to the lower Lyapunov exponent at $\Omega=1.0006$.
·············· : modal amplitude of the 70th mode ($\Omega=0.9832$); – · – · : straight line with slope (-0.2720) corresponding to the lower Lyapunov exponent at $\Omega=0.9832$.

To support the above argument, the correlation of the modal amplitude exponential decay rates and the lower Lyapunov exponents is studied. In the following analysis, mode shapes of the mistuned periodic structure with piezoelectric network are obtained by solving the eigenvalue problem of the system, the parameters of which are the same as those used for Fig. 7.2. For modal analysis, the mistuned periodic structure is assumed to have N=80 substructures. To demonstrate the exponential decay of the modal amplitudes, the substructure having the highest amplitude is chosen as the 1st substructure. Two examples are demonstrated in Fig. 7.3. Amplitudes of two modes are shown, the 70th mode with natural frequency of Ω=0.9832, which lies in the original passband 1 and the 93rd mode with natural frequency of Ω=1.0006, which lies in the original passband 2. The amplitudes of the 70th mode in logarithm scale are plotted with dotted line. It is shown that the modal amplitudes decay exponentially throughout the first half of the substructures. Actually, the exponential growth in the second half of the substructures can be seen as exponential decay at the same rate towards the other direction, due to the cyclic nature of the periodic structure.   At its natural frequency Ω=0.9832, the upper and lower Lyapunov exponents are computed to be 6.6969 and 0.2720 respectively. The dash-dotted line is a straight line with a slope of -0.2720, which is directly related to the lower Lyapunov exponent. It can be seen that the exponential decay rate for the 70th mode is well captured by this straight line, which means, the lower Lyapunov exponent can characterize the exponential decay of the modal amplitude.   Another example is the 93rd mode, with natural frequency of Ω=1.0006. At this frequency, the upper and lower Lyapunov exponents are calculated to be 5.8572 and 0.7202, respectively. The modal amplitudes in natural logarithm scale are plotted with solid line in Fig. 7.3. The dashed straight line has a slope of -0.7202, which is corresponding to the lower Lyapunov exponent at this frequency. Obviously, this straight line can approximate the spatial exponential decay of the mode amplitudes very well. Based on this correlation demonstration, we conclude that it is the lower Lyapunov exponent that can characterize the exponential decay of the mode localization for the mistuned system. Therefore, the lower Lyapunov exponent can serve as a measure to quantify the level of the modal localization.

7.4 PIEZOELETRIC TRANSDUCER CIRCUITRY NETWORK SYNTHESIS FOR VIBRATION MODE DELOCALIZATION

The preceding section provides the foundation for circuitry analysis in delocalization study. The focus of this section is on the actual implementation. A piezoelectric transducer circuitry is integrated to a scaled bladed-disk structure for experimental investigation. In addition, the negative capacitance idea described in Chapter 2 is incorporated to enhance the system performance.

7.4.1 Negative Capacitance for the Enhancement of Electro-mechanical Coupling

As can be seen, the proposed delocalization scheme is built upon the concept of energy propagation due to the additional electrical channel. Therefore, it is easy to envision that the generalized electro-mechanical coupling coefficient (ξ) plays an important role in the delocalization performance, since it characterizes the amount of energy that can be transformed from mechanical to electrical form. In Chapter 2, we have demonstrated that a negative capacitance circuit that is connected in series with the piezoelectric transducer can change the apparent piezoelectric capacitance and increase the coupling coefficient ξ (see Eqs. (2.14) and (2.17)). In this study, this negative capacitance idea is adopted to increase the generalized electro-mechanical coupling coefficient, for the purpose of enhancing the delocalization performance of the piezoelectric circuitry network.

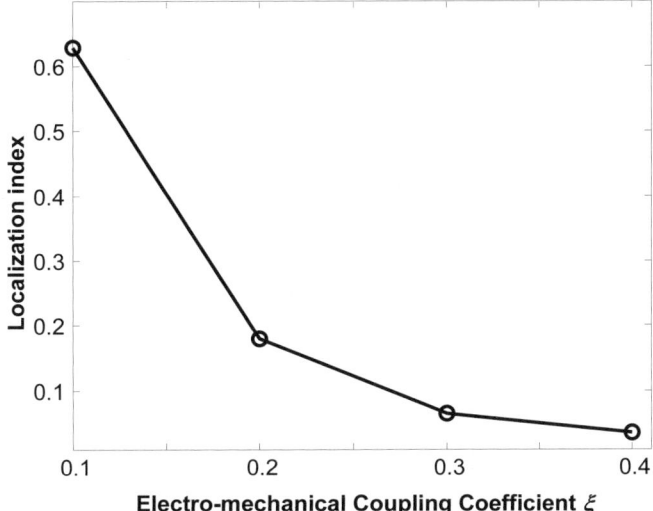

Fig. 7.4. Localization index versus electro-mechanical coupling coefficient ξ for the mistuned system ($R_a = 0.5$, $\delta = 0.5$, $R_c = 0.005$, $\sigma_m = 0.01$).

Based on the discussion in the preceding section, a mode localization index can be defined as the average of the lower Lyapunov exponents of the mistuned system within the frequency range where the original tuned system exhibits two separate passbands. The effect of negative capacitance on mode delocalization performance of the piezoelectric network can now be evaluated using the defined index. In the following parametric study, the average of the lower Lyapunov exponents is taken over fifty frequency points in each passband. As discussed earlier, the introduction of negative capacitance into the piezoelectric network can increase the generalized electro-mechanical coupling coefficient ξ. To examine the consequence of this coupling enhancement, a single case is first illustrated. In this case study, the system parameters are set to be R_a=0.5, δ=0.5, R_c=0.005, σ_m=0.01. It is shown in Fig. 7.4 that the localization index decreases as the electro-mechanical coupling coefficient is increased by adding the negative capacitance. Without negative capacitance ($\xi = 0.1$), the localization index is more than 0.6; with negative capacitance, as the electro-mechanical coupling coefficient ξ is increased, the localization index can be reduced to as low as values below 0.1. This reduction in localization index indicates that the level of localization is reduced by the introduction of the negative capacitance.

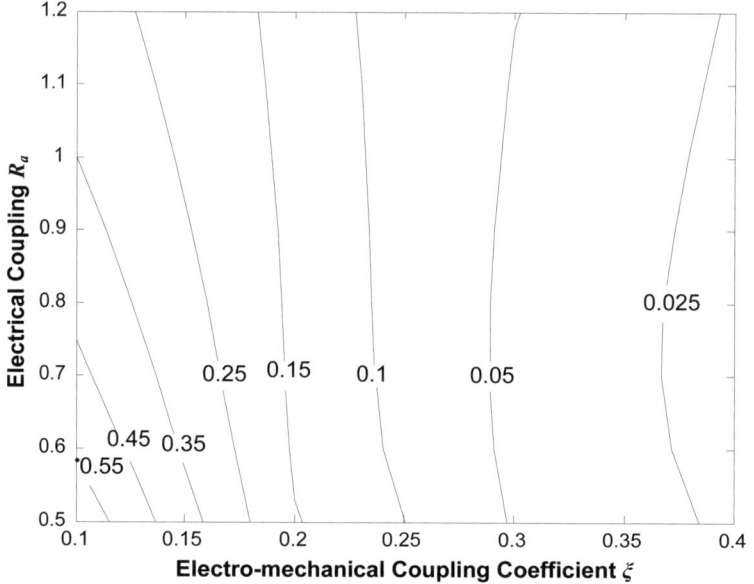

Fig. 7.5. Contour plot of localization index versus R_a and ξ for the mistuned system (R_c=0.005, δ =0.5, σ_m=0.01).

To gain more insights, a more extensive parametric study is carried out. The system parameters ξ, R_a and δ are varied within realistic application ranges, which cover the operation parameter region of the experimental study that follows. Figs. 7.5 and 7.6 show the contour plots of the localization index versus ξ and R_a (or δ). In Fig. 7.5, δ is fixed at 0.5, ξ and R_a are varied. In Fig. 7.6, R_a is fixed at 1.2, and ξ and δ are varied. As shown in both figures, the localization index tends to decrease as ξ increases. These results indicate that, as a consequence of increasing the electro-mechanical coupling coefficient by negative capacitance, the delocalization performance of the piezoelectric network can be improved. Physically, this is because by increasing the system electro-mechanical coupling coefficient one could increase the capability of energy transformation; hence more localized mechanical energy can be transferred into electrical form and propagate throughout the network.

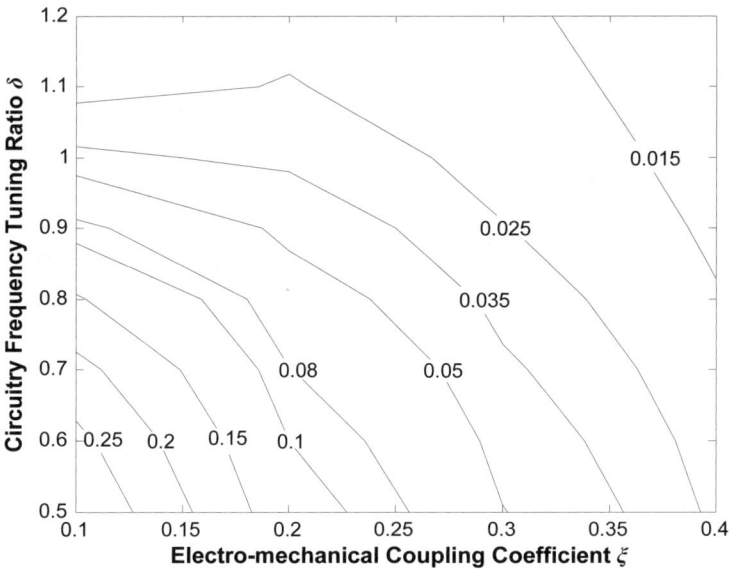

Fig. 7.6. Contour plot of localization index versus δ and ξ for the mistuned system (R_c=0.005, R_a =1.2, σ_m=0.01).

7.4.2 Experimental Setup

The experimental setup for delocalization study is shown in Fig. 7.7. The mistuned bladed-disk is vertically bolted at the hub disk center to a fixture mounted on an isolation table. A shaker is used to provide excitation at the disk hub. Tip displacements of the blades are measured by a laser vibrometer (OFV-303, Polytec Germany), which converts the displacement information into a voltage related output (calibrated in μm/Volt or 10^{-6} meter/Volt). This voltage related output is then recorded by an HP35665A analyzer. The resolution of the vibrometer could be as high as 0.5 μm/Volt. The laser vibrometer is mounted on two perpendicular stages (X-Y stages as shown in Fig. 7.7), which, controlled by LabVIEW programs, can precisely locate the measurement point on each blade tip.

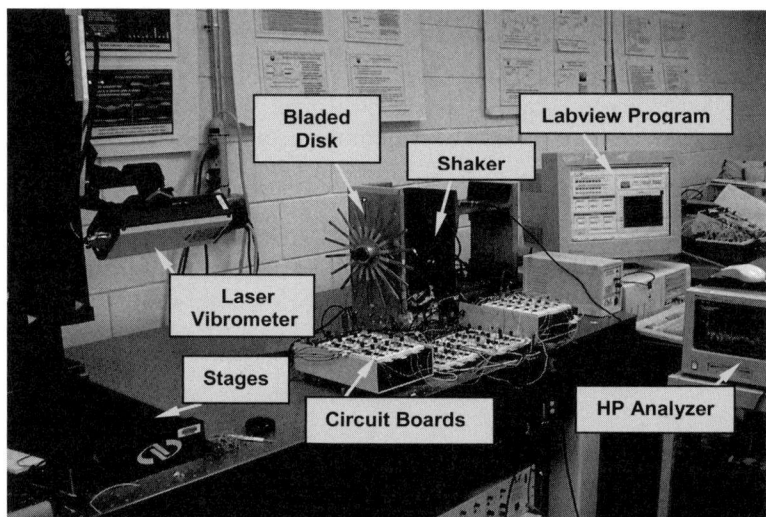

Fig. 7.7. Experimental setup.

The bladed-disk specimen with 18 equally spaced blades is fabricated from a single piece of aluminum alloy plate (see Table 7.1 for dimensions), which is mistuned in nature due to manufacturing tolerances. Each blade is bonded with an identical piezoelectric patch (type 5A, APCI, Ltd., dimensions are length 2.54 cm, width 0.76 cm, thickness 0.10 cm; material properties are shown in Table 7.2) at its root. The bonding of the piezoelectric patches also contributes to the mistuning of the bladed-disk. Therefore, the term *mistuned bladed-disk* in the following discussion refers to the bladed-disk with piezoelectric patches attached to it. Each piezoelectric patch has a negative electrode wrap-up design, which improves the bonding effectiveness and provides convenience in wiring. The patches are electrically insulated from the aluminum blades since subsequently the negative capacitance circuits will be inserted between the piezoelectric patches and the ground. The piezoelectric circuitry network is synthesized and integrated with the bladed-disk as shown in Fig. 7.1b. Each passive piezoelectric patch is connected in series with a synthetic inductor to form an *LC* resonant circuit.   Then these 18 resonant circuits are coupled through capacitors (C_a) each with a value of 2.2 nF (corresponding to R_a=1.2, as shown in Fig. 7.6).

The circuit diagrams of the synthetic inductor (Chen, 1986, Horowitz and Hill, 1989) and the negative capacitor are shown in Figs. 2.7 and 2.6b in Chapter 2. In order to increase the electro-mechanical coupling coefficient, each negative capacitance circuit is connected in series with each piezoelectric patch. Since the negative capacitance circuit needs to be grounded, it is inserted between the negative electrode of each piezoelectric patch and the ground.

Table 7.1. Dimensions of the bladed-disk model

Inner (hole) diameter of the hub disk	0.0381 m
Outer diameter of the hub disk	0.0889 m
Length of blade	0.1080 m
Width of blade	0.0077 m
Thickness of blade	0.0032 m

Table 7.2. Material properties of piezoelectric patches

Material type	Type 5A
Relative dielectric constant K^T	1750
Electro-mechanical coupling factor k_{31}	0.36
Piezoelectric charge constant	175×10^{-12} (m/V)
Young's modulus	6.3×10^{10} (N/m^2)
Capacitance (C_p)	3.3 nF

7.4.3 Experimental Results

A sample frequency response function (FRF) of the mistuned bladed-disk is shown in Fig. 7.8. The figure shows high modal density within the frequency range from 190 Hz to 250 Hz, which is a characteristic feature of mistuned periodic structures. Due to this feature, it is very difficult to obtain mode shapes using common modal analysis methods. Therefore, alternatively, the amplitudes of blade tips when the bladed-disk is under harmonic excitation at resonant frequencies are chosen as comparison objectives in evaluating the delocalization effect of the piezoelectric circuit network. To excite the bladed-disk at resonance, a series of frequencies within the range of 190 Hz to 250 Hz, most of which are around the FRF resonant peaks, are used for sine wave excitations.

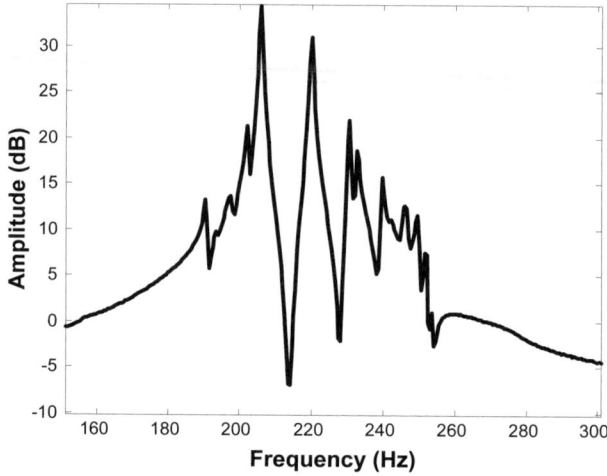

Fig. 7.8. Sample frequency response of the mistuned bladed-disk without circuit network.

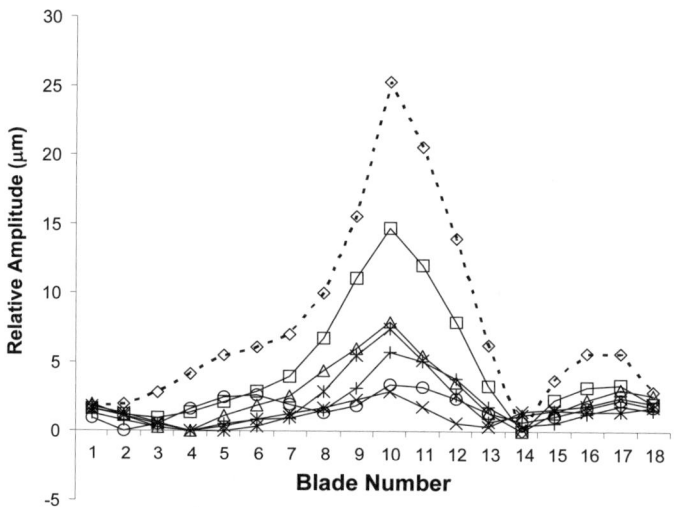

Fig. 7.9. Relative amplitudes of 18 blades for mistuned bladed-disk without network and with network at various circuit frequency tunings (f_e) from 193.5 Hz to 234 Hz. ············: (◊) without network; ———: (□) $f_e = 193.5$ Hz; (Δ) $f_e = $ 201 Hz; (×) $f_e = 206$ Hz; (*) $f_e = 212$ Hz; (○) $f_e = 222$ Hz; (+) $f_e = 234$ Hz.

In the preliminary test, it is found that under the excitation frequencies of 193.5 Hz and 202.3 Hz, the amplitude distributions show obvious localization phenomena, illustrated by the dotted lines in Figs. 7.9 and 7.11. In this investigation, we will thus focus our attention around these two resonant frequencies, and compare the amplitude distribution of the mistuned bladed disk with and without piezoelectric circuit network.

First, the mistuned bladed-disk without circuit network is excited at 193.5 Hz. The tip displacements of all 18 blades are measured with the laser vibrometer. Then the circuitry network is connected to the piezoelectric patches on the bladed-disk. The network has 18 synthetic inductors tuned to the same value. The circuit frequency, defined as $f_e = 1/\sqrt{LC_p}/2\pi$ (Hz), where C_p is the piezoelectric capacitance, can be tuned to different values by adjusting the synthetic inductance (L). Six different circuit frequency tunings are investigated, ranging from 193.5 Hz to 234 Hz (corresponding to the nondimensionalized parameter δ of range 0.88 ~ 1.06, which is covered in Fig. 7.6), as shown in Fig. 7.9. At each circuit frequency, the resonant response amplitudes of the blade tips are measured. These amplitudes are plotted in Fig. 7.9 for the mistuned bladed-disk with and without circuit network. The amplitude data shown in this figure reflect the *relative amplitude* (the relative amplitude of each blade is defined to be the difference between its absolute amplitude and the lowest amplitude among all the 18 blades), which also applies to all other figures showing amplitude hereafter. It should be noted that when the piezoelectric circuitry network is connected to the mistuned bladed-disk, the resonant frequencies are slightly changed. In order to maintain the resonant excitation, for each particular circuit frequency tuning (f_e), the new resonance of the system is identified and the structure is excited at the new resonant frequency. This also applies to the case when negative capacitance circuits are introduced into the network.

The dotted data line in Fig. 7.9 shows that the relative vibration amplitudes are high over a small region (only 4-5 blades) around blade number 10. Outside this region, amplitudes are relatively small. This indicates that vibration is highly localized in the mistuned bladed-disk. With the treatment, the solid lines corresponding to various circuit tunings show a more even distribution of the amplitudes over the 18 blades. This more even amplitude distribution indicates a reduction of the level of localization. As a tool to quantify how the amplitude data is distributed spatially, standard deviations are plotted in Fig. 7.10. However, it should be noted that the standard deviation is not used as an exact index for quantifying localization, but only as a measurement of the scatterness of the amplitude data distribution. As shown, the standard deviations for the mistuned bladed-disk with treatment are much smaller than that without treatment.

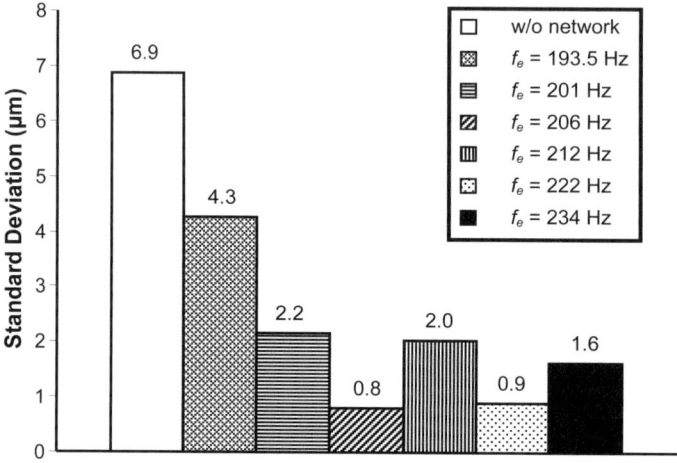

Fig. 7.10. Standard deviations of blade relative amplitudes for the system without network (1st column) and with network at various circuit frequency tunings (f_e).

Moreover, delocalization effect of the piezoelectric circuit network is also demonstrated when the mistuned bladed disk is excited under the resonant frequency of (and around) 202.3 Hz. Resonant response amplitudes for the mistuned bladed disk with and without piezoelectric network are plotted in Fig. 7.11. Without network, blade amplitudes are confined only in small regions around blade number 2, 3 and 7. With the piezoelectric circuit network (at three circuit frequency tunings: f_e= 201 Hz, 206 Hz, and 222 Hz), the blade amplitudes again become more evenly distributed over the 18 blades, which means the level of localization is reduced.

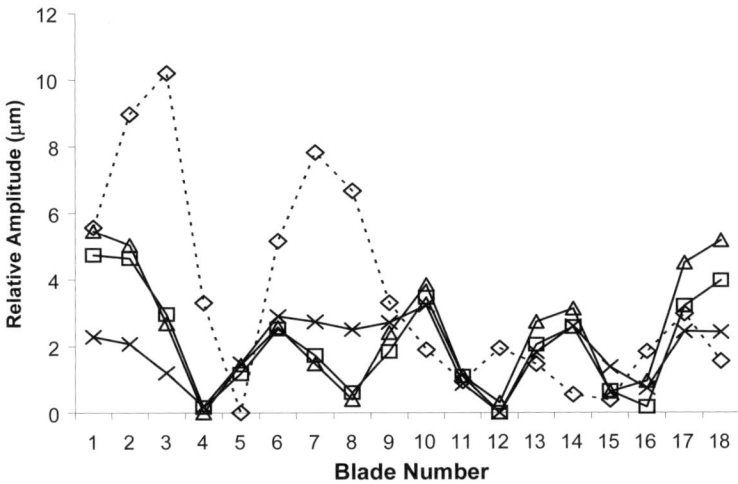

Fig. 7.11. Blade relative amplitudes distribution for the system at resonances around 202.3 Hz. •••••••••• : without network, ——————: with network. Legends: (◊) without network; (□) $f_e = 201$ Hz; (Δ) $f_e = 206$ Hz; (×) $f_e = 222$ Hz.

Next, negative capacitance circuits are incorporated into the piezoelectric network, and their effects on the delocalization performance are examined. The analytical results suggest that the electro-mechanical coupling coefficient (ξ) of the piezoelectric patch can be increased by the negative capacitance. As a result of this coupling enhancement, the overall delocalization effect of the piezoelectric circuit network can be further improved. To validate this prediction, negative capacitance circuits are built and connected in series with piezoelectric patches to the negative electrodes. The same experimental approach used in Chapter 2 is adopted here to measure the electro-mechanical coupling coefficient. The value of ξ is calculated according to $\xi = \sqrt{((\omega^D)^2 - (\omega^E)^2)/(\omega^D)^2}$, which is based on the resonant frequencies of the substructure under open circuit condition (ω^D) and short circuit condition (ω^E). Note that this formula is exactly the definition of the electro-mechanical coupling coefficient at the structural level. The original electro-mechanical coupling coefficient of the piezoelectric patch is measured to be $\xi = 0.1224$. With a negative capacitance of $C_n = -4.7$ nF, the coupling coefficient is measured to be $\xi = 0.1930$, which is a 57.7% increase. This range of ξ is covered in the analysis shown in Fig. 7.6.

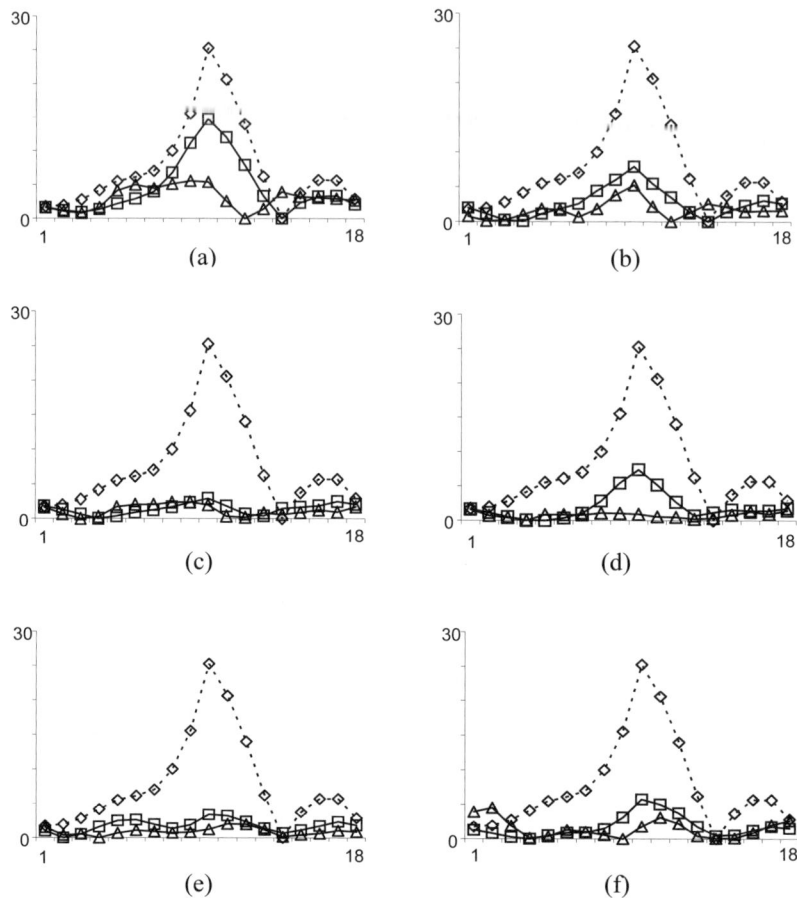

Fig. 7.12. Blade relative amplitudes of system without network (◊ in dotted line); with network (□ in solid line); and with network augmented by negative capacitance (Δ in solid line); (a) f_e = 193.5 Hz; (b) f_e = 201 Hz; (c) f_e = 206 Hz; (d) f_e = 212 Hz; (e) f_e = 222 Hz; (f) f_e = 234 Hz; Horizontal axis: blade number; Vertical axis: relative amplitude (unit: μm).

Eighteen negative capacitance circuits with the same value (C_n = -4.7 nF) are then built and integrated into the piezoelectric network. The network with negative capacitance is also referred to as the *augmented network*. Tip displacements of the 18 blades are re-measured for the bladed disk with this augmented network. Resonant response amplitudes are compared to previous results shown in Figs. 7.9 and 7.11 in order to evaluate the delocalization performance improvement. First, the responses at the resonance around 193.5 Hz are examined and amplitudes are plotted in Figs. 7.12 (a)-(f), with each corresponding to a circuit frequency tuning (f_e). It can be seen that for all of these six circuit frequency tunings, the vibration amplitude distributions with the augmented piezoelectric network become more uniform. This is expected, because with negative capacitance, the network has a larger electro-mechanical coupling (ξ). Therefore, the network is capable of transforming more mechanical energy into electrical form, which is then propagated throughout the network by the coupling capacitors. The standard deviations of the amplitudes are calculated and shown in Fig. 7.13 for the mistuned bladed disk without any treatment, with network but no negative capacitance, and with the augmented network, marked as Case 1, Case 2 and Case 3, respectively.

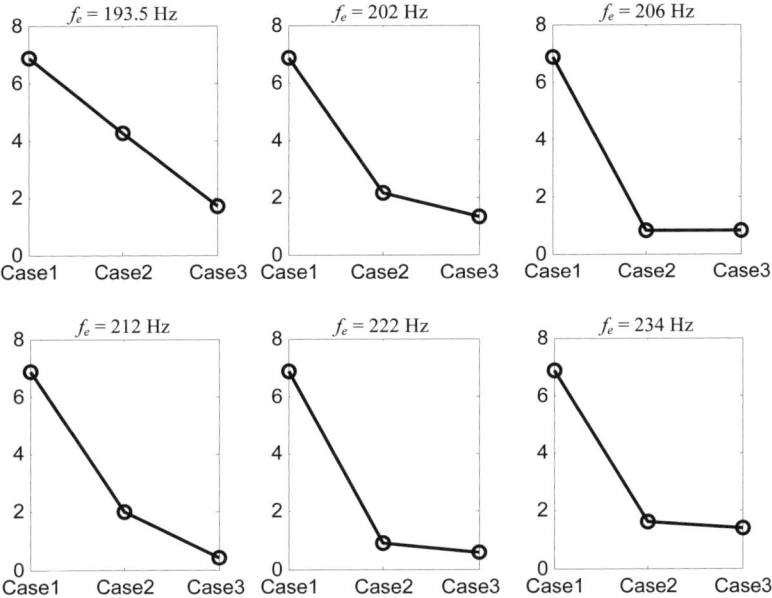

Fig. 7.13. Standard deviations of blade relative amplitudes for Case 1—mistuned bladed-disk without network; Case 2—with network; and Case 3—with network augmented by negative capacitance circuits. Vertical axis: standard deviation (unit: μm).

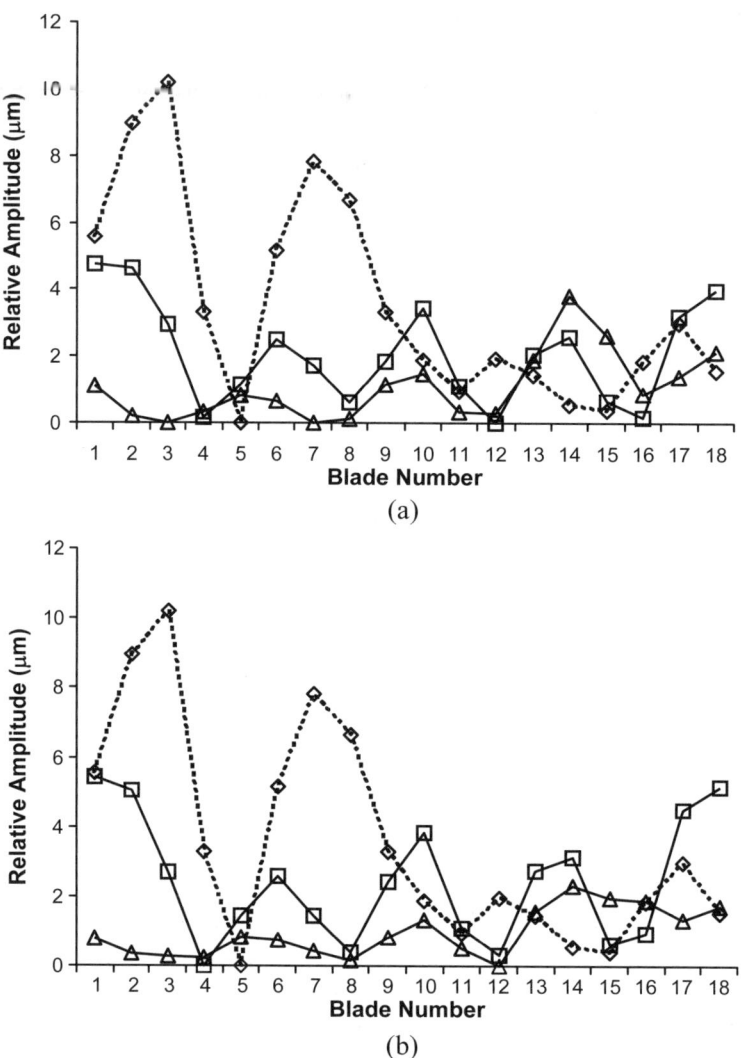

Fig. 7.14. Relative amplitudes distribution for the system without network ($\Diamond$ in dotted line), with network ($\Box$ in solid line) and with network augmented by negative capacitance (Δ in solid line). (a) $f_e = 201$ Hz; (b) $f_e = 206$ Hz.

In Fig. 7.13, the decreasing trend of the standard deviations from Case 1 to Case 3 indicates that vibration amplitudes become more and more evenly distributed. These results show that the delocalization ability of the piezoelectric network is improved by the negative capacitance circuits. Amplitudes of the mistuned bladed disk under resonant excitation frequencies of (and around) 202.3 Hz are also compared for Cases 1, 2 and 3, as shown in Figs. 7.14a and 7.14b, with $f_e = 201$ Hz and 206 Hz, respectively. Both figures show that amplitude localization is further reduced and the distribution is more even for Case 3, as compared to that of Case 2. The results again illustrate improvements in the delocalization performance when negative capacitance circuits are added.

7.5 VIBRATION SUPPRESSION IN PERIODIC STRUCTURES

In the previous section, the piezoelectric transducer circuitry network concept has been shown to be effective for vibration mode delocalization. While reducing vibration localization in mistuned periodic structures has practical significance, an equally important issue in such structures is the suppression of overall vibration under external disturbances such as the engine-order-excitation in bladed disks. In this section, we expand the investigation and present an approach of utilizing the piezoelectric circuitry network for vibration suppression of periodic structures under forced excitations. We start the discussion on perfectly periodic structure, and then incorporate mistuning into the analysis to examine the system behavior.

When excited by external vibratory loads, a periodic structure could have high strain in all of its substructures. Therefore, it is reasonable to apply vibration suppression devices on each of the substructures. One option is to develop piezoelectric damped vibration absorbers by using the configurations shown in Fig. 7.1b or 7.1c, with the addition of resistance elements. To maintain the design function of the original structure and for the sake of convenience in manufacturing, it is preferable to maintain system periodicity; that is, to implement identical piezoelectric absorbers onto each of the substructures. One of the major technical challenges here is that the piezoelectric absorber configuration and optimal tuning method discussed in Chapter 2 cannot be directly applied, due to the characteristics and complexity of periodic structures. Such phenomena and the proposed approach to effectively achieve vibration suppression are discussed in the following sections.

7.5.1 System Equations under External Excitation

We consider the circuitry configuration shown in Fig. 7.1b, whereas identical resistance R is now inserted into each local resonant shunt. The equations of motion are then given as

$$m\ddot{q}_j + c\dot{q}_j + (k + \Delta k_j)q_j + k_c(2q_j - q_{j-1} - q_{j+1}) + k_1 Q_j = f_j \quad (7.17a,b)$$

$$L\ddot{Q}_j + R\dot{Q}_j + k_2 Q_j + k_a(2Q_j - Q_{j-1} - Q_{j+1}) + k_1 q_j = 0$$

Here we include the mistuning effect in the system model. f_j is the external force applied to the blade. In this case, we assume engine-order-excitation force, which can be expressed as $f_j = F_0 e^{i(\omega t + \phi_j)}$. Here F_0 is the magnitude of the force. $i = \sqrt{-1}$. ϕ_j is the phase of the force, and $\phi_j = \dfrac{2\pi(E-1)(j-1)}{N}$, where E is the engine order number. This force expression is used to emulate the force that a bladed disk in a turbomachinery system experiences due to the aerodynamics of the fluid passing through the system. The force is harmonic in both time and spatial distribution. Its characteristic spatial distribution pattern is defined by the engine order number E. The phase difference in adjacent blades is $(E-1)\theta$, where θ is determined by the total number of blades, $\theta = 2\pi / N$.

Using a nondimensionalization procedure similar to that in section 7.1, we may obtain,

$$-\Omega^2 x_j + (1 + \Delta s_j)x_j + 2i\Omega\zeta_c x_j + R_c^2(2x_j - x_{j-1} - x_{j+1}) + \delta\xi y_j = \bar{f}_j \quad (7.18a,b)$$

$$-\Omega^2 y_j + \delta^2 y_j + 2i\Omega\delta\zeta_r y_j + R_a^2(2y_j - y_{j-1} - y_{j+1}) + \delta\xi x_j = 0$$

where, in addition to the nondimensionalization notations defined in Section 7.1, we define $\zeta_r = R / 2m\omega_c$, $\zeta_c = c / 2m\omega_m$ and $\bar{f}_j = f_j \sqrt{m/k}$. Similarly, the equation of motion for the original periodic structure without piezoelectric circuitry and its nondimensionalized version are expressed as, respectively,

$$m\ddot{q}_j + c\dot{q}_j + (k + \Delta k_j)q_j + k_c(2q_j - q_{j-1} - q_{j+1}) = f_j \tag{7.19}$$

$$-\Omega^2 x_j + (1 + \Delta s_j)x_j + 2i\Omega\zeta_c x_j + R_c^2(2x_j - x_{j-1} - x_{j+1}) = \overline{f}_j \tag{7.20}$$

7.5.2 Circuitry Synthesis

To design the absorber circuitry, we first consider the *tuned* systems in Eq. (7.19) and Eq. (7.17), *i.e.*, $\Delta k_j = 0$. Eqs. (7.19) and (7.17) can be written into matrix forms by grouping equations from $j = 1$ to N for a system with N subsystems (with or without circuits),

$$\mathbf{M}\ddot{\mathbf{z}} + \mathbf{C}\dot{\mathbf{z}} + \mathbf{K}\mathbf{z} = \mathbf{f} \tag{7.21}$$

where $\mathbf{M}$, $\mathbf{C}$, $\mathbf{K}$ are all circulant matrices (Davis, 1979),

$$\mathbf{M} = circ(\mathbf{a}_m, 0, ..., 0) = \begin{bmatrix} \mathbf{a}_m & & \\ & \ddots & \\ & & \mathbf{a}_m \end{bmatrix} \tag{7.22}$$

$$\mathbf{C} = circ(\mathbf{a}_c, 0, ..., 0) = \begin{bmatrix} \mathbf{a}_c & & \\ & \ddots & \\ & & \mathbf{a}_c \end{bmatrix} \tag{7.23}$$

$$\mathbf{K} = circ(\mathbf{a}_k, \mathbf{b}_k, 0, ..., 0, \mathbf{b}_k) = \begin{bmatrix} \mathbf{a}_k & \mathbf{b}_k & 0 & ... & 0 & \mathbf{b}_k \\ \mathbf{b}_k & \mathbf{a}_k & \mathbf{b}_k & 0 & ... & 0 \\ & \ddots & \ddots & \ddots & & \\ & & \ddots & \ddots & \ddots & \\ 0 & ... & 0 & \mathbf{b}_k & \mathbf{a}_k & \mathbf{b}_k \\ \mathbf{b}_k & 0 & ... & 0 & \mathbf{b}_k & \mathbf{a}_k \end{bmatrix} \tag{7.24}$$

For the original mechanical system without the piezoelectric transducer circuitry, *i.e.*, for Eq. (7.19), the displacement vector is of size $1 \times N$,

$\mathbf{z} = [q_1, q_2, ..., q_N]^T$. The elements in those circulant matrices $\mathbf{M}$, $\mathbf{C}$, $\mathbf{K}$ are scalars listed below:

$$\mathbf{a}_m = m, \quad \mathbf{a}_c = c, \quad \mathbf{a}_k = k + 2k_c, \quad \mathbf{b}_k = -k_c \quad\quad (7.25)$$

For the mechanical system integrated with the piezoelectric circuitry network, *i.e.*, for Eq. (7.17), the displacement vector is of size $1 \times 2N$, $\mathbf{z} = [q_1, Q_1, q_2, Q_2, ..., q_N, Q_N]^T$. Parameters in these circulant matrices are 2×2 matrices shown below; thus $\mathbf{M}$, $\mathbf{C}$, $\mathbf{K}$ are also called *block-circulant* matrices.

$$\mathbf{a}_m = \begin{bmatrix} m & \\ & L \end{bmatrix}, \quad \mathbf{a}_c = \begin{bmatrix} c & \\ & R \end{bmatrix} \quad\quad (7.26)$$

$$\mathbf{a}_k = \begin{bmatrix} k + 2k_c & k_1 \\ k_1 & k_2 + 2k_a \end{bmatrix}, \quad \mathbf{b}_k = \begin{bmatrix} -k_c & 0 \\ 0 & -k_a \end{bmatrix}$$

The force vector for mechanical system is:

$$\mathbf{f} = F_0 e^{i\omega t} [e^0, e^{i\phi_2}, ..., e^{i\phi_j}, ..., e^{i\phi_N}]^T \qu\quad (7.27)$$

For the mechanical system with integrated piezoelectric networks, since we assume no external voltage source is applied, the forcing vector is:

$$\mathbf{f} = F_0 e^{i\omega t} [e^0, 0, e^{i\phi_2}, 0, ..., e^{i\phi_j}, 0, ..., e^{i\phi_N}, 0]^T \qu\quad (7.28)$$

The U transform (Tang and Wang, 1999) can be applied to diagonalize (or block-diagonalize) the circulant matrices (or block-circulant matrices) in Eq. (7.21). For the mechanical structure alone, the U transform matrix is an $N \times N$ matrix, denoted as $\mathbf{U}_m$, whose (p,q)th element is defined as:

$$(\mathbf{U}_m)_{pq} = \frac{1}{\sqrt{N}} e^{i\theta(p-1)(q-1)} \qu\quad (7.29)$$

For the mechanical structure with the circuitry, the U transform matrix is a $2N \times 2N$ matrix, denoted as $\mathbf{U}_{me}$, which is an expansion of $\mathbf{U}_m$, i.e., $\mathbf{U}_{me} = \mathbf{U}_m \otimes \mathbf{I}_2$, where $\otimes$ is the Kronecker tensor product.

We define

$$\mathbf{z} = \mathbf{U}\mathbf{v} \tag{7.30}$$

Substituting Eq. (7.30) into Eq. (7.21) and pre-multiplying by $\mathbf{U}^*$, we obtain

$$\mathbf{U}^*\mathbf{M}\mathbf{U}\mathbf{v} + \mathbf{U}^*\mathbf{C}\mathbf{U}\mathbf{v} + \mathbf{U}^*\mathbf{K}\mathbf{U}\mathbf{v} = \mathbf{U}^*\mathbf{f} \tag{7.31}$$

where $\mathbf{U}^*$ is the complex conjugate transpose of $\mathbf{U}$. The system equations in Eq. (7.21) are transformed into the spatial harmonic space, where the originally coupled stiffness matrix $\mathbf{K}$ can be decoupled so that the stiffness matrix becomes diagonal (for mechanical system without circuit) or block-diagonal (for mechanical system with circuit).

To design the absorber circuitry, let us consider Eq. (7.17a) with $\Delta k_j = 0$, and neglect the structural damping c since bladed-disk systems usually have very light damping. After the U transformation Eq.(7.17) becomes:

$$m\ddot{v}_{mj} + [k + 2k_c(1 - \cos((j-1)\theta))]v_{mj} + k_1 v_{ej} = h_j \tag{7.32a,b}$$

$$L\ddot{v}_{ej} + R\dot{v}_{ej} + [k_2 + 2k_a(1 - \cos((j-1)\theta))]v_{ej} + k_1 v_{mj} = 0$$

where $h_j = \sqrt{N}F_0 e^{i\omega t}$. Note that in Eq. (7.17), j denotes the jth spatial harmonic. The non-zero h_j exists only when the spatial harmonic equals to the engine order number, i.e., $j=E$, otherwise, it will be zero. In this case, the state vectors are $\mathbf{z} = [q_1, Q_1, q_2, Q_2, ..., q_N, Q_N]^T$ and $\mathbf{v} = [v_{m1}, v_{e1}, ..., v_{mN}, v_{eN}]^T$.

Assuming harmonic motion, the frequency response function between the blade motion and the force can be obtained from Eq. (7.32):

$$v_{mj}/h_j = \cfrac{1}{\alpha - \cfrac{k_1^2}{\beta}}$$

(7.33)

where

$$\alpha = -\omega^2 m + k + 2k_c(1 - \cos((j-1)\theta))$$

$$\beta = -\omega^2 L + i\omega R + k_2 + 2k_a(1 - \cos((j-1)\theta))$$

(7.34a,b)

First we examine the traditional absorber design without the network capacitance, *i.e.*, $k_a=0$. By following the procedure in (Tang and Wang, 1999) for vibration absorber design, one can find the optimal inductance L,

$$L_{opt} = \frac{mk_2}{k + 2k_c(1 - \cos((j-1)\theta)}$$

(7.35)

In this case, it is obvious that the optimal tuning is dependent on the spatial harmonic number j. In other words, the non-networked traditional absorber can only be optimally designed to suppress a specific spatial harmonic excitation.

By applying the coupling capacitance and forming the circuitry, one can derive the optimal inductance to be:

$$L_{opt} = \frac{m[k_2 + 2k_a(1 - \cos((j-1)\theta)]}{k + 2k_c(1 - \cos((j-1)\theta)}$$

(7.36)

While the expression in Eq. (7.36) is still j-dependent, by properly tuning the coupling capacitance k_a, one can design an L_{opt} that is independent of the spatial harmonic number j. This can be accomplished by letting

$$k_a = \frac{k_2 k_c}{k}$$

(7.37)

Then the optimal L will become,

$$L^* = \frac{mk_2}{k} \qquad (7.38)$$

It is thus obvious that this expression is no longer j-dependent, meaning it will be effective for all spatial harmonic excitations.

For resistance tuning, it is found that the system performance is not very sensitive to small perturbation in the resistance, thus a single resistance value is used for all spatial harmonics by taking $j=1$. Thus,

$$R^* = \frac{k_1}{k}\sqrt{2mk_2} \qquad (7.39)$$

The nondimensionalized optimal parameters corresponding to Eqs. (7.38) and (7.39) are:

$$R_a = R_c, \quad \delta^* = 1, \quad \zeta_r^* = \xi/\sqrt{2} \qquad (7.40)$$

7.5.3 Vibration Suppression Performance Illustration: Comparison with Traditional Absorber

It should be noted that the optimal absorber design in Eq. (7.40) is derived in the context of perfectly tuned periodic structure system (*i.e.*, tuned bladed disk in this case). As discussed earlier in this chapter, in reality, periodic structures such as bladed-disks in turbo-engine are often mistuned due to factors such as manufacturing tolerance and in-service degradation. Mistuning in bladed disks can drastically change the system dynamic characteristics and increase the maximum forced response compared to the ideally tuned case. Thus mistuned bladed-disk system is considered next to examine the performance of the optimal network.

First, we verify the above theory on optimal network using single random mistuning. In Fig. 7.15, we compare the vibration suppression effects of the traditional absorber and the optimal network for bladed-disk under summation of engine order excitations. Shown in this figure are the maximum blade responses versus frequency. Here, the mistuned system is realized by generating a random mistuning set for the mechanical stiffness matrix. In this case, the nondimensionalized system equations are used. The mechanical parameters used in the simulation are: R_c=0.5, N=10. The random mistuning follows normal distribution, with a standard deviation of $\sigma_m = 0.05$. The optimal network is designed according to Eq. (7.40) with a default electro-mechanical coupling coefficient ξ=0.1. The non-dimensionalized equivalence to the traditional absorber design in Eq. (7.36) is $\delta_{opt} = \sqrt{1 + 2R_c^2(1 - \cos(j - 1)\theta)}$. For the traditional absorber, harmonic number j is arbitrarily picked to be j=2 in the above equation. In Fig. 7.15, the maximum blade response is plotted against frequency. One may see that the traditional absorber can only effectively suppress a few frequency-response peaks and loses its effectiveness on others. Nevertheless, the optimal network can effectively suppress the vibration at all peaks. This is because the traditional absorber can only be optimally tuned to a specific harmonic, while the optimal network, through networking, can be tuned to suppress all spatial harmonics (*i.e.*, harmonic-independent).

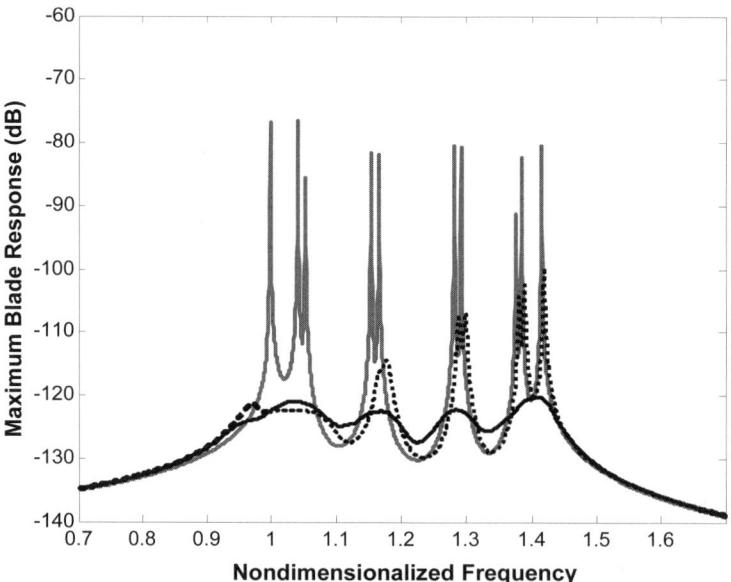

Fig. 7.15. Comparison of suppression effectiveness between traditional absorber and optimal network.———: without control; ———: with optimal network; ·········· : with traditional absorber.

7.5.4 Monte Carlo Simulation

In this section, we examine the optimal network's vibration suppression performance on mistuned bladed-disks from a statistical point of view, from which we can establish a more rigorous conclusion.

In this study, Monte Carlo simulation is used to study the network's vibration suppression performance. Due to the random nature of mistuning, for a given standard deviation σ_m, there are virtually infinite number of sets of mistuning realizations that follow the same random distribution pattern. For each realization of mistuned system, it may yield different maximum forced response. Hence, one has to simulate a large number of the mistuning realizations in order to draw a solid conclusion. Monte Carlo simulation is thus well-suited for this study.

For a given standard deviation σ_m of the random mistuning, a large number (e.g., P=500 sets) of mistuning realizations are generated according to the distribution pattern, in this case, normal distribution. Correspondingly, there are P mistuned bladed-disk assemblies. For each mistuned assembly, its maximum blade amplitude is calculated using frequency sweep under all possible engine order excitations (E=1~N). One engine order is considered at each calculation. The maximum blade response for this specific realization of mistuned system is obtained from all blades, at all frequencies and under all engine orders. The same process is applied to the mistuned bladed disk incorporated with the piezoelectric network and the maximum blade response is obtained in the same way. A ratio r is then defined based on Eq. (7.41).

$$r = \frac{\max(A_{me})}{\max(A_m)} \tag{7.41}$$

where, $\max(A_{me})$ is the maximum forced response of the mistuned system with networks, and $\max(A_m)$ is the maximum forced response of the mistuned system without networks.

For all P mistuned bladed-disk realizations, one will get P ratios. Since the ratio r is between the maximum forced response of mistuned system with and without network, smaller r means better suppression performance. Then, the vibration suppression performance index of the circuitry network is defined as the 95th percentile value of the P ratios generated by the Monte Carlo simulation. The 95th percentile is the statistical value from a set of data that only 5 percent of the data lies above.

7.5.5 Effects of Random Mistuning in Structure

In this section, the effectiveness of the network against mistuning level is studied using Monte Carlo simulation and the performance index defined above. As mentioned earlier, the networked absorber design in Eq. (7.40) is derived from tuned bladed-disk system where mistuning is not considered. In reality, the bladed disk is often mistuned. Here, we perform the study within a reasonable range of the standard deviation σ, 0~0.8, where σ =0 corresponds to the tuned system. The random mistuning is assumed to be normally distributed and only exists in stiffness (mass mistuning can be tackled in the same way). For Monte Carlo simulation, the nondimensionalized equations of motion in Eq. (7.18) are used. The parameters used in Fig. 7.16 are: R_c=0.05, N=20, P=500, damping in the structure is assumed to be very small, with damping ratio $\zeta_c = 0.001$. Network is optimally tuned according to Eq. (7.40). Here for the original piezoelectric transducers that are used on/in the bladed disk, we assume the electro-mechanical coupling coefficient to be ξ=0.1, which will be the default value without negative capacitance. As we discussed in the previous sections, this coefficient is physically determined by the property of the material and the host structure and is difficult to alter by passive means. However, one can use a negative capacitance treatment to increase this coefficient, the effect of which will be discussed later.

Observing Fig. 7.16, when σ =0, i.e., when the bladed-disk system is tuned, the performance index is r=0.025, which means that with the network, the maximum blade response is reduced by 97.5%. As the mistuning level increases, the performance degrades, as seen in the increasing performance index. However, even with σ =0.08, which is considered quite large a mistuning level, the performance index is still smaller than 0.12. This means the maximum blade amplitude is reduced by approximately 88%. Such a result shows that our piezoelectric circuitry network, although designed based on a tuned system, performs very well for mistuned systems. The same approach is carried out to analyze the system with random mistuning in electrical elements and similar results can be concluded. That is, the system's performance is quite robust against differences among the electrical circuitry parameters as well as that among the mechanical structural parameters.

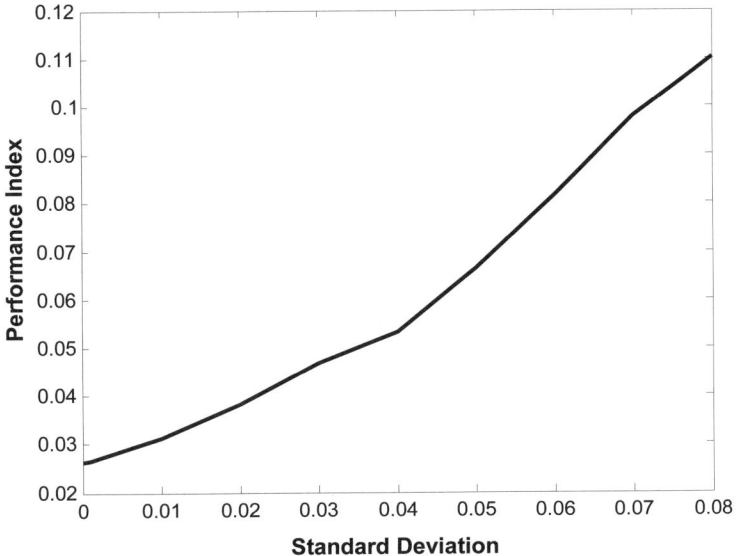

Fig. 7.16 Performance index versus standard deviation of mistuning.

7.5.6 Effects of Circuitry Detuning

In this section, the performance robustness of the network in terms of circuitry detuning is investigated. Detuning means that the local circuitry parameter nominal values are uniformly off-tuned by the same amount. Such detuning could be results of modeling or design errors, where the relative errors could be larger than those caused by random mistuning. In Fig. 7.17, the detuning effect on δ is examined with parameters R_c=0.05, N=20, ζ_c=0.001, and engine order E=11. Here, a set of random mistuning is used instead of using the Monte Carlo simulation. The frequency response with maximum blade amplitude is illustrated for comparison. The circuit frequency tuning ratio δ is detuned from its optimal value 1.0 to 0.95 (-5% detuning) and 0.9 (-10% detuning). Fig. 7.17 shows the maximum blade response in dB. With -5% detuning and -10% detuning in the optimal circuitry, the circuit becomes non-optimal and the maximum blade response is increased by 7 dB and 11 dB. However, the overall amplitude reduction is still significant (over 40 dB reduction) compared to the maximum response of the original mechanical system without piezoelectric circuitry network. Case studies involving same detuning in other circuitry parameters show similar results: the maximum blade response is slightly increased when compared with the perfectly tuned case but is still significantly lower than that of the original mechanical system without treatment. In other words, the network is quite robust against moderate detuning in the circuitry parameters. However, as can be seen, the vibration suppression performance degrades when the detuning increases. In this case, the negative capacitance can be included to further improve the robustness of the network, as will be discussed later.

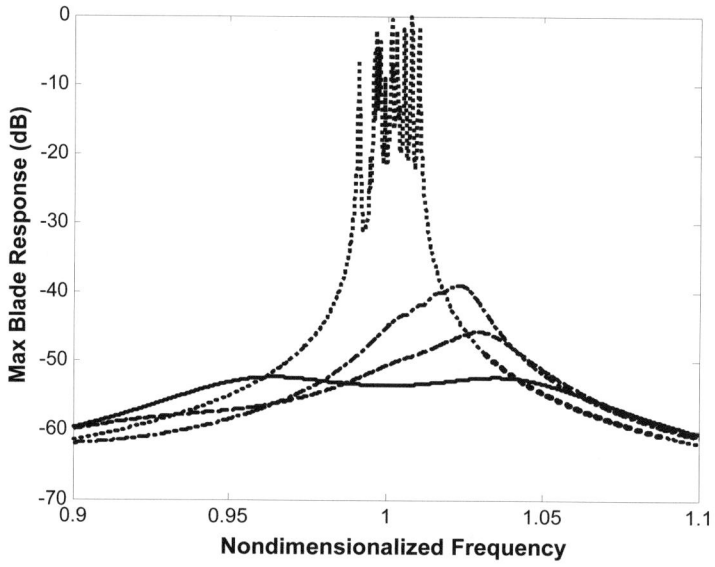

Fig. 7.17. Maximum blade response versus frequency. ⋯⋯⋯⋯ : original mechanical system without network; ───── : system with optimal tuning (δ=1.0); ─ ─ ─ ─ : system with -5% detuning (δ=0.95); ─ · ─ · ─ : system with -10% detuning (δ=0.9).

A more detailed study is conducted via Monte Carlo simulation using the performance index. Fig. 7.18 shows the effect of detuning in the circuit frequency tuning ratio on the performance of the network. Parameters used in generating Fig. 7.18 are: R_c=0.05, N=20, P=500, with all engine orders (E=1~N) and optimal circuit tuning as in Eq. (7.40). The optimal tuning is δ =1.0, corresponding to the lowest point on the curve. Then δ is detuned from 0.5 (-50% detuning) to 1.5 (+50% detuning). In either direction departing from the optimal tuning, the performance index increases, indicating performance degradation. This range of detuning is quite large, emulating possible worst cases. However, even with this large range, the performance index is still less than 0.45, which means that 55%

reduction can be achieved in the maximum blade response. With smaller detuning range, *e.g.*, 20% detuning, the performance index is smaller than 0.1, indicating that 90% of reduction. Therefore, it can be concluded that the network performance is quite robust under moderate detuning.

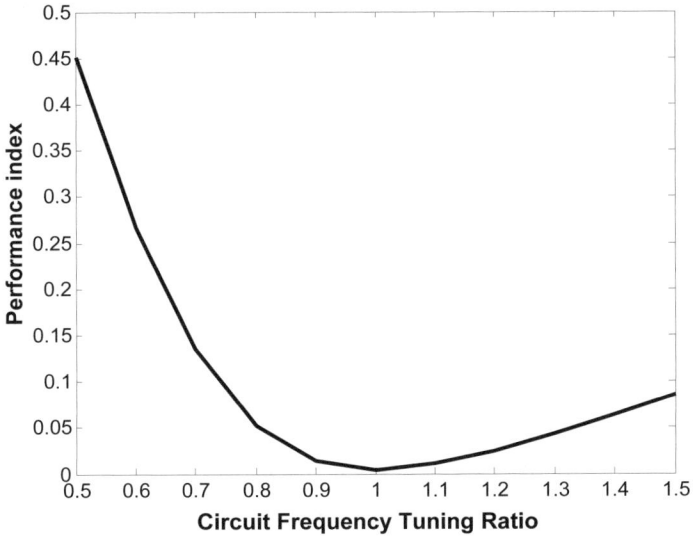

Fig. 7.18. Effect of detuning in circuit frequency tuning ratio δ on the network performance for $\xi = 0.1$.

The effects of detuning of the other two circuit parameters, circuit resistive modal damping ratio ζ_r and coupling capacitance R_a, are shown in Fig. 7.19 and Fig. 7.20, respectively. Parameters used in Fig. 7.19 are the same as in Fig. 7.18 except that δ is kept at optimum and only ζ_r is detuned. It can be seen from Fig. 7.19 that the network performance is not so sensitive to detuning in ζ_r. Here ζ_r is detuned from the optimal value 0.0707 to 0.02 (-72% detuning) and 0.15 (+112% detuning), the maximum performance index is still below 0.06, meaning 94% reduction in maximum blade response. Secondly, it is seen that detuning toward lower damping ratio degrades the performance faster than detuning toward larger damping ratio. In Fig. 7.20, to better illustrate the detuning effect, $R_c=0.5$ is used, other parameters are the same as in Fig. 7.18 except R_a (the only detuned parameter). The optimal $R_a=0.5$ is detuned within a range from 0.3 (-40% detuning) to 0.7 (+40% detuning). One can see that if the detuning is within roughly a smaller 20% range, the performance index can be kept below 0.1. As the detuning goes larger in the range shown, the performance index can go up to as high as 0.6.

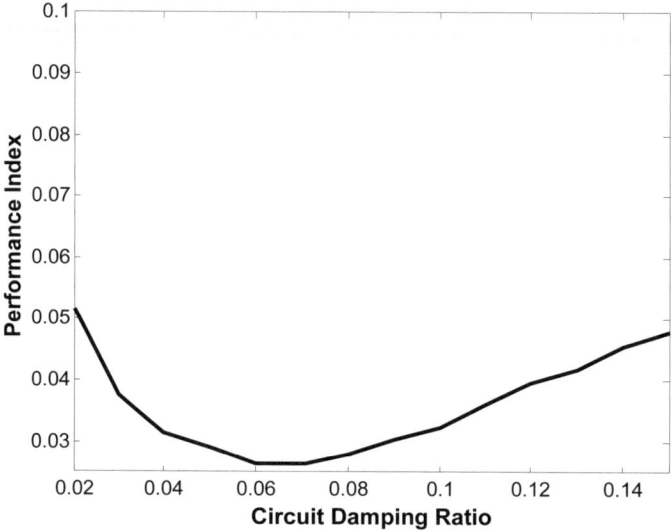

Fig. 7.19. Effects of detuning in circuit damping ratio on network performance for $\xi=0.1$ (optimal $\zeta_r=0.0707$).

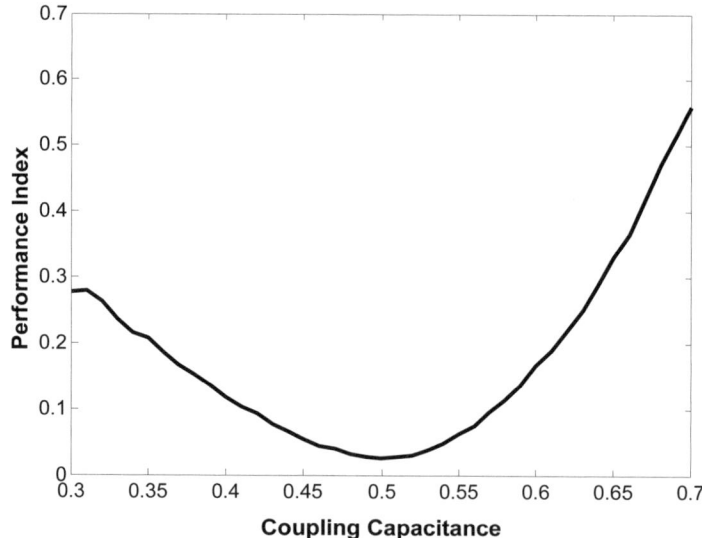

Fig. 7.20. Effect of detuning in R_a for $\xi=0.1$ (optimal $R_a=0.5$).

7.5.7 System Enhancement via Negative Capacitance

The preceding analyses indicate that the system performance is quite robust under moderate level of random mistuning and electrical element detuning. In this section, we present an approach, by incorporating negative
capacitances into the network, to further improve the system robustness if
needed (*i.e.*, if the mistuning or detuning level is higher, or better vibration
suppression performance is required).

As discussed in Section 7.4, negative capacitance can increase the generalized electro-mechanical coupling coefficient and greatly improve the
delocalization effect. Here, by increasing the electro-mechanical coupling,
more mechanical vibration energy can be transformed into the electrical
form and be dissipated by the resistance. Further reduction in the maximum blade amplitude can be achieved by this enhancement, thus improving the performance of the network. Here in our simulation, we consider
two cases of increased electro-mechanical coupling coefficient after negative capacitance is used: $\xi=0.2$ and $\xi=0.3$. The Monte Carlo simulation
is performed for these two cases for scenarios shown in Fig. 7.16 and Figs.
7.18-7.20. The results are shown in Figs. 7.21-7.24.

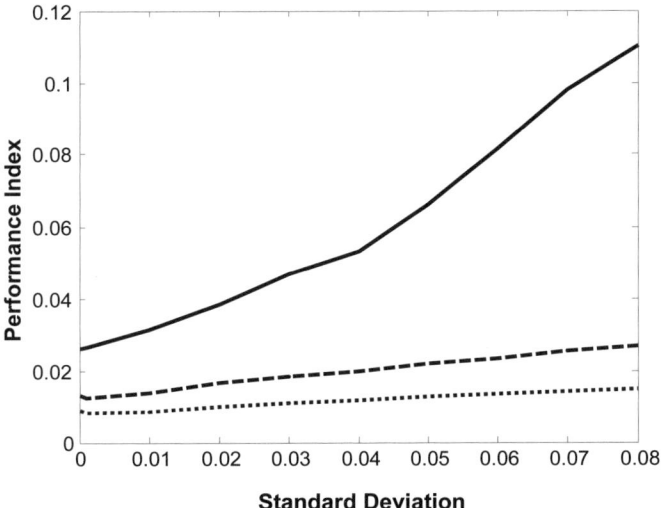

Fig. 7.21. Performance index versus standard deviation comparison between without negative capacitance case (———— for $\xi=0.1$) and with negative capacitance
case (▬ ▬ ▬ ▬ for $\xi=0.2$ and ⋯⋯⋯⋯for $\xi=0.3$).

It can be seen from Fig. 7.21 (corresponding to Fig. 7.16) that with negative capacitance (thus higher ξ), the performance index is lowered throughout the entire range of stiffness mistuning level. For example, with ξ =0.2, for mistuning level as large as σ =0.08, the performance index is below 0.02, meaning 98% reduction in the maximum forced response. Higher ξ (ξ = 0.3) yields even better vibration suppression results.

When circuitry parameter detuning is considered, results with negative capacitance are shown in Figs. 7.22-7.24. In these figures, with negative capacitance (thus higher electro-mechanical coupling ξ), the performance index can be further reduced, not only making the network more robust against detuning around optimal tuning, but also making the network capable of tolerating a wider range of detuning. For example, if the performance satisfaction threshold is set at performance index = 0.05 in Fig. 7.22 with detuning in δ, then without negative capacitance the detuning tolerance range in δ is roughly [0.8, 1.3]. On the other hand, with negative capacitance, for ξ=0.2, δ can be detuned within a range [0.6, 1.5], and for ξ=0.3, the entire range [0.5, 1.5] can be used. Similar results can be seen in Figs. 7.23 and 7.24.

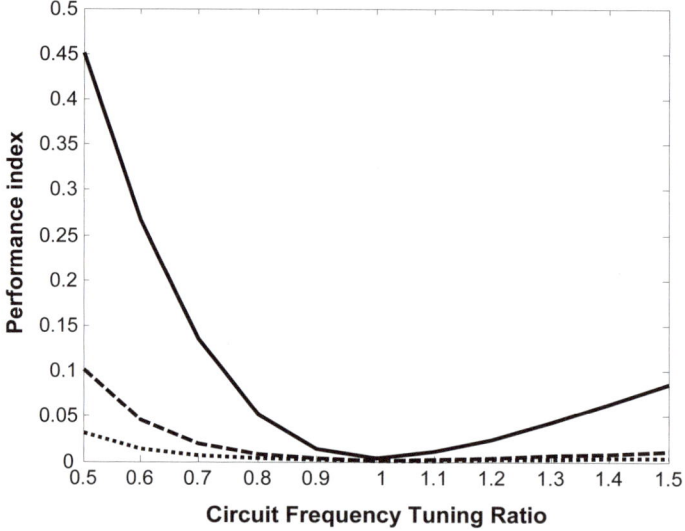

Fig. 7.22. Detuning effect of circuit frequency tuning ratio δ on the performance, without negative capacitance (———— for ξ= 0.1) and with negative capacitance (− − − − for ξ= 0.2 and ·············· for ξ= 0.3).

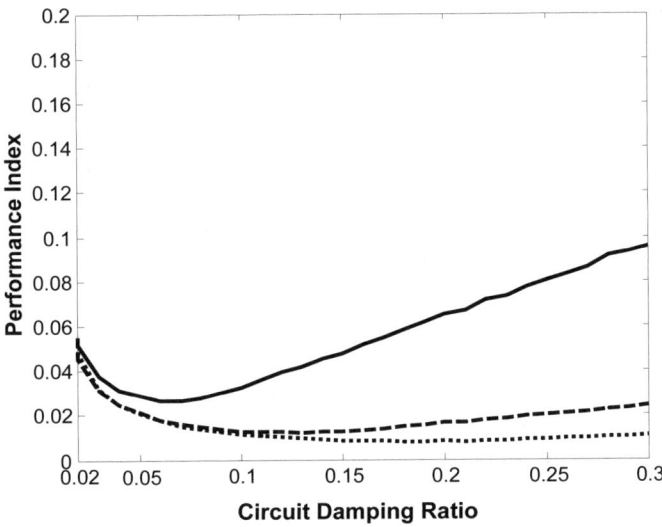

Fig. 7.23. Detuning effect of circuit damping ratio ζ_r on the performance, without negative capacitance (———— for $\xi = 0.1$, optimal $\zeta_r = 0.0707$) and with negative capacitance (– – – – for $\xi = 0.2$, optimal $\zeta_r = 0.1414$, and ············ for $\xi = 0.3$, optimal $\zeta_r = 0.2121$).

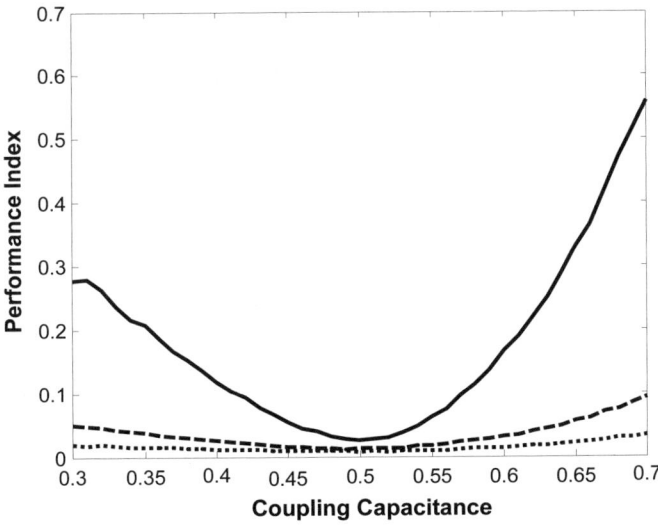

Fig. 7.24. Detuning effect of coupling capacitance R_a on performance, without negative capacitance (———— for $\xi = 0.1$) and with negative capacitance (– – – – for $\xi = 0.2$ and ············ for $\xi = 0.3$).

7.6 CONCLUDING REMARKS

This chapter presents the design of piezoelectric transducer circuitry networks that can achieve effective vibration delocalization and vibration suppression in mistuned periodic structures.

In the delocalization study, identical piezoelectric local inductive circuits are applied to all substructures and connected to each other with capacitive elements. With this design, the otherwise localized vibration energy can be transferred into electrical form and propagate throughout the integral system by way of the strong electrical coupling. The proposed scheme retains the nominal periodicity of the system. Lyapunov exponent is used to characterize the level of localization. It is found that with proper selection of circuitry elements, the localization level of the nearly periodic structure can be significantly reduced. The proposed concept is experimentally verified, and the negative capacitance idea is evaluated to further enhance the delocalization performance.

Piezoelectric networking for vibration suppression of mistuned periodic structures, such as bladed disks, is also investigated in this research. An optimal network design that is independent of spatial harmonics is derived using proper selection of the coupling capacitance. The optimal network's performance is examined through Monte Carlo simulation. Simulation results show that the optimal network, although designed based on the perfectly tuned periodic structure assumption, is very effective on reducing the maximum forced response of mistuned bladed disk with a large range of mistuning level. The network performance is also shown to be robust against moderate detuning in the circuitry parameters. Negative capacitance is introduced to further improve the performance and robustness of the network.

8 Structural Damage Identification Enhancement

Because of its importance for various engineering systems, research on structural health monitoring has been performed extensively in recent years (Sohn *et al.*, 2003). Among the many different damage detection and identification approaches, the vibration-based methods (Doebling *et al.*, 1996; Doebling *et al.*, 1998; Cornwell *et al.*, 1998) have been quite popular. The basic idea of vibration-based damage identification is that damage in the structure will change the structural properties (mass, stiffness or damping) and these changes will result in changes in the global vibratory characteristics of the system, such as natural frequencies, damping ratio and mode shapes. Since the measurement of natural frequencies and frequency response functions is quite straightforward (Dascotte, 1990; Friswell and Penny, 1997), the damage detection schemes that require only the measurement of natural frequencies (hereafter referred to as the frequency-shift-based methods (Salawu, 1997)) are considered to be the easiest to implement, which is a significant advantage especially for complex structures.

The current practice of frequency-shift-based damage identification methods, nevertheless, has severe limitations. For example, one problem is that the natural frequencies can be relatively insensitive to damage occurrence. To address this issue, Ray and Tian (1999) introduced the concept of sensitivity-enhancing control to increase modal frequency sensitivity to damage. They illustrated such idea using a single-degree-of freedom (SDOF) system analysis. The feedback control law was designed by placing the closed-loop modal frequencies at appropriate locations so that their sensitivities toward mass or stiffness damage could be enhanced. Later, this concept was experimentally validated on a cantilevered beam (Ray *et al.*, 2000) and extended to multi-input systems (Koh and Ray, 2004). Jiang *et al.* (2007) further advanced the sensitivity-enhancing feedback control for damage identification of multiple-degrees-of-freedom (MDOF) systems by developing a new eigenstructure assignment-based approach. Eigenvalue sensitivity analysis of a MDOF system indicated that the closed-loop eigenvalue sensitivity depends on both the closed-loop eigenvalues and eigenvectors. Therefore, optimal assignment of both eigenval-

K.W. Wang and J. Tang, *Adaptive Structural Systems with Piezoelectric Transducer Circuitry*, doi: 10.1007/978-0-387-78751-0_8,
© Springer Science + Business Media, LLC 2008

ues and eigenvectors were performed to achieve the best sensitivity enhancement result, where the eigenstructure assignment algorithm essentially followed the approach presented in Chapter 6. Another serious limitation of the traditional frequency-shift-based damage identification method is that the number of natural frequencies that can be accurately measured is normally much smaller than the number of system parameters required to completely characterize the damage. To address this issue, several methods have been proposed in the literature. Cha and Gu (2000) and Nalitolela *et al.* (1992) introduced a mass/stiffness addition technique to enrich the modal information measurement. However, the direct addition of mass/stiffness to a structure might be difficult to implement for many practical applications. To overcome this difficulty, Lew and Juang (2002) proposed an active control approach by using a virtual passive controller to enrich modal frequency measurement. Specifically, different feedback controllers are incorporated to generate additional closed-loop natural frequencies. Koh and Ray (2004) also addressed this issue by utilizing multiple closed-loop systems that can lead to a much enlarged dataset of frequency measurement. An alternative approach utilizing piezoelectric transducer circuitry with tunable inductance to enrich frequency measurement data has been proposed by Jiang *et al.* (2006, 2008). The underlying idea is to use a tunable piezoelectric transducer circuitry coupled to the mechanical structure to favorably alter the dynamics of the electromechanical integrated system. First, the circuitry can be tailored to change the system frequency/modal distribution by introducing additional resonant frequencies. Second, through tuning the circuitry elements (*i.e.*, the inductors), one can obtain a much enlarged dataset consisting of a family of frequency response functions (under different circuitry tunings) as compared to the original single frequency response of the mechanical structure without circuit. This method can directly address the aforementioned second drawback of frequency-shift-based damage detection approach, and could more completely and accurately characterize the variation of system dynamic response due to damage.

8.1 PROBLEM STATEMENT AND OBJECTIVE

In this chapter, the enhancement of frequency-shift-based damage detection using tunable piezoelectric circuitry is systematically introduced. The identification of structural damage is realized by an inverse-sensitivity algorithm. The tunable inductance can be easily realized by utilizing synthetic inductor as mentioned in Chapter 2 (Fig. 2.7). In such approach for damage identification enhancement, the synthetic inductors entertain advantages including small size, low power, high accuracy and wide range of tuning. This approach is analogues to adding extra mechanical spring-mass elements to the structure (Nalitolela, *et al.*, 1992; Cha and Gu, 2000). However, electrical tailoring with variable circuitry is much easier to implement than mechanical tailoring in real systems. Meanwhile, compared with the schemes based on active feedback control (Koh and Ray, 2004; Lew and Juang, 2002), although the tunable circuitry approach (Jiang *et al.*, 2006, 2008) also requires attachment of additional physical element (*i.e.*, piezoelectric transducers) to the structure, it does not require a complex sensor-actuator-controller architecture and significant external energy source. The objective of this chapter is to highlight the following issues:

(a) Why the tunable piezoelectric circuitry can benefit the mission of structural damage detection enhancement?
(b) How to optimally tune the circuitry elements, so the damage identification performance can be best improved?

The overall concept of using tunable piezoelectric transducer circuitry to enhance damage detection is outlined in 8.2. The tuning methodology is presented in 8.3, followed by analysis results detailed in 8.4.

8.2 DAMAGE IDENTIFICATION USING TUNABLE PIEZOELECTRIC CIRCUITRY

When a piezoelectric transducer circuitry with tunable inductors is integrated to a structure to form an electro-mechanical integrated system, the discretized equations of motion of the integrated system can be written as:

$$\begin{bmatrix} \mathbf{M}_s & 0 \\ 0 & \mathbf{L} \end{bmatrix} \begin{Bmatrix} \ddot{\mathbf{q}} \\ \ddot{\mathbf{Q}} \end{Bmatrix} + \begin{bmatrix} \mathbf{C}_d & 0 \\ 0 & \mathbf{R} \end{bmatrix} \begin{Bmatrix} \dot{\mathbf{q}} \\ \dot{\mathbf{Q}} \end{Bmatrix} + \begin{bmatrix} \mathbf{K}_s & \mathbf{K}_c \\ \mathbf{K}_c^T & \mathbf{K}_p \end{bmatrix} \begin{Bmatrix} \mathbf{q} \\ \mathbf{Q} \end{Bmatrix} = \begin{Bmatrix} \mathbf{F}_d \\ 0 \end{Bmatrix} \tag{8.1}$$

where $\mathbf{q}$ is the generalized displacement vector of the structure, $\mathbf{Q}$ is the electrical charge flow vector in the circuit, $\mathbf{M}_s$, $\mathbf{C}_d$, and $\mathbf{K}_s$ are the mass, damping and stiffness matrices of the mechanical structure, respectively, $\mathbf{L}$, $\mathbf{R}$, and $\mathbf{K}_p$ are the inductance, resistance, and inverse capacitance matrices of the circuit, respectively, $\mathbf{K}_c$ is the coupling term between the mechanical and electrical fields, and $\mathbf{F_d}$ is the external disturbance/excitation. The generalized mass and stiffness matrices of the electro-mechanical integrated system can be written as

$$\tilde{\mathbf{M}} = \begin{bmatrix} \mathbf{M_s} & \mathbf{0} \\ \mathbf{0} & \mathbf{L} \end{bmatrix}, \quad \tilde{\mathbf{K}} = \begin{bmatrix} \mathbf{K_s} & \mathbf{K_c} \\ \mathbf{K_c^T} & \mathbf{K_p} \end{bmatrix} \tag{8.2a,b}$$

Neglecting damping, we can obtain the eigenvalue problems of the integrated system associated with the undamaged (healthy) structure and the damaged structure, respectively, as following

$$\left(-\lambda_i \tilde{\mathbf{M}} + \tilde{\mathbf{K}} \right) \boldsymbol{\varphi}_i = \mathbf{0} \tag{8.3}$$

$$\left(-\lambda_i^d \tilde{\mathbf{M}}^d + \tilde{\mathbf{K}}^d \right) \boldsymbol{\varphi}_i^d = \mathbf{0} \tag{8.4}$$

where λ_i and $\boldsymbol{\varphi}_i$ are the ith eigenvalue and eigenvector of the undamaged system, respectively, and λ_i^d, $\boldsymbol{\varphi}_i^d$ are the ith eigenvalue and eigenvector of the damaged system, respectively. In this chapter, for the purpose of illustration without loss of generosity, we are concerned with the structural damage that only causes the change of structural stiffness. Thus,

$$\tilde{\mathbf{M}}^d = \tilde{\mathbf{M}}, \quad \tilde{\mathbf{K}}^d = \tilde{\mathbf{K}} + \delta\tilde{\mathbf{K}} \tag{8.5a,b}$$

Assuming that the damage results in eigenvalue change $\delta\lambda_i$ and eigenvector change $\delta\boldsymbol{\varphi}_i$, we then have

$$\lambda_i^d = \lambda_i + \delta\lambda_i, \quad \boldsymbol{\varphi}_i^d = \boldsymbol{\varphi}_i + \delta\boldsymbol{\varphi}_i \tag{8.6a,b}$$

Substituting Eqs. (8.5a,b) and (8.6a,b) into Eq. (8.4) and neglecting the high-order terms, we obtain a first-order approximation,

$$\delta\lambda_i = \left(\boldsymbol{\varphi}_i \right)^T \delta\tilde{\mathbf{K}} \left(\boldsymbol{\varphi}_i \right) \tag{8.7}$$

In a finite element model, the change of the global stiffness matrix can be expressed as the summation of elemental stiffness matrix changes,

$$\delta \tilde{\mathbf{K}} = \sum_{j=1}^{n_e} \delta \alpha_j \cdot \tilde{\mathbf{K}}_j^e \tag{8.8}$$

where $\tilde{\mathbf{K}}_j^e$ is the elemental stiffness matrix of the jth element positioned within the generalized stiffness matrix of the integrated system $\tilde{\mathbf{K}}$, and $\delta \alpha_j = \alpha_j^d - \alpha_j^h$ denotes the damage-induced stiffness parameter variation of the jth element.

Combining the eigenvalue change relation shown in Eq. (8.7) together for all measured (available) eigenvalues, we can formulate a first-order sensitivity based equation relating the vector of damage-induced eigenvalue changes ($\delta \lambda$) to the vector of damage-induced stiffness parameter variation ($\delta \alpha$) as follows,

$$\delta \lambda = \mathbf{S} \cdot \delta \alpha \tag{8.9}$$

where $\delta \lambda = \begin{bmatrix} \delta \lambda_1 & \delta \lambda_2 & \cdots & \delta \lambda_m \end{bmatrix}^T$ in which m is the number of measured natural frequencies of the integrated system, $\delta \alpha = \begin{bmatrix} \delta \alpha_1 & \delta \alpha_2 & \cdots & \delta \alpha_{n_e} \end{bmatrix}^T$ in which n_e is the number of structural elements in the finite element model, and $\mathbf{S} \in \Re^{m \times n_e}$ is the first-order sensitivity matrix whose elements can be calculated by using the following equation

$$\mathbf{S}(i,j) = (\boldsymbol{\varphi}_i)^T \tilde{\mathbf{K}}_j^e (\boldsymbol{\varphi}_i) \tag{8.10}$$

Since in general, the number of measured frequencies (m) is much smaller than the number of structural elements (n_e), Eq. (8.9) is a significantly underdetermined problem.

One of the key features of integrating tunable piezoelectric transducer circuitry to the structure is that we can obtain multiple frequency response functions (with different inductances) and their changes due to (the same) damage, which leads to a much enlarged dataset for damage detection and identification. Let the inductances in the piezoelectric transducer circuitry network be tuned to form a sequence $\mathbf{L}^{(i)}\left(i=1,2,\cdots,n\right)$

$$
\begin{bmatrix} \mathbf{L}^{(1)} \\ \mathbf{L}^{(2)} \\ \vdots \\ \mathbf{L}^{(n)} \end{bmatrix} = \begin{bmatrix} L_1^{(1)} & L_2^{(1)} & \cdots & L_p^{(1)} \\ L_1^{(2)} & L_2^{(2)} & \cdots & L_p^{(2)} \\ \vdots & \vdots & \vdots & \vdots \\ L_1^{(n)} & L_2^{(n)} & \cdots & L_p^{(n)} \end{bmatrix} \tag{8.11}
$$

where the number of columns, p, represents the number of tunable piezo-electric transducer circuitries integrated to the mechanical structure, and the number of rows, n, represents the number of inductance tuning sets. A set of simultaneous equations similar to Eq. (8.9) can then be obtained and these equations can be written in the matrix form as follows

$$
\begin{bmatrix} \delta\lambda(\mathbf{L}^{(1)}) \\ \delta\lambda(\mathbf{L}^{(2)}) \\ \vdots \\ \delta\lambda(\mathbf{L}^{(n)}) \end{bmatrix} = \begin{bmatrix} \mathbf{S}(\mathbf{L}^{(1)}) \\ \mathbf{S}(\mathbf{L}^{(2)}) \\ \vdots \\ \mathbf{S}(\mathbf{L}^{(n)}) \end{bmatrix} \cdot \delta\boldsymbol{\alpha} \tag{8.12}
$$

This formulation clearly illustrates the advantage of using tunable piezo-electric circuitry. We can now significantly increase the number of measurements in eigenvalue changes (or frequency shifts) and thus increase the number of simultaneous equations that characterize the damage features. In other words, we can make the problem much less underdetermined.

The main limitation of the above first-order approximation based algorithm is that the information regarding the change in eigenvectors (mode shapes) is not included, which could deteriorate the accuracy of damage identification. Wong et al. (2004) developed a general high-order perturbation expression for eigenvalue problem with changes in stiffness, and the perturbation method was used iteratively in conjunction with an optimization method to identify the stiffness parameters of the structure. In what follows, we use a second-order perturbation to describe the changes of eigenvalues, which incorporates the prediction of changes of mode shapes in the derivation. The changes of the kth eigenvalue of the integrated system after damage occurrence can be expressed as

$$\delta\lambda_k = \sum_{i=1}^{n_e}\frac{\partial\lambda_k}{\partial\alpha_i}\delta\alpha_i + \sum_{i=1}^{n_e}\sum_{j=1}^{n_e}\frac{\partial^2\lambda_k}{\partial\alpha_i\partial\alpha_j}\delta\alpha_i\delta\alpha_j = \mathbf{S}_k^{(1)}\delta\boldsymbol{\alpha} + \delta\boldsymbol{\alpha}^T\mathbf{S}_k^{(2)}\delta\boldsymbol{\alpha} \tag{8.13}$$

where $\mathbf{S}_k^{(1)}$ and $\mathbf{S}_k^{(2)}$ are the coefficients of the first and second order perturbations for the kth eigenvalue, respectively,

$$\mathbf{S}_k^{(1)}(i) = \left(\boldsymbol{\varphi}_k\right)^T \tilde{\mathbf{K}}_i^e \left(\boldsymbol{\varphi}_k\right) \tag{8.14}$$

$$\mathbf{S}_k^{(2)}(i,j) = \frac{1}{2!}\left(\boldsymbol{\varphi}_k\right)^T \left\{\tilde{\mathbf{K}}_i^e\mathbf{D}_k^{(1)}(j) + \tilde{\mathbf{K}}_j^e\mathbf{D}_k^{(1)}(i)\right\} \tag{8.15}$$

In Eq. (8.15), where $\mathbf{D}_k^{(1)}(i)$ is the coefficient vector of the first-order perturbation for the kth mass-normalized eigenvector

$$\delta\boldsymbol{\varphi}_k = \sum_{i=1}^{n_e}\frac{\partial\boldsymbol{\varphi}_k}{\partial\alpha_i}\delta\alpha_i = \sum_{i=1}^{n_e}\mathbf{D}_k^{(1)}(i)\delta\alpha_i \tag{8.16}$$

This $\mathbf{D}_k^{(1)}(i)$ and can be calculated as follows,

$$\mathbf{D}_k^{(1)}(i) = \sum_{j=1}^{m}\mathbf{P}_k^{(1)}(i,j)\boldsymbol{\varphi}_j \tag{8.17}$$

$$\mathbf{P}_k^{(1)}(i,j) = \begin{cases} 0 & j=k; \\ \dfrac{1}{\lambda_k - \lambda_j}\left(\boldsymbol{\varphi}_j\right)^T\tilde{\mathbf{K}}_i^e\left(\boldsymbol{\varphi}_k\right) & j\neq k; \end{cases} \tag{8.18}$$

For a given set of inductance tuning, $\mathbf{L}^{(0)} = \begin{bmatrix} L_1^{(0)} & L_2^{(0)} & \cdots & L_p^{(0)} \end{bmatrix}$, a second-order perturbation based equation can be obtained as

$$\delta\lambda(\mathbf{L}^{(0)}) = \begin{Bmatrix} \delta\lambda_1(\mathbf{L}^{(0)}) \\ \delta\lambda_2(\mathbf{L}^{(0)}) \\ \vdots \\ \delta\lambda_k(\mathbf{L}^{(0)}) \end{Bmatrix} = \begin{bmatrix} \mathbf{S}_1^{(1)} \\ \mathbf{S}_2^{(1)} \\ \vdots \\ \mathbf{S}_k^{(1)} \end{bmatrix} \cdot \delta\boldsymbol{\alpha} + \begin{bmatrix} (\delta\boldsymbol{\alpha})^T \mathbf{S}_1^{(2)} (\delta\boldsymbol{\alpha}) \\ (\delta\boldsymbol{\alpha})^T \mathbf{S}_2^{(2)} (\delta\boldsymbol{\alpha}) \\ \vdots \\ (\delta\boldsymbol{\alpha})^T \mathbf{S}_k^{(2)} (\delta\boldsymbol{\alpha}) \end{bmatrix} \tag{8.19}$$

$$= \mathbf{S}^{(1)}(\mathbf{L}^{(0)}) \cdot \delta\boldsymbol{\alpha} + \mathbf{P}\{\mathbf{S}^{(2)}(\mathbf{L}^{(0)}), \delta\boldsymbol{\alpha}\}$$

When we tune the inductances to form a sequence as $\mathbf{L}^{(i)}\,(i=1,2,\cdots,n)$ and perform frequency (eigenvalue) measurements, correspondingly we may obtain a series of eigenvalue change equations, which, collectively, lead to a set of equations in the following form,

$$\begin{bmatrix} \delta\lambda(\mathbf{L}^{(1)}) \\ \delta\lambda(\mathbf{L}^{(2)}) \\ \vdots \\ \delta\lambda(\mathbf{L}^{(n)}) \end{bmatrix} = \begin{bmatrix} \mathbf{S}^{(1)}(\mathbf{L}^{(1)}) \\ \mathbf{S}^{(1)}(\mathbf{L}^{(2)}) \\ \vdots \\ \mathbf{S}^{(1)}(\mathbf{L}^{(n)}) \end{bmatrix} \cdot \delta\boldsymbol{\alpha} + \begin{bmatrix} \mathbf{P}\{\mathbf{S}^{(2)}(\mathbf{L}^{(1)}), \delta\boldsymbol{\alpha}\} \\ \mathbf{P}\{\mathbf{S}^{(2)}(\mathbf{L}^{(2)}), \delta\boldsymbol{\alpha}\} \\ \vdots \\ \mathbf{P}\{\mathbf{S}^{(2)}(\mathbf{L}^{(n)}), \delta\boldsymbol{\alpha}\} \end{bmatrix} \tag{8.20}$$

For the nonlinear equation (8.20), a constrained optimization can be formulated to find the approximate solution of $\delta\boldsymbol{\alpha}$. The objective of the constrained optimization is to minimize the norm of the difference between the actual eigenvalue change $\delta\lambda^{actual}$ (from the frequency response measurement) and $\delta\lambda(\delta\boldsymbol{\alpha})$ (which is the eigenvalue change due to the estimated stiffness parameter variation $\delta\boldsymbol{\alpha}$) in the following manner:

$$\text{Minimize} \left\| \delta\lambda(\delta\boldsymbol{\alpha}) - \delta\lambda^{actual} \right\| \tag{8.21}$$

Subject to $-1 < \delta\alpha_j \le 0$, when $j = 1,2,\cdots,n_e$.

8.3 FORMULATION OF FAVORABLE INDUCTANCE TUNING

A fundamental issue of the tunable piezoelectric transducer circuitry concept is how to tune the inductances to optimize the improvement of the performance of damage identification. In this section, we present the guidelines of forming a favorable inductance tuning sequence based on the analysis on how the inductance tuning affects the characteristics of system dynamics and the damage-induced eigenvalue changes. First, a benchmark beam structure integrated with a single tunable piezoelectric transducer circuitry is used to obtain the fundamental understandings of the effects of inductance tuning. Then, a more complicated plate structure integrated with multiple piezoelectric transducer circuitries is studied to verify and extend the observations to multiple inductance tunings.

8.3.1 Integrated System with Single Tunable Piezoelectric Circuitry

We first study a benchmark example of cantilevered beam integrated with a single tunable piezoelectric transducer circuitry. A circuit with tunable inductor is integrated to a homogenous cantilevered beam through the piezoelectric transducer, which is bonded on the upper surface of the beam from x_1 to x_2. Using Galerkin's method to discretize the partial differential equations, we can obtain a set of ordinary differential equations in the form of Eq. (8.1).

8.3.1.1 Analysis of 2-DOF simplified system

If only the first mode is used in the Galerkin's method, the integrated system can be modeled as a 2-DOF system and the eigenvalue problem of this simplified system is given as

$$\left(\begin{bmatrix} k_{11} & k_c \\ k_c & k_p \end{bmatrix} - \lambda_i \begin{bmatrix} m_{11} & 0 \\ 0 & L \end{bmatrix} \right) \{ \varphi_i \} = \begin{Bmatrix} 0 \\ 0 \end{Bmatrix} \qquad (8.22)$$

which yields two eigenvalues,

$$\lambda_{1,2} = \frac{(k_{11}L + k_p m_{11}) \pm \sqrt{(k_{11}L - k_p m_{11})^2 + 4m_{11}Lk_c^2}}{2m_{11}L} \qquad (8.23)$$

The difference between these two eigenvalues is

$$\lambda_1 - \lambda_2 = \frac{\sqrt{(k_{11}L - k_p m_{11})^2 + 4m_{11}Lk_c^2}}{m_{11}L} \qquad (8.24)$$

From Eq. (8.24) we can derive and conclude that the difference of the two eigenvalues reaches its minimum or, in other words, the two eigenvalues are the closest, when the inductance takes the value of

$$L^* = \frac{k_p m_{11}}{k_{11}} \left(\frac{1}{1 - 2k_c^2 / (k_{11}k_p)} \right) \qquad (8.25)$$

This is referred to as the *critical* inductance value in this chapter. If we assume that $k_c^2 \leq k_{11}k_p$, Eq. (8.25) can be simplified as

$$L^* \approx \frac{k_p m_{11}}{k_{11}} \qquad (8.26)$$

For analysis purpose, we let the left and right ends of the piezoelectric transducer be $x_1 = 0.04184\,\text{m}$ and $x_2 = 0.08368\,\text{m}$. All other system parameters of this illustrative case are specified in Table 8.1. Relevant notations are listed in Nomenclature. The variation of the two system eigenvalues with respect to the inductance tuning is plotted in Fig. 8.1. In this figure, the horizontal axis corresponds to the normalized inductance ξ, which is defined as the ratio between the actual inductance to the critical inductance L^* given in Eq. (8.26),

$$\xi = \frac{L}{L^*} = \frac{L}{\left(k_p \cdot m_{11}\right)/k_{11}} \tag{8.27}$$

From Fig. 8.1, we can see that the loci of the two eigenvalues approach each other in the first stage, and then diverge abruptly when we continuously increase the inductance. This phenomenon of rapid changes of system eigenvalues with respect to the system parameter (inductance L) indicates the occurrence of eigenvalue curve veering (Leissa, 1974; Kutter and Sigillito, 1981; Perkins and Mote, 1986).

Table 8.1. Parameters for the integrated beam system

$\rho_b = 2700\,\mathrm{kg/m^3}$	$L_b = 0.4184\,\mathrm{m}$	$h_b = 3.175\,\mathrm{mm}$
$b = 0.0381\,\mathrm{m}$	$E_b = 7.1\times10^{10}\,\mathrm{N/m^2}$	$\rho_p = 7800\,\mathrm{kg/m^3}$
$E_p = 6.6\times10^{10}\,\mathrm{N/m^2}$	$h_p = 0.191\,\mathrm{mm}$	$\beta_{33} = 7.1445\times10^7\,\mathrm{V\cdot m/C}$
$h_{31} = 1.0707\times10^9\,\mathrm{N/C}$		

In this chapter, we are concerned with the sensitivity of damage-induced eigenvalue change with respect to inductance tuning. A favorable tuning of the inductance value should yield a system in which the eigenvalue change due to the damage occurrence (causing stiffness parameter change) is sensitive to the inductance variations. This way we can enrich the data set more effectively. This sensitivity can be expressed as the second-order derivatives of the system eigenvalues with respect to stiffness parameter ($G = E_b I_b$) and inductance (L),

$$\frac{\partial(\delta\lambda_{1,2})}{\partial L} = \frac{\partial}{\partial L}\left(\frac{\partial\lambda_{1,2}}{\partial G}\delta G\right) = \frac{\partial^2\lambda_{1,2}}{\partial L\partial G}\delta G \tag{8.28}$$

The derivatives of the eigenvalues with respect to stiffness parameter change can be solved as

$$\frac{\partial\lambda_{1,2}}{\partial G} = \frac{\partial\lambda_{1,2}}{\partial k_{11}}\cdot\frac{\partial k_{11}}{\partial G} = \frac{1}{2m_{11}}\left[1\pm\frac{k_{11}L - k_p m_{11}}{\sqrt{(k_{11}L - k_p m_{11})^2 + 4m_{11}Lk_c^2}}\right]\cdot\frac{\partial k_{11}}{\partial G} \tag{8.29}$$

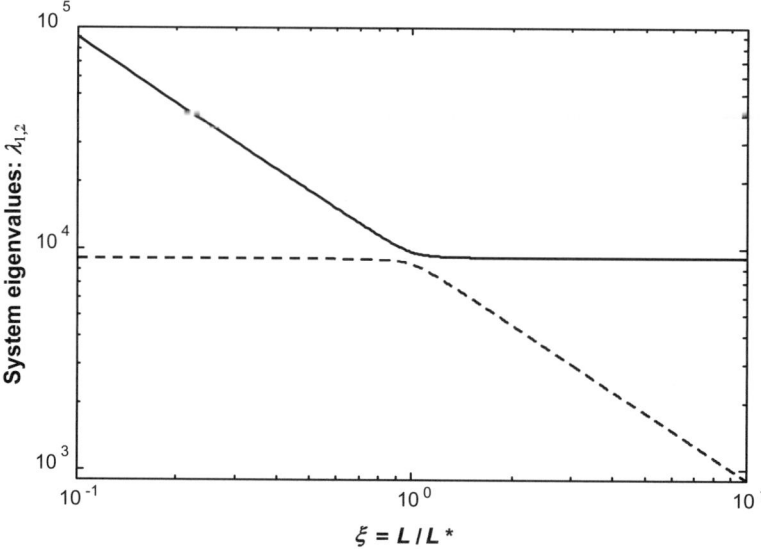

Fig. 8.1. System eigenvalues versus the normalized inductance when using the simplified 2-DOF system model for the integrated beam system. ———— λ_1, ---- λ_2.

Substituting Eq. (8.29) into Eq. (8.28), we can obtain the derivative of damage-induced eigenvalue changes ($\delta\lambda_1$ and $\delta\lambda_2$) with respect to inductance (L) as

$$\frac{\partial(\delta\lambda_{1,2})}{\partial L} = \pm \frac{Lk_{11}k_c^2 + m_{11}k_p k_c^2}{\left[(k_{11}L - k_p m_{11})^2 + 4m_{11}Lk_c^2\right]^{\frac{3}{2}}} \cdot \left(\frac{\partial k_{11}}{\partial G}\delta G\right) \tag{8.30}$$

For a specific damage scenario, $(\partial k_{11}/\partial G)\delta G$ on the right hand side of Eq. (8.30) is fixed, and the inductance value which yields the maximum of $\partial(\delta\lambda_{1,2})/\partial L$ can be easily solved by letting the derivative of the fraction part on the right hand side of Eq. (8.30) with respect to the inductance (L) be zero,

$$L^* = \frac{m_{11}k_p}{2k_{11}} \cdot \left[\left(9 - \frac{10k_c^2}{k_{11}k_p} + \frac{k_c^4}{k_{11}^2k_p^2} \right)^{\frac{1}{2}} - \left(1 - \frac{k_c^2}{k_{11}k_p} \right) \right] \tag{8.31}$$

If we assume that $k_c^2 \leq k_{11}k_p$, Eq. (8.31) can be simplified as

$$L^* \approx \frac{k_p m_{11}}{k_{11}} \tag{8.32}$$

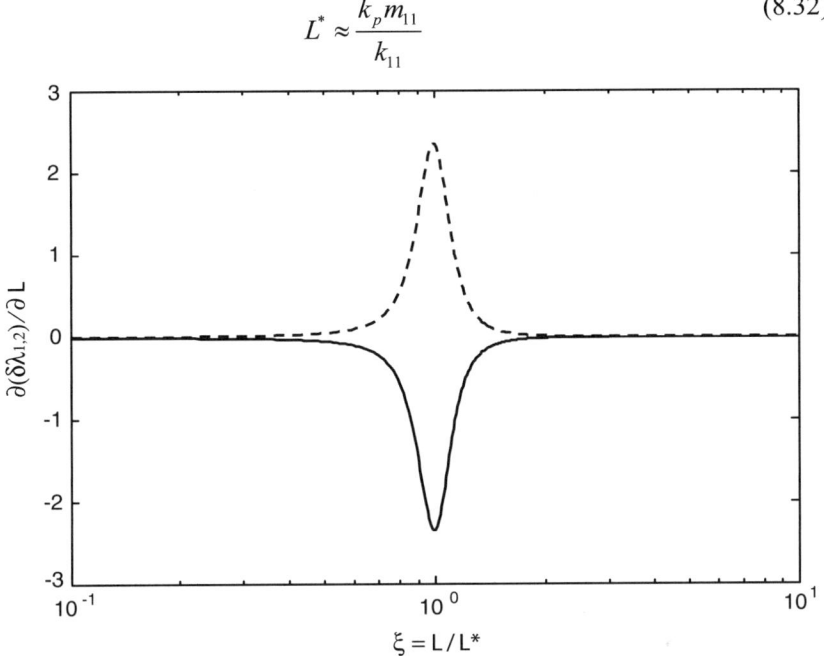

Fig. 8.2. Sensitivities of the damage-induced eigenvalue changes with respect to the normalized inductance when using the simplified 2-DOF system model for the integrated beam system. ——— $\partial(\delta\lambda_1)/\partial L$, – – – – $\partial(\delta\lambda_2)/\partial L$.

Assuming that the damage-induced stiffness reduction is $\delta G = -0.15G$, we can plot the derivative of the damage-induced eigenvalue changes ($\delta\lambda_1$ and $\delta\lambda_2$) with respect to inductance (L), as shown in Fig. 8.2. From this figure, we can see that the sensitivities of the two eigenvalue changes with respect to inductance reach their maximal absolute values when the inductance is tuned near the critical value, $L^* = k_p \cdot m_{11}/k_{11}$, and the sensitivities decrease dramatically when the inductance is tuned away from this critical value. Recall Eq. (8.26) and note that eigenvalue curve veering occurs when the inductance is tuned around $L^* = k_p \cdot m_{11}/k_{11}$ (Fig. 8.1). We can conclude that the occurrence of eigenvalue curve veering produces an inductance tuning range in which high sensitivity of damaged-induced eigenvalue changes with respect to inductance tuning can be expected. Therefore, when the inductance is tuned near L^*, multiple sets of frequency-shift measurements with different sensitivity relations to the potential damage can be obtained. This can greatly enrich the frequency data available for damage identification and help to more completely capture the information about the damage occurrence.

8.3.1.2 Multiple-DOF system analysis

The above observations for the 2-DOF system are based on the analytical sensitivity analysis of the damage-induced eigenvalue changes with respect to inductance tuning. We now extend this sensitivity analysis to MDOF systems. The beam is analyzed using finite element analysis, which is evenly discretized into 10 elements. The piezoelectric transducer is bonded on the upper surface of the beam from the second element to the fourth element. The relevant system parameters are specified in Table 8.1.

First, we examine how the inductance tuning alters the characteristics of system dynamics (i.e., system eigenvalues). Fig. 8.3 shows the variations of the first four system eigenvalues with respect to inductance tuning. From this figure, we can see that eigenvalue curve veering occurs between each two consecutive system eigenvalues from low frequency mode to high frequency mode when the inductance is tuned from $1500\,\mathrm{H}$ down to $0.3\,\mathrm{H}$. In each eigenvalue curve veering, only the two associated system eigenvalues change dramatically with respect to inductance tuning, while other system eigenvalues are hardly affected. Meanwhile, as a well-known phenomenon associated with the curve veering, during the eigenvalue curve veering the eigenvectors corresponding to the veering eigenvalues will interchange in a rapid but continuous way (Leissa, 1974; Kutter and Sigillito, 1981).

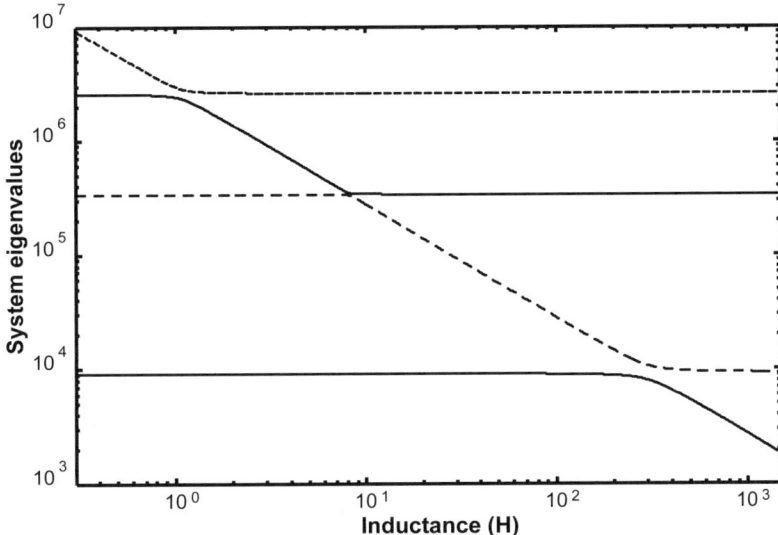

Fig. 8.3. System eigenvalues versus inductance when using the multiple-DOF system model for the integrated beam system. ——— λ_1, – – – – λ_2, ——— λ_3, – – – – λ_4.

It has been shown that for the 2-DOF system (Figs. 8.1 and 8.2) the occurrence of eigenvalue curve veering is related to an inductance tuning range with high sensitivity of damage-induced eigenvalue changes. In order to examine the case of a MDOF system, we calculate the sensitivities of the damage-induced eigenvalue changes with respect to inductance L, as plotted in Fig. 8.4 where (a)-(c) correspond to the first, second and third eigenvalue changes, respectively. The structural damage is assumed to be on the second beam element and the damage causes a 25% stiffness reduction. As shown in Fig. 8.4(a), there is only one peak region indicating high sensitivity of the first eigenvalue change with respect to inductance tuning, and this region corresponds to eigenvalue curve veering between the first and second system eigenvalues. In Fig. 8.4(b), two peak regions with high sensitivity of eigenvalue change with respect to inductance tuning are found, and it is easy to verify that these two regions correspond to the eigenvalue curve veering between two pairs of system eigenvalues (the first and second, and the second and third), respectively. Similar conclusion can be drawn in Fig. 8.4(c), where two peak regions with high sensitivity are achieved when the third system eigenvalue has curve veering with the second and fourth system eigenvalues.

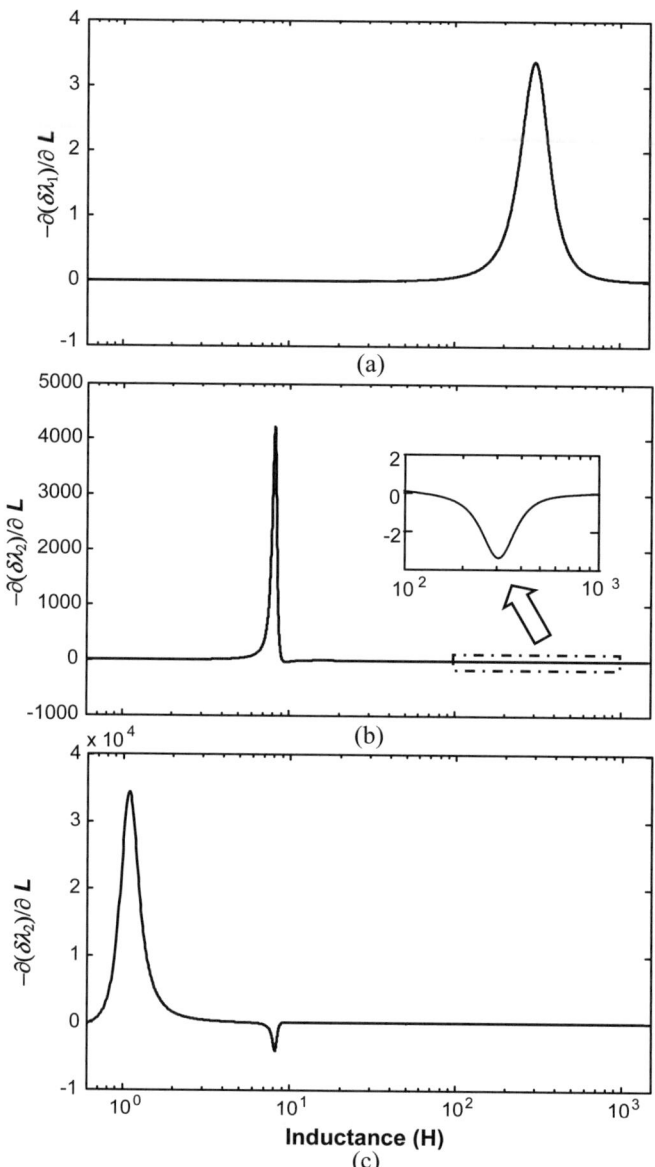

Fig. 8.4. Sensitivities of the damage-induced eigenvalue changes with respect to inductance when using the MDOF system model for the integrated beam system: (a) Sensitivity of the first eigenvalue change; (b) Sensitivity of the second eigenvalue change; (c) Sensitivity of the third eigenvalue change.

Fig. 8.5 shows the variation of the damage-induced eigenvalue changes with respect to inductance tuning when the second element is damaged with a 25% stiffness reduction. If the inductance is tuned around each of the eigenvalue curve veering value, the damage-induced changes of the two system eigenvalues associated with the two loci in that curve veering vary significantly with respect to inductance. Therefore, if the inductance is tuned around values corresponding to eigenvalue curve veering, multiple sets of frequency-shift measurement with dramatically different sensitivity relations to the damage can be obtained. Again this will greatly enrich the modal data measurement available for damage identification.

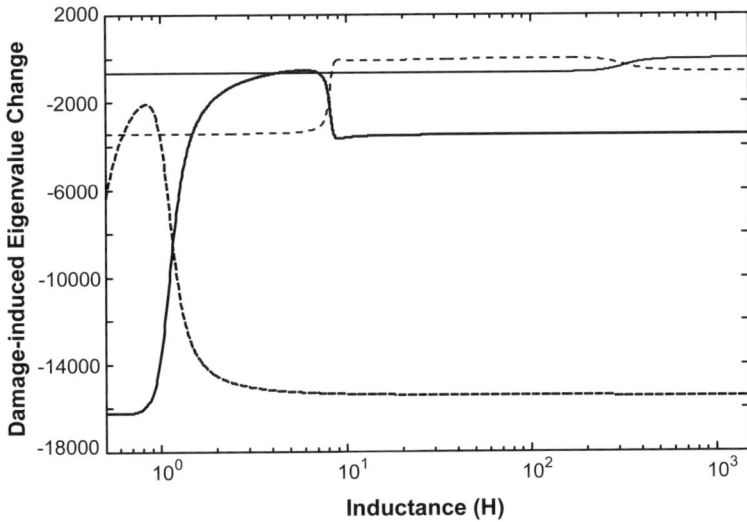

Fig. 8.5. Variation of the damage-induced eigenvalue changes with respect to inductance when using the multiple-DOF system model for the integrated beam system. ——— $\delta\lambda_1$, ———— $\delta\lambda_2$, ▬▬▬ $\delta\lambda_3$, ▬ ▬ ▬ $\delta\lambda_4$.

8.3.2 Integrated System with Multiple Tunable Piezoelectric Circuitries

8.3.2.1 General methodology

The preceding example concerns the integration of a single tunable piezo-electric circuitry onto a beam structure. For more complicated structures, it can be envisioned that multiple tunable circuitries could be more beneficial for damage identification. In this section, we use a plate structure, as shown in Fig. 8.6, to explore the tuning of multiple tunable piezoelectric transducer circuitries. For this benchmark plate, the left edge of the plate is clamped and the other three edges have free boundary conditions. The plate is discretized into 25 elements (with uniform element size $0.05\text{m} \times 0.05\text{m}$), and the element numbers are labeled as shown in the figure. Three piezoelectric transducers are bonded onto the 7th, 13th and 19th elements, respectively. Each piezoelectric transducer patch has a size of $0.05\text{m} \times 0.05\text{m}$. Three piezoelectric circuitries with tunable inductances (L_1, L_2 and L_3) are integrated to the plate structure through three piezoelectric transducers, respectively. The parameters of the system are listed in Table 8.2.

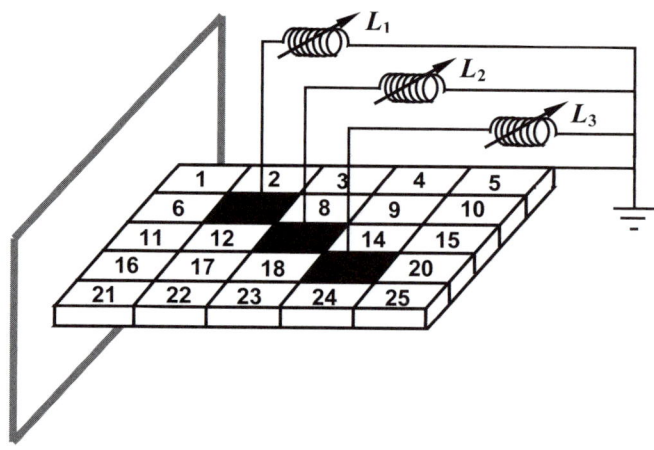

Fig. 8.6. Configuration of a cantilevered plate integrated with multiple tunable piezoelectric transducer circuitries.

Table 8.2. Parameters for the integrated plate system

$\rho_s = 2700\,\text{kg/m}^3$	$a = 0.25\,\text{m}$	$b = 0.25\,\text{m}$
$h_s = 5\,\text{mm}$	$E_s = 30 \times 10^9\,\text{N/m}^2$	$\rho_p = 7800\,\text{kg/m}^3$
$E_p = 6.9 \times 10^{10}\,\text{N/m}^2$	$h_p = 0.25\,\text{mm}$	$\beta_{33} = 7.1445 \times 10^7\,\text{V}\cdot\text{m/C}$
$h_{31} = 7.664 \times 10^8\,\text{N/C}$		

In the case of single piezoelectric circuitry, only one eigenvalue curve veering can be achieved under certain inductance tuning, and all other eigenvalues are much less affected. On the other hand, multiple piezoelectric transducer circuitries make it possible for multiple pairs of system eigenvalues to achieve curve veering simultaneously. Such multiple curve-veering phenomena could further enhance the system response sensitivity with respect to inductance tunings. When multiple piezoelectric circuitries are integrated with the mechanical structure, these circuitries are not only directly coupled with the mechanical structure, but also coupled indirectly with other circuitries through energy exchange within the entire electro-mechanical system. In other words, when tuning the inductance in one circuitry, the interactions between the mechanical structure and other circuitries are also affected even if the inductances in those circuitries remain the same. Therefore, it is not feasible to tune each of the inductances separately. We need to tune these inductances simultaneously to achieve the desired set of eigenvalue curve veering concurrently. Since there are three tunable piezoelectric transducer circuitries in this plate example, these circuitries can be tuned to accomplish at least three eigenvalue curve veering. In each curve veering, one additional resonance frequency, due to the dynamics of one piezoelectric circuitry, is introduced into the frequency response function near the structural resonance frequency corresponding to that veering.

Typically, clustered eigenvalues or close natural frequencies in a dynamic system are related to eigenvalue curve veering (Leissa, 1974; Kutter and Sigillito, 1981; Perkins and Mote, 1986; Pierre, 1988; Liu, 2002). Indeed, the occurrence of close eigenvalues is an indication of eigenvalue curve veering and thus can be used as the criterion for tuning the inductances to realize multiple eigenvalue curve-veering phenomena. In this research, an optimization scheme is formulated to find the critical values of three inductances $\langle L_1^*, L_2^*, L_3^* \rangle$ which yield three pairs of close eigenvalues. That is, three eigenvalue curve veerings are realized simultaneously when the inductances are tuned around their respective critical values. The objective function to be minimized is defined as the summation of the difference between each pair of system eigenvalues that are targeted for eigenvalue curve veering.

8.3.2.2 Option 1 of inductance tuning

The integrated system has a large number of eigenvalues/natural frequencies, which can all be potential candidates for eigenvalue curve veering.

We first formulate an optimization problem to find the critical values of three inductances to simultaneously achieve eigenvalue curve veering between the 1st and 2nd eigenvalues, the 3rd and 4th eigenvalues, and the 5th and 6th eigenvalues respectively. The objective function is defined as

$$J_1 = \sum_{i=1}^{3} \left| \lambda_{2i-1} - \lambda_{2i} \right| \qquad (8.33)$$

By using the standard constrained minimum subroutine, FMINCON, provided by MATLAB, the minimization of the above objective function yields the following critical inductance values,

$$L_1^* = 84.6\,\text{H}, \qquad L_2^* = 1.91\,\text{H}, \qquad L_3^* = 14.5\,\text{H} \qquad (8.34)$$

Utilizing the above critical values as center values and expanding the inductance on both sides by $20.0\,\text{H}$, $0.5\,\text{H}$ and $4.0\,\text{H}$, respectively, the tuning ranges for the three inductances can be formulated as

$$L_1 \in [64.6\,\text{H}, 104.6\,\text{H}], \ L_2 \in [1.41\,\text{H}, 2.41\,\text{H}], \ L_3 \in [10.5\,\text{H}, 18.5\,\text{H}] \qquad (8.35)$$

Fig. 8.7 shows the variation of the first six eigenvalues with respect to inductance tuning. The horizontal axis only shows the change of the inductance L_3, while it should be noted that inductances L_1 and L_2 are actually tuned synchronously when tuning L_3 (This statement holds for all the following figures in the same category). It is shown in this figure that three eigenvalue curve veerings occur between the 1st and 2nd eigenvalues, the 3rd and 4th eigenvalues, and the 5th and 6th eigenvalues, respectively and simultaneously. Fig. 8.8 shows the variation of the damage-induced eigenvalue changes with respect to inductance tuning inside the proposed inductance tuning ranges. The structural damages are assumed to result in 20%, 30%, 10%, and 20% of stiffness parameter reductions on the 6th, 11th, 12th, and 16th elements, respectively. It can be easily observed from the figure that the damage-induced changes of the three eigenvalue pairs vary significantly around the curve veering values of the inductances.

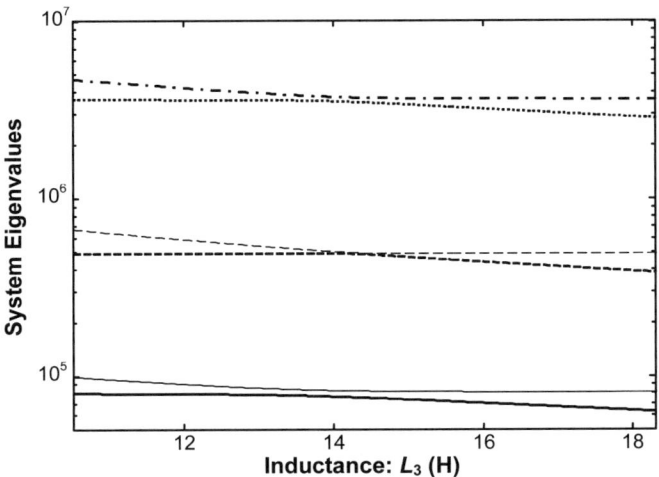

Fig. 8.7. System eigenvalues versus inductance L_3 for the integrated plate system when using option 1 of inductance tuning. ━━━ λ_1, ━━━ λ_2, ▬▬▬ λ_3, ━ ━ ━ λ_4, ▪▪▪▪▪▪▪▪ λ_5, ▬▪▬▪▬ λ_6.

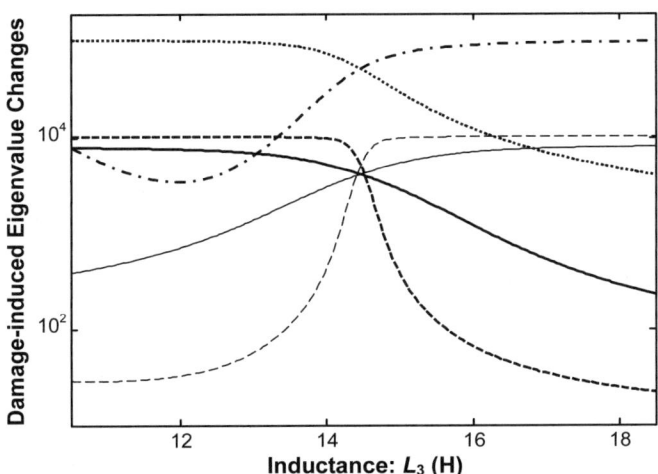

Fig. 8.8. Variation of the damage-induced eigenvalue changes with respect to inductance L_3 for the integrated plate system when using option 1 of inductance tuning. ━━━ $\left(-\delta\lambda_1\right)$, ━━━ $\left(-\delta\lambda_2\right)$, ▬▬▬ $\left(-\delta\lambda_3\right)$, ━ ━ ━ $\left(-\delta\lambda_4\right)$, ▪▪▪▪▪▪▪▪ $\left(-\delta\lambda_5\right)$, ▬▪▬▪▬ $\left(-\delta\lambda_6\right)$.

8.3.2.3 Option 2 of inductance tuning

Here we investigate an alternative option of inductance tuning, *i.e.*, we aim at achieving eigenvalue curve veering between the 3rd and 4th eigenvalues, the 5th and 6th eigenvalues, and the 7th and 8th eigenvalues, respectively and simultaneously. The objective function is then defined as

$$J_2 = \sum_{i=1}^{3} |\lambda_{2i+1} - \lambda_{2i+2}| \tag{8.36}$$

The numerical minimization of the above objective function yields

$$L_1^* = 0.96\,\text{H}\,, \qquad L_2^* = 1.25\,\text{H}\,, \qquad L_3^* = 1.91\,\text{H} \tag{8.37}$$

The tuning ranges for three inductances can be determined as

$$L_1 \in [0.51\,\text{H}, 1.41\,\text{H}]\,, \; L_2 \in [0.65\,\text{H}, 1.85\,\text{H}]\,, \; L_3 \in [1.16\,\text{H}, 2.66\,\text{H}] \tag{8.38}$$

The variation of the system eigenvalues with respect to inductance tuning inside the proposed tuning ranges are shown in Fig. 8.9. The eigenvalue curve veering are achieved not only between those desired pairs of system eigenvalues (the 3rd and 4th, the 5th and 6th, and the 7th and 8th), but also between the 4th and 5th, and the 6th and 7th system eigenvalues. The reason is that the three circuitry modes are tuned to accomplish curve veerings with three twisting-bending modes (3rd-5th) of the plate structure, and the resonance frequencies of these twisting-bending modes are close to each other, which makes the veering more sensitive to inductance tuning. As a result, it now becomes possible for one eigenvalue to veer with its two adjacent eigenvalues successively inside the tuning ranges.

For the same damage scenario as used in Fig. 8.8, the variations of the damage-induced eigenvalue changes with respect to inductance tuning inside the tuning ranges are plotted in Fig. 8.10. The variation of the damage-induced eigenvalue changes is more noticeable than that in Fig. 8.8 (produced under tuning option 1), and the reason is that additional eigenvalue veerings occur under tuning option 2, as shown in Fig. 8.9. Obviously, by integrating multiple tunable piezoelectric transducer circuitries, the frequency measurement data available for damage identification can be further enriched by formulating inductance tuning sequence to accomplish eigenvalue curve veering between different pairs of system eigenvalues.

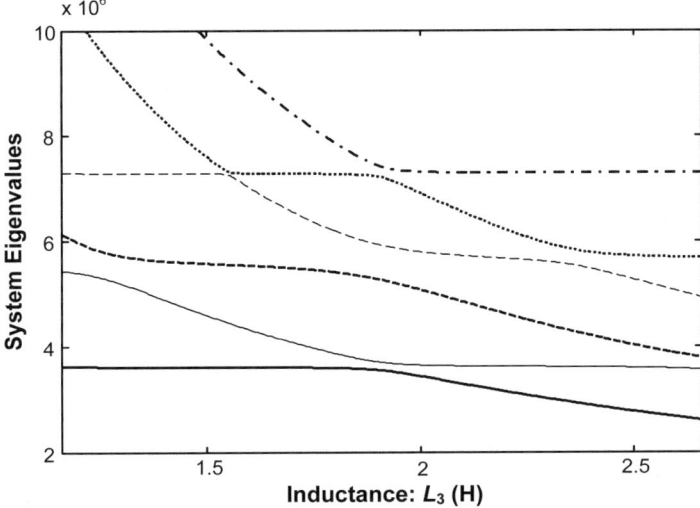

Fig. 8.9. System eigenvalues versus inductance L_3 for the integrated plate system when using option 2 of inductance tuning. ▬▬ λ_3, ▬▬ λ_4, ▬▬▬ λ_5, ▬ ▬ ▬ λ_6, ▪▪▪▪▪▪▪▪ λ_7, ▬▪▬▪▬ λ_8.

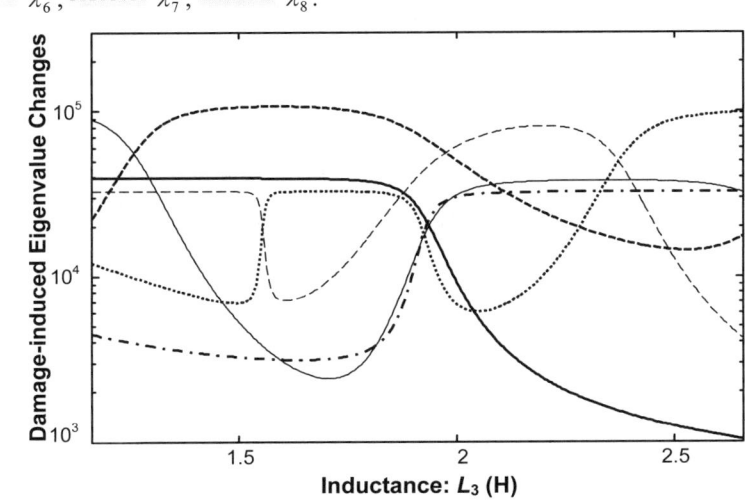

Fig. 8.10 Variation of the damage-induced eigenvalue changes with respect to inductance L_3 for the integrated plate system when using option 2 tuning. ▬▬▬ $\left(-\delta\lambda_1\right)$, ▬▬▬ $\left(-\delta\lambda_2\right)$, ▪▬▪▬ $\left(-\delta\lambda_3\right)$, ▬ ▬ ▬ $\left(-\delta\lambda_4\right)$, ▪▪▪▪▪▪▪▪ $\left(-\delta\lambda_5\right)$, ▬▪▬▪▬ $\left(-\delta\lambda_6\right)$.

8.4 DAMAGE IDENTIFICATION ANALYSES AND CASE STUDIES

The preceding sections have outlined the basis of achieving eigenvalue curve veering under inductance tuning and illustrated such phenomena. In this section, the performance improvements on damage identification in beam and plate structures caused by the tunable piezoelectric transducer circuitry are discussed. Here we utilize the favorable inductance tuning results obtained in Section 8.3.

8.4.1 Damage Identification in Cantilevered Beam with Single Tunable Piezoelectric Circuitry

In order to illustrate the performance improvement due to tunable piezo-electric transducer circuitry network and verify the guidelines on favorable inductance tuning obtained in Section 8.3, here we compare the damage identification results under three different approaches: (1) the traditional method (iterative second-order perturbation based algorithm without integration of tunable piezoelectric circuitry); (2) the enhanced integrated-system approach (iterative second-order perturbation based algorithm with integration of tunable piezoelectric circuitry) with *ad hoc* inductance tuning; and (3) the enhanced integrated-system approach with the proposed favorable inductance tuning method. For a fair comparison, we assume that only the frequencies below 1000 rad/sec are available for accurate measurement. This means that only the first two natural frequencies can be accurately measured for damage identification when we use the traditional method without tunable piezoelectric circuitry, and only the first three modal frequencies, of which the additional one comes from the dynamics of the piezoelectric circuitry, are available when using the enhanced integrated-system approach. According to the analysis and results given in Section 8.3.1, a favorable inductance tuning sequence can be selected as follows

$$\mathbf{L} = \begin{bmatrix} 7.50 & 7.75 & 8.00 & 8.25 & 8.50 & 225 & 250 & 275 & 300 & 325 \end{bmatrix} \mathrm{H} \ (8.39)$$

In the above tuning sequence, the first 5 values are selected from the curve veering between the 2nd and 3rd system eigenvalues, and the last 5 values are selected from the curve veering between the 1st and 2nd eigenvalues.

The first case we examine is to identify single element damage. The damage in the beam is assumed to be on the 2nd element and results in a 25% stiffness reduction. Fig. 8.11 shows the predictions of structural damage by using the traditional method and the enhanced integrated-system approach with *ad hoc* and *favorable* inductance tuning. From this figure, we can see that the prediction using the traditional method has significant error because the major damage is predicted to be on the 3rd element and the predicted damage severity of the 2nd element is much less than the actual value. When the enhanced integrated-system approach with tunable piezoelectric circuitry is used, no obvious improvement is observed in the case of *ad hoc* inductance tuning ($\tilde{L}_i = (100 + 10 \times i)$ H; $i = 1, 2, \cdots, 10$), while a much more accurate prediction is achieved when using the *favorable* inductance tuning. This clearly demonstrates the necessity of employing the proposed tuning methodology.

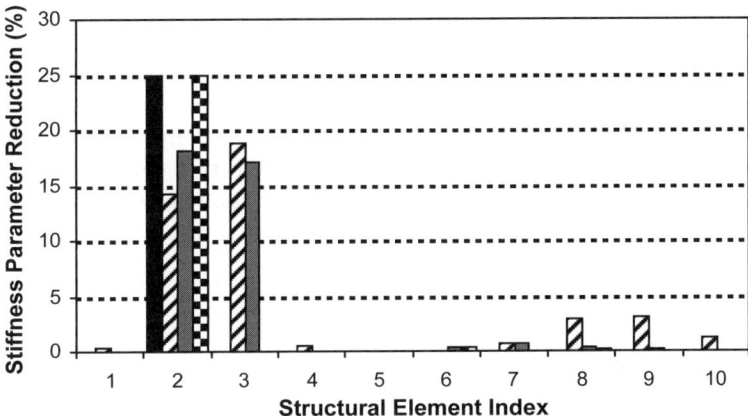

Fig. 8.11. Identification of structural damage on the second element of the beam structure. ■ Actual stiffness parameter reduction, ▨ Prediction using the traditional method, ■ Prediction using the enhanced integrated-system approach with *ad hoc* inductance tuning, ▨ Prediction using the enhanced integrated-system approach with *favorable* inductance tuning.

In order to quantitatively evaluate the performance of damage identification using different methods, we define a performance metric as the root-mean-square difference (RMSD) between the predicted stiffness parameter reduction using a specific method ($\delta\boldsymbol{\alpha}^p$) and the actual damage-induced stiffness parameter reduction ($\delta\boldsymbol{\alpha}^a$),

$$\text{RMSD(\%)} = \sqrt{\frac{\left(\delta\boldsymbol{\alpha}^{p} - \delta\boldsymbol{\alpha}^{a}\right)^{T}\left(\delta\boldsymbol{\alpha}^{p} - \delta\boldsymbol{\alpha}^{a}\right)}{\left(\delta\boldsymbol{\alpha}^{a}\right)^{T}\left(\delta\boldsymbol{\alpha}^{a}\right)}} \times 100 \tag{8.40}$$

For the example shown in Fig. 8.11, the prediction obtained from the traditional method results in a RMSD error of 92%, and the predictions obtained from the enhanced integrated-system approach with *ad hoc* inductance tuning and *favorable* inductance tuning have RMSD errors of 73% and 2%, respectively.

In the second example, we detect multiple element damages, and the result is shown in Fig. 8.12, where the actual structural damages are assumed to be on the 2nd and 5th elements with 25% and 15% damage-induced stiffness reductions, respectively. Neither the traditional method nor the enhanced integrated-system approach with *ad hoc* inductance tuning gives an accurate prediction of the actual damage situation. When the enhanced integrated-system approach with the *favorable* inductance tuning is used, both the locations and severities of the two damages are accurately predicted. The prediction obtained from the traditional method results in a RMSD error of 60%, and the predictions obtained from the enhanced integrated-system approach with *ad hoc* inductance tuning and *favorable* inductance tuning have RMSD errors of 45% and 5%, respectively.

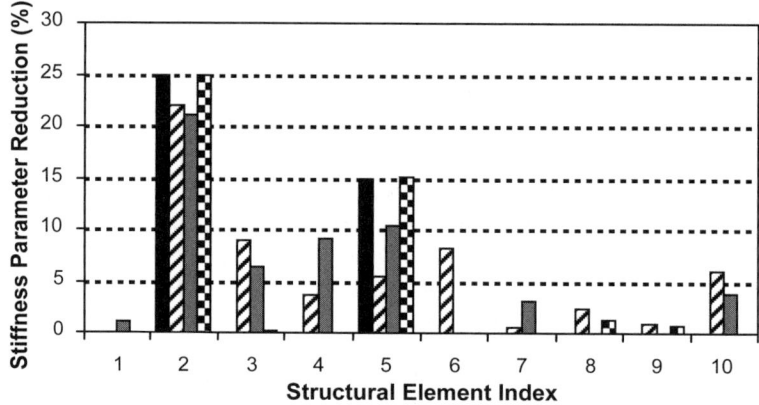

Fig. 8.12. Identification of structural damages on the second and fifth elements of the beam structure. ■ Actual stiffness parameter reduction, ▨ Prediction using the traditional method, ▨ Prediction using the enhanced integrated-system approach with *ad hoc* inductance tuning, ▨ Prediction using enhanced integrated-system approach with *favorable* inductance tuning.

Since measurement noise is inevitable in practical applications, it is necessary to investigate the effects of measurement noise on the damage identification performance of the proposed approach. The effects of measurement noise on the measured eigenvalues (natural frequencies) can be simulated using the following equation

$$\tilde{\lambda}^d = \lambda^d + v\mathbf{R}\lambda^d \qquad (8.41)$$

where $\tilde{\lambda}^d$ is the noise-contaminated eigenvalues, λ^d is the noise-free eigenvalue, $\mathbf{R}$ is a diagonal matrix whose diagonal entries are independently, uniformly distributed random numbers in the interval of $[-1,1]$, and v represents the noise level. The noise-contaminated eigenvalue measurement with two different noise levels, $v = 1.0\%$ and $v = 2.0\%$, are used to identify the same damage scenario as given in Fig. 8.12, and the damage identification results are shown in Fig. 8.13. Compared with the result using noise-free eigenvlaue measurement, the inclusion of measurement noise in the measured eigenvalues only leads to small variations in the damage identification results. In other words, the overall damage characteristics (locations and severities) are still successfully identified.

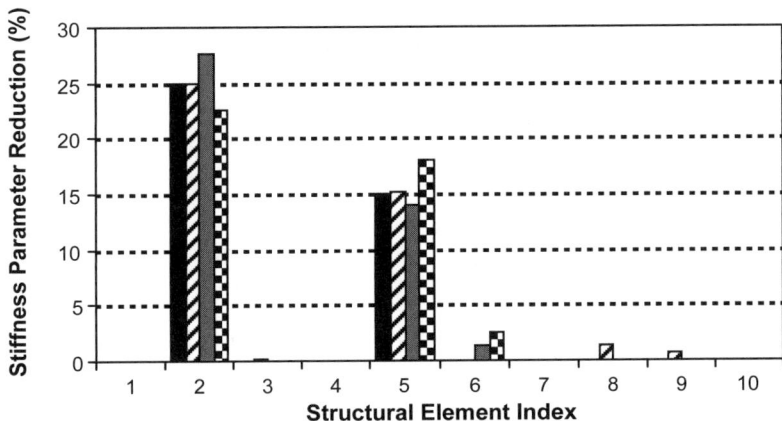

Fig. 8.13. Comparison of damage identification results using noise-free and noise-contaminated eigenvalues of the integrated beam system. The damages are assumed to cause 25% and 15% stiffness reduction on the second and fifth elements, respectively. ■ Actual stiffness parameter reduction, ▨ Prediction using noise-free eigenvalues, ■ Prediction using noise-contaminated eigenvalues ($v = 1.0\%$), ▨ Prediction using noise-contaminated eigenvalues ($v = 2.0\%$).

8.4.2 Damage Identification in Cantilevered Plate with Multiple Tunable Piezoelectric Circuitries

In this second case study, we examine the performance improvement of using multiple tunable piezoelectric circuitries to detect damages in the more complicated plate structure, and we also compare the damage detection performance by using different tuning options. In this example, we assume that only the frequencies under 4000rad/sec are available for accurate measurement. Therefore, only the first five natural frequencies can be used for damage identification when we use the traditional method without tunable piezoelectric circuitry, and only the first eight modal frequencies, of which the additional three come from the inductance circuit connected to the piezoelectric transducer, are available when using the enhanced method with integration of tunable piezoelectric circuitry. According to the two tuning options derived in the previous section, the first set of inductance tuning sequence can be selected from the inductance tuning ranges obtained for tuning option 1,

$$
\begin{Bmatrix} \mathbf{L}^{(1)} \\ \mathbf{L}^{(2)} \\ \mathbf{L}^{(3)} \\ \mathbf{L}^{(4)} \\ \mathbf{L}^{(5)} \end{Bmatrix} = \begin{bmatrix} L_1^{(1)} & L_2^{(1)} & L_3^{(1)} \\ L_1^{(2)} & L_2^{(2)} & L_3^{(2)} \\ L_1^{(3)} & L_2^{(3)} & L_3^{(3)} \\ L_1^{(4)} & L_2^{(4)} & L_3^{(4)} \\ L_1^{(5)} & L_2^{(5)} & L_3^{(5)} \end{bmatrix} = \begin{bmatrix} 78.6 & 1.51 & 14.00 \\ 81.6 & 1.61 & 14.15 \\ 84.6 & 1.71 & 14.30 \\ 87.6 & 1.81 & 14.45 \\ 90.6 & 1.91 & 14.60 \end{bmatrix} \text{H} \tag{8.42}
$$

A second set of inductance tuning sequence is selected from the inductance tuning ranges obtained for tuning option 2

$$
\begin{Bmatrix} \mathbf{L}^{(1)} \\ \mathbf{L}^{(2)} \\ \mathbf{L}^{(3)} \\ \mathbf{L}^{(4)} \\ \mathbf{L}^{(5)} \end{Bmatrix} = \begin{bmatrix} L_1^{(1)} & L_2^{(1)} & L_3^{(1)} \\ L_1^{(2)} & L_2^{(2)} & L_3^{(2)} \\ L_1^{(3)} & L_2^{(3)} & L_3^{(3)} \\ L_1^{(4)} & L_2^{(4)} & L_3^{(4)} \\ L_1^{(5)} & L_2^{(5)} & L_3^{(5)} \end{bmatrix} = \begin{bmatrix} 0.51 & 0.65 & 1.16 \\ 0.71 & 0.95 & 1.56 \\ 0.91 & 1.25 & 1.96 \\ 1.16 & 1.55 & 2.36 \\ 1.41 & 1.85 & 2.66 \end{bmatrix} \text{H} \tag{8.43}
$$

As showing in Fig. 8.7 and Fig. 8.9, respectively, each set of the inductance tuning sequence formulated above achieves one set of eigenvalue curve veering between different pairs of system eigenvalues, and the ability of accomplishing multiple sets of eigenvalue curve veering is exactly the additional benefit offered by integrating multiple piezoelectric circuitries. Therefore, utilizing multiple sets of eigenvalue curve veering to further enrich the frequency measurement data for damage detection, the inductances should be tuned according to a combined tuning sequence, which can be obtained by assembling the two sets of inductance tuning sequences in Eqs. (8.42) and (8.43),

$$
\begin{bmatrix} L_1^{(1)} & L_1^{(2)} & \cdots & L_1^{(10)} \\ L_2^{(1)} & L_2^{(2)} & \cdots & L_2^{(10)} \\ L_3^{(1)} & L_3^{(2)} & \cdots & L_3^{(10)} \end{bmatrix} =
$$

$$
\begin{bmatrix} 78.6 & 81.6 & 84.6 & 87.6 & 90.6 & 0.51 & 0.71 & 0.91 & 1.16 & 1.41 \\ 1.51 & 1.61 & 1.71 & 1.81 & 1.91 & 0.65 & 0.95 & 1.25 & 1.55 & 1.85 \\ 14.00 & 14.15 & 14.30 & 14.45 & 14.60 & 1.16 & 1.56 & 1.96 & 2.36 & 2.66 \end{bmatrix} H
$$

$$(8.44)$$

Unless otherwise specified, hereafter this combined inductance tuning sequence is used as the favorable tuning sequence in the proposed new method for damage identification.

Table 8.3. System natural frequencies (rad/sec) under inductance tunings

ω_1	ω_2	ω_3	ω_4	ω_5	ω_6	ω_7	ω_8
281.46	299.38	698.57	704.98	1900.0	2132.4	2385.8	2700.2
280.70	294.64	697.85	702.14	1898.9	2070.5	2381.2	2700.2
279.34	290.76	694.98	701.43	1896.6	2014.0	2378.4	2700.0
277.11	288.05	692.10	700.71	1891.0	1964.7	2376.8	2700.0
273.99	286.44	688.48	700.71	1874.8	1930.3	2375.3	2700.0
283.23	699.86	1901.6	2331.1	2471.8	2700.9	3279.3	3689.9
283.23	699.86	1900.0	2101.9	2358.0	2698.3	2725.6	3130.8
283.23	699.86	1866.8	1912.1	2295.4	2434.7	2695.7	2771.6
283.23	699.86	1710.6	1901.1	2107.1	2373.0	2460.3	2701.5
283.23	699.86	1612.5	1887.1	1948.6	2215.6	2383.3	2700.7

The variation of the first eight natural frequencies with respect to inductance tuning sequence in Eq. (8.44) is shown in Table 8.3. In the first 5 rows in this table, the first 6 natural frequencies vary significantly with respect to inductance tuning. This is because the corresponding inductance tuning sequence is selected from tuning option 1 where eigenvalue curve veering occur between the 1st and 2nd, the 3rd and 4th, and the 5th and 6th system eigenvalues. In the last 5 rows, the 3rd through the 8th natural frequencies vary significantly with respect to inductance tuning because the inductance tuning sequence is selected from tuning option 2 where eigenvalue curve veering occur between the 3rd through the 8th eigenvalues.

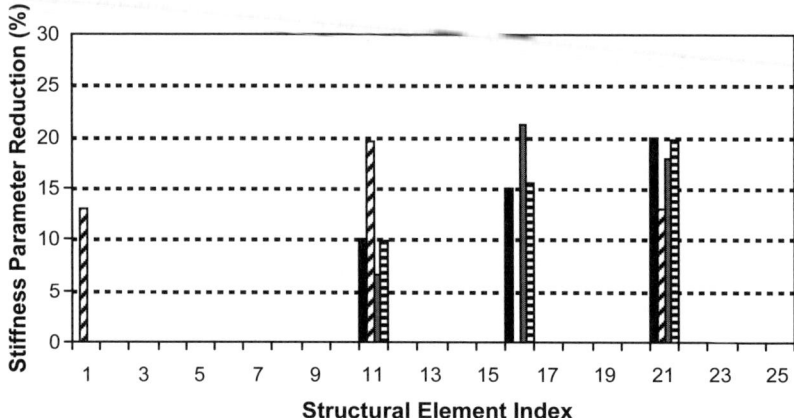

Fig. 8.14. Identification of damages on the 11th, 16th and 21st elements of the plate structure with 10%, 15% and 20% stiffness parameter reductions, respectively. ■ Actual stiffness parameter reduction, ◩ Prediction using the traditional method, ▨ Prediction using the enhanced integrated-system approach with *ad hoc* inductance tuning, ▤ Prediction using the enhanced integrated-system approach with *favorable* inductance tuning.

The first example of damage detection is shown in Fig. 8.14. The actual structural damages, denoted by the solid black bars in this figure, are assumed to be on the 11th, 16th, and 21st elements with 10%, 15%, and 20% stiffness reductions, respectively. The variation of damage-induced frequency shift with respect to inductance tuning sequence in Table 8.3 is shown in Table 8.4. Compared to only one set of frequency-shifts when using the traditional method without tunable piezoelectric circuitry, we can now obtain as many sets of frequency-shift measurement as the number of tuned inductance sequence. As shown in Fig. 8.14, the prediction by using the traditional method has significant RMSD error of 86%, which is not acceptable. Although the enhanced integrated-system approach with *ad hoc* inductance tuning ($\tilde{L}_1^{(i)} = (i)$ H, $\tilde{L}_2^{(i)} = (2 \times i - 1)$ H, $\tilde{L}_3^{(i)} = (3 \times i - 2)$ H, where $i = 1, 2, \cdots, 10$) successfully locate all three damaged elements, the predicted damage severities are not accurate and have a RMSD error of 28%. When the enhanced integrated-system approach with *favorable* inductance tuning is used, the three locations of the structural damages are exactly identified, and the predicted damage severities are very close to the actual stiffness parameter reductions. The RMSD error of the prediction by using the *favorable* inductance tuning is only 2%.

Table 8.4. Damage-induced natural frequency changes (rad/sec) of the integrated plate system with respect to inductance tuning. The damages cause 10%, 15% and 20% stiffness reductions on the 11th, 16th and 21st elements, respectively.

$\delta\omega_1$	$\delta\omega_2$	$\delta\omega_3$	$\delta\omega_4$	$\delta\omega_5$	$\delta\omega_6$	$\delta\omega_7$	$\delta\omega_8$
-7.62	-0.67	-8.64	-0.71	-23.57	-0.94	-7.56	-24.55
-7.18	-1.12	-7.93	-1.43	-23.58	-0.97	-7.78	-24.55
-6.34	-1.95	-5.05	-4.29	-23.08	-1.49	-7.79	-24.37
-4.97	-3.35	-2.89	-6.45	-21.00	-3.06	-8.22	-24.37
-3.29	-5.00	-0.73	-8.62	-14.46	-9.87	-8.01	-24.56
-8.28	-9.28	-24.08	-6.66	-1.62	-24.55	-0.15	-0.27
-8.28	-9.28	-23.83	-0.48	-7.65	-23.82	-1.28	-0.32
-8.28	-9.28	-8.86	-14.96	-4.36	-4.11	-23.85	-1.08
-8.28	-9.28	-0.29	-23.29	-0.95	-6.75	-1.22	-24.36
-8.28	-9.27	-0.31	-19.17	-5.14	-0.68	-7.78	-24.55

In order to further examine the effects of inductance tuning sequence on the performance of damage detection, the predictions using different sequences are compared in Fig. 8.15. The first set of inductance tuning sequence is selected from tuning option 1, as shown in Eq. 8.42, the second set is selected from tuning option 2, as shown in Eq. 8.43, and the third set is the combination of the first two sets, as given by Eq. 8.44. As shown in Fig. 8.15, the locations of the three damaged elements are accurately predicted by using any set of the tuning sequence. However, the predicted severities of the three damages have a RMSD error of 22% when using the tuning option 1. When tuning option 2 is used, the predicted severities are much more accurate and the prediction error in terms of RMSD decreases to 7%. The performance improvement is not only owing to that the eigenvalue curve veering is achieved for high order system eigenvalues which are usually more sensitive to structural damage, but also caused by the more noticeable variation of damage-induced eigenvalue change with respect to inductance tuning as shown in Fig. 8.10. Finally, when the combined tuning sequence is used, the accuracy of the predicted damage severities is further improved with a prediction error of 2%.

The above case study demonstrates the merits of using multiple piezoelectric transducer circuitries to detect damage. To best benefit the damage identification process, the *favorable* inductance tuning sequence can be formed by accomplishing a 'comprehensive' set of eigenvalue curve veering in the sense that each of the measurable modal frequencies is under curve veering at least once.

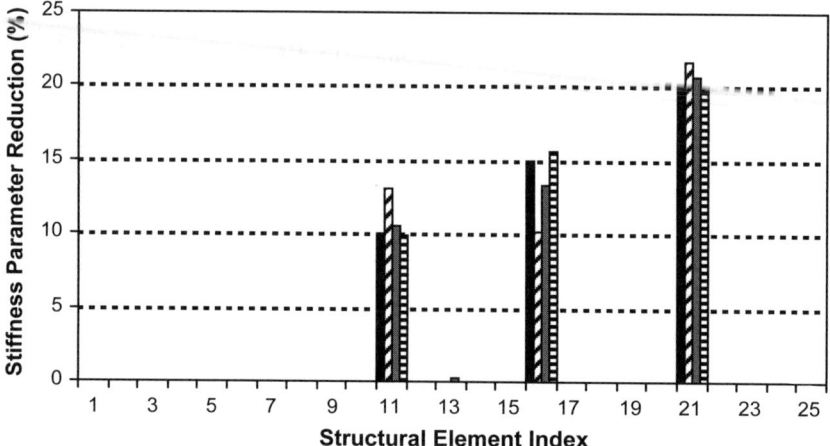

Fig. 8.15. Effects of different inductance tuning sequences on the performance of damage identification for the plate structure. ■ Actual stiffness parameter reduction, ▨ Prediction using tuning option 1, ■ Prediction using tuning option 2, ▤ Prediction using the combined tuning option.

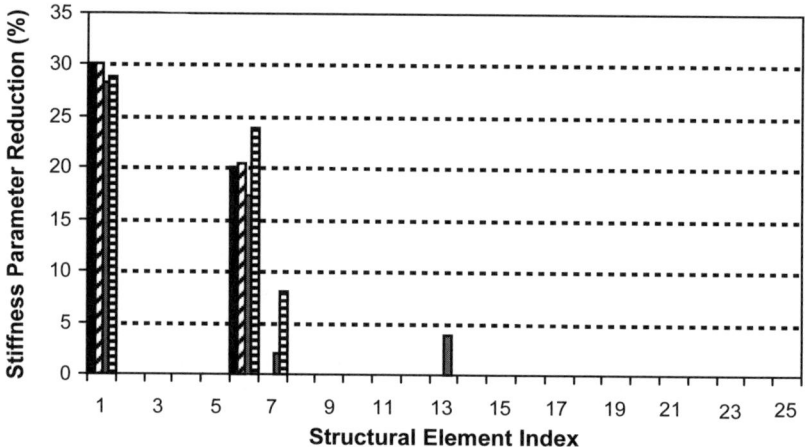

Fig. 8.16 Comparison of damage identification results using noise-free and noise-contaminated eigenvalues of the integrated plate system. The damages are assumed to cause 30% and 20% stiffness reduction on the first and sixth elements, respectively. ■ Actual stiffness parameter reduction, ▨ Prediction using noise-free eigenvalues, ■ Prediction using noise-contaminated eigenvalues ($v = 0.5\%$), ▤ Prediction using noise-contaminated eigenvalues ($v = 1.0\%$).

Another case study is performed to illustrate the effects of measurement noise on the performance of damage identification. As shown in Fig. 8.16, the noise-free and noise contaminated eigenvalues with noise levels of $v = 0.5\%$ and $v = 1.0\%$, are used to identify 30% stiffness reduction in the 1st element and 20% stiffness reduction in the 6th element. The prediction of damages using noise-free eigenvalues matches very well with the actual damage scenario. When the noise-contaminated eigenvalues are used to perform damage identification, the two damages are also successfully located and the damage severities are well predicted. Therefore, even in the presence of measurement noise, the proposed damage identification method may still provide a satisfactory prediction of the damage scenario.

8.5 CONCLUDING REMARKS

In this chapter, the method of using tunable piezoelectric transducer circuitry to enhance frequency-shift based damage identification is presented. Such an approach can introduce additional resonant peaks into the system frequency response function, and these additional peaks can be placed/adjusted over the frequency band by tuning the inductance. As a result, a significantly increased amount of frequency-shift information can be obtained to reflect the damage effect.

To take full advantage of the frequency measurement enrichment feature of the circuitry, an inductance tuning criterion is established. Analysis shows that in the region of eigenvalue curve veering due to inductance change, high sensitivity of damaged-induced eigenvalue changes with respect to inductance tuning can be expected. By tuning the inductance around the curve veering region, one may obtain a family of frequency response functions that could reflect the damage occurrence effectively. When multiple tunable piezoelectric transducer circuitries are integrated to the mechanical structure, multiple eigenvalue curve veering can be simultaneously achieved, and a series of inductance tunings can be formed by accomplishing curve veering between different pairs of system eigenvalues. To best characterize the damage occurrence, the favorable inductance tuning sequence should be selected as that leads to a comprehensive set of eigenvalue curve veering, i.e., all measurable resonant frequencies undergo curve veering at least once. A series of case studies have been performed, which demonstrate the tuning criterion and the performance improvement in damage identification practices.

Nomenclature

a : plate length

A_b : cross sectional area of beam

A_c and A_p : cross sectional area of piezoelectric transducer

b : beam and plate width

b_s: width of beam and piezoelectric transducer

c_b : uniform beam damping constant

C_{11}^D : elastic modulus of piezoelectric transducer - open circuit

C_a : coupling capacitance

$\mathbf{C_d}$: damping matrix

$C_p{}^S$: capacitance of the piezoelectric transducer

D: electrical displacement

D_a: electrical displacement of piezoelectric actuator

D_s: electrical displacement of piezoelectric sensor

E_b: Young's modulus of beam

E_p: Young's modulus of piezoelectric transducer

E_r: Young's modulus of rod

E_s: Young's modulus of plate

E_v: electrical field

E_{va}: electrical field of piezoelectric actuator

E_{vs}: electrical field of piezoelectric sensor

$E[\ ]$: expected value operator

f_o: nominal frequency of chirp signal

$\dot{f}$: frequency rate of change of the chirp signal

F_m: generalized external disturbance force

F_p: first moment of area

$\hat{\mathbf{F}}$: disturbance distribution vector

$\tilde{\mathbf{F}}$: disturbance distribution vector in modal space

g: beam equivalent damping

G_{ac} : factor indicating apparent coupling increase

h_{31} : piezoelectric constant

h_b: distance from neutral axis of beam to outside surface of piezoelectric transducer

h_s: distance from neutral axis of beam to outside surface of piezoelectric transducer

h_i: filter coefficient

h_p: thickness of piezoelectric transducer

I_b : moment of inertia of beam

I_c: moment of inertia of piezoelectric transducer

I_p : moment of inertia of piezoelectric transducer

J_c: $b_s(h_s{}^2 - h_b{}^2)/2$

k : equivalent stiffness

k_1 : electro-mechanical coupling

k_2 : inverse of capacitance of piezoelectric transducer

k_{1s}: spring constant of spring at excitation end

k_{2s}: spring constant of spring at attenuated end

k_a : inverse of coupling capacitance

k_c : coupling stiffness

$\tilde{k}_2$: inverse of negative capacitance

$\hat{k}_2$: inverse of overall circuit capacitance

$\tilde{k}_j$: stiffness of the j-th beam

K_{ij}: generalized coupling coefficient

$\mathbf{K}$: stiffness matrix

$\mathbf{K^c}$: control gain

$\mathbf{K^D}$: open-circuit stiffness matrix

$\mathbf{K_C}$: electro-mechanical coupling matrix

$\mathbf{K_p}$: inverse capacitance matrix of the piezoelectric circuitry network

$\mathbf{K_s}$: stiffness matrix of the mechanical structure before damage

$\mathbf{K_s^d}$: stiffness matrix of the mechanical structure after damage

$\tilde{\mathbf{K}}_C$: electro-mechanical coupling matrix in modal space

$\tilde{\mathbf{K}}$: generalized stiffness matrix of the electro-mechanical integrated system before damage

$\tilde{\mathbf{K}}^d$: generalized stiffness matrix of the electro-mechanical integrated system after damage

$\tilde{\mathbf{K}}_j^e$: elemental stiffness matrix of the jth element positioned within the global stiffness matrix $\tilde{\mathbf{K}}$

l_b : length of beam

l_p : length of piezoelectric transducer

L: inductance

L_a: variable active inductance L_a

L_b : length of beam

L_j : circuit inductance (j=1,2,3,4)

L_p: passive inductance

L_r: thickness of ring

L_t: total inductance

L : inductance matrix of the piezoelectric circuitry network

L* : critical inductances to accomplish pairs of close eigenvalues

m: equivalent mass

M: mass matrix

M$_s$: mass matrix of the mechanical structure only

M̃ : generalized mass matrix of the electro-mechanical integrated system

p: generalized electrical displacement

P$_l$: solution of the Lyapunov equation

P$_r$: solution of the Riccati equation

q: generalized mechanical displacement

q_j: generalized mechanical displacement of the j-th beam

q : displacement vector of mechanical structure

Q : electrical charge of piezoelectric transducer

Q_a: electrical charge of piezoelectric actuator

Q_i : charge input

Q_j : electrical charge of the j-th piezoelectric transducer

Q_s: electrical charge of piezoelectric sensor

Q : electrical charge flow vector in the circuit

r : circuit resistance tuning ratio

R: resistance

R_a : electrical coupling ratio

R_c : mechanical coupling ratio

R_i: radius of rod

R_j : circuit resistance (j=1,2,3,4)

R_o: radius of ring

t_b : thickness of beam

t_p : thickness of piezoelectric transducer

V_a: voltage across piezoelectric transducer

V_{aj} : voltage across the jth piezoelectric transducer

V_c: active voltage input

V_t : active voltage input

V_L: voltage across inductor

V_S: piezoelectric sensor voltage

x_1: left end of piezoelectric transducer

x_2: right end of piezoelectric transducer

x_l : left end of piezoelectric transducer

x_j : nondimensionalized mechanical displacement

x_r : right end of piezoelectric transducer

w : transversal displacement

$w_d(t)$: displacement at point of interest

w_b : width of beam

w_p : width of piezoelectric transducer

$\mathbf{W}$: modal transformation vector

y_j : nondimensionalized electrical displacement

α_j : stiffness parameter of the jth element

β_{33} : dielectric constant

δ : inductance tuning ratio

$\delta\alpha_j$: stiffness parameter reduction of the jth element

δ_{opt} : optimal inductance tuning

$\delta\boldsymbol{\alpha}$: vector of elemental stiffness parameter reduction

$\delta\lambda_i$: change of the ith system eigenvalue due to the structural damage

$\delta\boldsymbol{\varphi}_i$: change of the ith system eigenvector due to the structural damage

Δ: bandwidth of frequency variation of the chirp signal

Δk_j : stiffness mistuning

Δs_j : mistuning ratio

ε : strain

ζ: damping ratio

ζ_{cl}: closed-loop damping ratio

λ_i : the ith eigenvalue of the electro-mechanical integrated system before damage

λ_i^d : the ith eigenvalue of the electro-mechanical integrated system after damage

ξ : generalized electro-mechanical coupling

ρ_b : density of beam

ρ_p : density of piezoelectric transducer

ρ_r: density of rod

ρ_s : density of plate

τ : stress

ϕ : assumed mode for mechanical displacement

$\boldsymbol{\varphi}_i$: the ith eigenvector of the healthy integrated electro-mechanical system

$\boldsymbol{\varphi}_i^d$: the ith eigenvector of the damaged integrated electro-mechanical system

ψ : assumed mode for electrical displacement

ω_a: natural frequency of the circuit

ω_c : natural frequency of the circuit

ω_e: excitation frequency

ω_m : beam (first) natural frequency

ω_n: open-circuit modal frequency

ω_n^D: open-circuit natural frequency of the system

ω_n^E: short-circuit natural frequency of the system

Ω : nondimensionalized harmonic frequency

References

Abuelma'atti, MT, Khan, MH (1995) Current-controlled OTA-based single-capacitor simulations of grounded inductors. International Journal of Electronics 78: 881-885.

Agnes, GS (1994) Active/passive piezoelectric vibration suppression. Proceedings of SPIE, Smart Structures and Materials, 2193, pp 24-34.

Agnes, GS (1995) Development of a modal model for simultaneous active and passive piezoelectric vibration suppression. Journal of Intelligent Material Systems and Structures 6: 482-487.

Agnes, GS (1999) Piezoelectric coupling of bladed-disk assemblies. Proceedings of SPIE, Smart Structures and Materials 3672, pp. 94-103.

Allaei, D, Tarnowski, DJ (1997) Enhancing the performance of constrained layer damping confining vibrational energy. Proceedings of ASME, Active/Passive Vibration Control and Nonlinear Dynamics of Structures, DE-Vol.95/AMD-Vol.223, pp 31-46.

Andry, AN Jr., Shapiro, EY, and Chung, JC (1983) Eigenstructure assignment for linear systems. IEEE Transactions on Aerospace and Electronic Systems AES-19: 711-729.

Anton, SR, Sodano, HA (2007) A review of power harvesting using piezoelectric materials (2003-2006), Smart Materials and Structures 16: R1-R21.

Arora, JS (1989) Introduction to optimum design, McGraw Hill Inc.

Belasco D, Wang, KW (2003) Structurally integrated piezoelectric stack absorbers for nonstationary disturbance rejection. Fifth International Conference on Intelligent Materials, State College, PA.

Bondoux, D (1996) Piezo-damping: a low power consumption technique for semi-active damping of light structures. Proceedings of SPIE, Smart Structures and Materials 2779, pp 694-699.

Bouzit, D, Pierre, C (1992) Vibration confinement phenomena in disordered, mono-coupled, multi-span beams. ASME Journal of Vibration and Acoustics 114: 521-530.

Bruneau, H, Le Letty, R, Claeyssen, F, Barillot, F, Lhermet, N, Bouchilloux, P (1999) Semi-passive and semi-active vibration control using new amplified piezoelectric actuators. Proceedings of SPIE, Smart Structures and Materials 3668, pp 814-821.

Castanier, MP, Pierre, C (1995) Lyapunov exponents and localization phenomena in multi-coupled nearly-periodic systems. Journal of Sound and Vibration 183: 493-515.

Castanier, MP, Pierre, C (2002) Using intentional mistuning in the design of turbomachinery rotors. AIAA Journal 40: 2077-2086.

Cha, PD, Gu, W (2000) Model updating using an incomplete set of experimental modes. Journal of Sound and Vibration 233: 587–600.

Cha, PD, Pierre, C (1991) Vibration localization by disorder in assemblies of monocoupled, multimode component systems. ASME Journal of Applied Mechanics 58: 1072-1081.

Chang, T, Sun, X (2001) Analysis and control of monolithic piezoelectric nano-actuator. IEEE Transactions on Control Systems Technology 9: 69-75.

Chen, CT (1984) Linear system theory and design. Holt, Rinehart and Winston, Inc., NY.

Chen, W (1986) Passive and active filters, theory and implementation. John Wiley & Sons, New York.

Chen, P, Montgomery, S (1980) A macroscopic theory for the existence of the hysteresis and butterfly loops in ferroelectricity. Ferroelectrics 23: 199-207.

Cheng, CC, and Liu, IM (1999) Design of MIMO integral variable structure controllers. Journal of The Franklin Institute 336: 1119-1134.

Chern, TL, Wu, YC (1991) Design of integral variable structure controller and application to electrohydraulic velocity servosystems. IEE Proceedings-D 138: 439-444.

Choi, GS, Lim, YA, Choi, GH (2002) Tracking position control of piezoelectric actuators for periodic reference inputs. Mechatronics 12: 669-684.

Chonan, S, Jiang, Z, Yamamoto, T (1996) Nonlinear hysteresis compensation of piezoelectric ceramic actuators. Journal of Intelligent Material Systems and Structures 7: 150-156.

Choura, S (1995) Control of flexible structures with the confinement of vibrations. ASME Journal of Dynamic Systems, Measurement, and Control 117: 155-164.

Choura, S, Yigit, AS (1995) Vibration confinement in flexible structures by distributed feedback. Computers and Structures 54: 531-540.

Cicekoglu, MO (1998) Active simulation of grounded inductors with CCII+s and grounded passive elements. International Journal of Electronics 85: 455-462.

Corr, LR, Clark, WW (1999) Active and passive vibration confinement using piezoelectric transducers and dynamic vibration absorbers. Proceedings of SPIE, Smart Structures and Materials 3668, pp 747-758.

Cornwell, PJ, Bendiksen, OO (1989) Localization of vibrations in large space reflectors. AIAA Journal 27: 219-226.

Cornwell, P, Kan, M, Carlson, B, Hoerst, LB, Doebling, SW, Farrar, CR (1998) Comparative study of vibration-based damage identification algorithms. Proceedings of the 16th International Modal Analysis Conference, Part 2, Santa Barbara, CA, pp 1710-1716.

Cox, AM, Agnes, GS (1999) A statistical analysis of space structure mode localization. Proceedings of the 1999 AIAA/ASME/ASCE/AHS /ASC Structures, Structural Dynamics, and Materials Conference and Exhibit, V4, pp 3123-3133.

Croft, D, Devasia, S (1998) Hysteresis and vibration compensation for piezoactuators. Journal of Guidance, Control, and Dynamics 21: 710-717.

Cunningham, TB (1980) Eigenspace selection procedures for closed-loop response shaping with modal control. Proceedings of the American Control Conference, pp 178-186.

Dascotte, E (1990) Practical application of finite element tuning using experimental modal data. Proceedings of the 8th International Modal Analysis Conference, Kissimmee, FL, pp 1032–1037.

Datta, BN, Elhay, S, Ram, YM, Sarkissian, DR (2000) Partial eigenstructure assignment for the quadratic pencil. Journal of Sound and Vibration 230: 101-110.

Davis, CL, Lesieutre, GA (2000) An actively tuned solid-state vibration absorber using capacitive shunting of piezoelectric stiffness. Journal of Sound and Vibration 232: 601-617.

Davis, PL (1979) Circulant Matrices, Wiley, New York.

Davison, DJ, Goldenberg, A (1975) Robust control of general servomechanism problem: the servo compensator. Automatica 11: 461-471.

Damjanovic, D (1997) Stress and frequency dependence of the direct piezoelectric effect in ferroelectric ceramics. Journal of Applied Physics 82: 1788-1797.

Den Hartog, JP (1956) Mechanical Vibrations. McGraw-Hill, NY, pp 87-105.

Devonshire, AF (1954) Theory of ferroelectrics, Advances in Physics 3: 85-130.

Doebling, SW, Farrar, CR, Prime, MR (1998) A summary review of vibration-based damage identification methods. Shock and Vibration Digest 30: 91-105.

Doebling, SW, Farrar, CR, Prime, MB, Shevitz, DW (1996) Damage identification and health monitoring of structural and mechanical systems from changes in their vibration characteristics: a literature review, Los Alamos National Laboratory Report LA-13070-MS.

Edberg, DL, Bicos, AS, Fuller, CM, Tracy, JJ, Fechter, JS (1992) Theoretical and experimental studies of a truss incorporating active members. Journal of Intelligent Material Systems and Structures 3: 333-347.

Fleming, AJ, Behrens, S, Moheimani, SOR (2002) Optimization and implementation of multimode piezoelectric shunt damping systems. IEEE/ASME Transactions on Mechatronics 7: 87-94.

Friswell, MI, Penny, JET (1997) The practical limits of damage detection and location using vibration data. Proceedings of the 11th VPI&SU Symposium on Structural Dynamics and Control, Blacksburg, VA, pp 31–40.

Fuller, CR, Elliott, SJ, Nelson, PA (1996) Active control of vibration. Academic Press, London.

Ge, P, Jouaneh, M (1995) Modeling hysteresis in piezoceramic actuators. Precision Engineering 17: 211-221.

Ge, P, Jouaneh, M (1996) Tracking control of a piezoceramic actuator. IEEE Transactions on Control Systems Technology 4: 209-216.

Ge, P, Jouaneh, M (1997) Generalized Preisach model for hysteresis nonlinearity of piezoceramic actuators. Precision Engineering 20: 99-111.

Gift, SJG (2004) New simulated inductor using operational conveyors. International Journal of Electronics 91: 477-483.

Goldfarb, M, Celanovic, N (1997a) A lumped parameter electrochemical model for describing the nonlinear behavior of piezoelectric actuators. ASME Journal of Dynamic Systems, Measurement, and Control, 119: 478-485.

Goldfarb, M, Celanovic, N (1997b) Modeling piezoelectric stack actuator for control of micromanipulation. IEEE Control System Magazine, 17: 69-79.

Gordon, RW, Hollkamp, JJ (2000) An experimental investigation of piezoelectric coupling in jet engine fan blades. Collection of Technical Papers – AIAA/ASME/ASCE/AHS/ASC Structures, Structural Dynamics and Materials Conference. v 1, n III, pp 1975-1983.

Hagood, NW, Crawley, EF (1991) Experimental investigation of passive enhancement of damping for space structures. Journal of Guidance, Control, and Dynamics 14: 1100-1109.

Hagood NW, von Flotow A (1991) Damping of structural vibrations with piezoelectric materials and passive electrical networks. Journal of Sound and Vibration 146: 243-268

Handel, P, Tichavsky, P (1994) Adaptive estimation for periodic signal enhancement and tracking. International Journal of Adaptive Control and Signal Processing, 8: 447-456.

Heverly, D, Wang, KW, Smith, EC (2002) Optimal actuator placement and active structure design for control of helicopter airframe vibrations. Proceedings of the 58th American Helicopter Society Forum, V2, pp. 2149-2166.

Hodges, CH (1982) Confinement of vibration by structural irregularity. Journal of Sound and Vibration 82: 411-424.

Hodges, CH, Woodhouse, J (1983) Vibration isolation from irregularity in a nearly periodic structure: theory and measurements. Journal of Acoustical Society of America. 74: 894-905.

Hodges, CH, Woodhouse, J (1989) Confinement of vibration by one-dimensional disorder, I: theory of ensemble averaging. Journal of Sound and Vibration 130: 253-268.

Hollkamp, JH (1994) Multimodal passive vibration suppression with piezoelectric materials and passive electrical networks. Journal of Intelligent Material Systems and Structures 5: 49-57.

Hollkamp, JH, Starchville, Jr. TF (1994) A self-tuning piezoelectric vibration absorber. Journal of Intelligent Material Systems and Structures 5: 559-566.

Horowitz, P, Hill W (1989) The art of electronics, second edition, Cambridge University Press.

Hung, JY, Gao, W, Hung, JC (1993) Variable structural control: a survey. IEEE Transactions on Industrial Electronics 40: 2-22.

Inman, D. J. (1989) Vibrations with control, measurement, and stability, Prentice Hall.

Institute of Electrical and Electronics Engineers (1988) IEEE Standard on Piezoelectricity (ANSI/IEEE Std. 176-1987), IEEE, New York.

Jiang, LJ, Tang, J, Wang, KW (2006) An enhanced frequency-shift based damage identification method using tunable piezoelectric transducer circuitry. Smart Materials and Structures 15:799-808.

Jiang, LJ, Tang, J, Wang, KW (2007) Optimal sensitivity-enhancing feedback control via eigenstructure assignment for structural damage detection. ASME Journal of Vibration and Acoustics 129:771-783.

Jiang, LJ, Tang, J, Wang, KW (2008), On the tuning of variable piezoelectric transducer circuitry network for structural damage identification. Journal of Sound and Vibration, 309:695-717.

Kahlil, HK (1996) Nonlinear systems. Prentice-Hall, NJ.

Kahn SP, Wang KW (1994) Structural vibration controls via piezoelectric materials with active-passive hybrid networks. Proceedings of ASME IMECE DE-75, pp 187-194.

Kahn, SP, Wang, KW (1995) On the simultaneous design of active-passive hybrid control actions for structures with piezoelectrical networks. ASME Paper # 95-WA/AD5.

Keane, AJ, (1995) Passive vibration control via unusual geometries: the application of genetic algorithm optimization to structural design. Journal of Sound and Vibration 185: 441-453.

King, TG, Preston, ME, Murphy, BJM,, Cannel, DS (1990) Piezoelectric ceramic actuators: a review of machinery applications. Precision Engineering 12: 131-136.

Kim, HS, Kim, Y (1999) Partial eigenstructure assignment algorithm in flight control system design. IEEE Transactions on Aerospace and Electronic Systems 35: 1403-1408.

Kissel, GJ (1991) Localization factor for multichannel disordered systems. Physical Review A 44:1008-1014.

Klema VC, Laub, AJ (1980) The singular value decomposition: its computation and some applications. IEEE Transactions on Automatic Control AC-25: 164-176.

Koh, BH, Ray, LR (2004) Feedback controller design for sensitivity-based damage localization. Journal of Sound and Vibration 273: 317-335.

Kuo, BC (1995) Automatic control systems. Prentice-Hall, Englewood Cliffs, NJ.

Kung, YS, Fung, RF (2004) Precision control of a piezoceramic actuator using neutral networks. ASME Journal of Dynamic Systems, Measurement, and Control 126: 235-238.

Kutter, JR, Sigillito, VG (1981) On curve veering. Journal of Sound and Vibration 75: 585-588.

Kwakernaak H, Sivan R (1972) Linear optimal control systems, John Wiley and Sons, Inc.

Kwon, BH, Youn, MJ (1987) Eigenvalue-generalized eigenvector assignment by output feedback. IEEE Transactions on Automatic Control AC-32: 417-421.

Lee, SH, Royston, TJ (2000) Modeling piezoceramic transducer hysteresis in the structural vibration control problem. Journal of Acoustic Society of America, 108: 2843-2855.

Leissa, AW (1974) On a curve veering aberration, Journal of Applied Mathematics and Physics (ZAMP) 25: 99-111.

Lesieutre, GA (1998) Vibration damping and control using shunted piezoelectric materials. The Shock and Vibration Digest 30: 187-195.

Lesieutre, GA, Davis, CL (1997) Can a coupling coefficient of a piezoelectric device be higher than those of its active material? Journal of Intelligent Material Systems and Structures 8:859-867.

Lew, J-S, Juang, JN (2002) Structural damage detection using virtual passive controllers. Journal of Guidance, Control, and Dynamics 25: 419-424.

Lewis, FL, Syrmos, VL (1995) Optimal Control, 2nd ED., Wiley, New York.

Liu, XL (2002) Behavior of derivatives of eigenvalues and eigenvectors in curve veering and mode localization and their relation to close eigenvalues. Journal of Sound and Vibration 256: 551-564.

Main, JA, Garcia, E, Newton, DV (1995) Precision position control of piezoelectric actuators using charge feedback. Journal of Guidance, Control, and Dynamics 18: 1068-1073.

Main, JA, and Garcia, E (1997) Piezoelectric stack actuators and control system design: strategies and pitfalls. Journal of Guidance, Control, and Dynamics 20: 479–485.

Mayergoyz, I (1991) Mathematical Models of Hysteresis, Springer-Verlag, New York.

Mead, DJ (1975a) Wave propagation and natural modes in periodic systems: I. Mono-coupled systems. Journal of Sound and Vibration 40: 1-18.

Mead, DJ (1975b) Wave propagation and natural modes in periodic systems: II. Multi-coupled systems, with and without damping. Journal of Sound and Vibration 40: 19-39.

Meirovitch, L (1990) Dynamics and Control of Structures, John Wiley & Sons, Inc., New York.

Mester, SS, Benaroya, H (1995) Periodic and near-periodic structures. Shock and Vibration 2: 69-95.

Miller, LR, Ahmadian, M, Nobles, CM, Swanson, DA (1995) Modeling and performance of an experimental active vibration isolator. ASME Journal of Vibration and Acoustics 117: 272-278.

Moheimani, SOR, Behrens, S (2004) Multi-mode piezoelectric shunt damping with a highly resonant impedance. IEEE Transactions on Control Systems Technology 12: 484-491.

Moore, BC (1976) On the flexibility offered by state feedback in multivariable systems beyond closed-loop eigenvalue assignment. IEEE Transactions on Automatic Control AC-21: 689-692.

Morgan, R, Wang, KW (2002a) An active-passive piezoelectric vibration absorber for structural control under harmonic excitations with time-varying frequency, Part 1: algorithm development and analysis. ASME Journal of Vibration and Acoustics 124: 77-83.

Morgan, R, Wang, KW (2002b) An active-passive piezoelectric vibration absorber for structural control under harmonic excitations with time-varying frequency, Part 2: experimental validation and parametric study. ASME Journal of Vibration and Acoustics 124: 84-89.

Morgan, R, Wang, KW (2002c) Active-passive piezoelectric absorbers for suppression of multiple non-stationary harmonic excitations. Journal of Sound and Vibration 255: 685-700.

Morgan, RA, Wang, KW, Tang, J (2000) Active tuning and coupling enhancement of piezoelectric vibration absorbers for variable-frequency harmonic excitations in multiple degrees of freedom mechanical systems. Proceedings of the SPIE, Smart Materials and Structures 3985, pp. 497-509.

Nalitolela, NG, Penny, JET, Friswell, MI (1992) Mass or stiffness addition technique for structural parameter updating, Modal Analysis: The International Journal of Analytical and Experimental Modal Analysis 7: 157–168.

Niezrecki, C, Cudney, HH (1994) Improving the power consumption characteristics of piezoelectric actuators. Journal of Intelligent Material Systems and Structures 5: 522-529.

Panza, MJ, McGuire, DP, Jones, PJ, (1997) Modeling, actuation, and control of an active fluid vibration isolator. ASME Journal of Vibration and Acoustics 119: 52-59.

Perkins, NC, Mote, CD (1986) Comments on curve veering in eigenvalue problems. Journal of Sound and Vibration 106:451-463.

Pierre, C (1988) Mode localization and eigenvalue loci veering phenomena in disordered structures. Journal of Sound and Vibration 126: 485-502.

Pierre, C (1990) Weak and strong vibration localization in disordered structures. Journal of Sound and Vibration 126: 485-502.

Pierre, C, Castanier, MP, Chen, WJ (1996) Wave localization in multi-coupled periodic structures: application to truss beams. Applied Mechanics Review 49: 65-86.

Pierre, C, Cha, PD (1989) Strong mode localization in nearly periodic disordered structures. AIAA Journal 27: 227-241.

Pierre, C, Dowell, EH (1987) Localization of vibrations by structural irregularity. Journal of Sound and Vibration 114: 549-564.

Ray, LR, Tian, L (1999) Damage detection in smart structures through sensitivity enhancing feedback control. Journal of Sound and Vibration 227: 987–1002.

Ray, LR, Koh, BH, Tian, L (2000) Damage detection and vibration control in smart plates: towards multifunctional smart structures. Journal of Intelligent Material Systems and Structures 11: 725–739.

Salawu, OS (1997) Detection of structural damage through changes in frequency: a review. Engineering Structures 19: 718-723.

Sciulli, D, Inman, DJ (1996) Comparison of single- and two-degree-of-freedom models for passive and active vibration isolation design. Proceedings of SPIE, Smart Materials and Structures 2720, pp 293-304.

Senani, R (1980) New tunable synthetic floating inductors, Electronics Letters 16: 382-283.

Senani, R (1987) Generation of new two-amplifier synthetic floating inductors. Electronics Letters 23: 1202-1203.

Shelley, FJ, Clark, WW (1996) Eigenvector scaling for mode localization in vibrating systems. Journal of Guidance, Control, and Dynamics 19: 1342-1348.

Shelley, FJ, Clark, WW (2000a) Experimental application of feedback control to localize vibration. ASME Journal of Vibration and Acoustics 122: 143-150.

Shelley, FJ, Clark, WW (2000b) Active mode localization in distributed parameter systems with consideration of limited actuator placement, part 1: theory. ASME Journal of Vibration and Acoustics 122: 160-164.

Shelly, FJ, Clark, WW (2000c) Active mode localization in distributed parameter systems with consideration of limited actuator placement, part 2: simulations and experiments. ASME Journal of Vibration and Acoustics 122: 165-168.

Sievers, LA, von Flotow, AH (1992) Comparison and extensions of control methods for narrow-band disturbance rejection. IEEE Transactions on Signal Processing. 40: 2377-2391.

Sirohi, J, Chopra, I (2000) Fundamental understanding of piezoelectric strain sensors. Journal of Intelligent Material Systems and Structures 11: 246-257.

Slater, JC, Minkiewicz, GR, Blair, AJ (1999) Forced response of bladed disk assemblies – a survey. The Shock and Vibration Digest 31: 17-24.

Sodano, HA, Inman, DJ, Park, G (2004) A review of power harvesting from vibration using piezoelectric materials. The Shock and Vibration Digest 36: 197-205.

Sohn, H., Farrar, CR, Hemez, FM, Shunk, DD, Stinemates, DW, Nadler BR (2003), A review of structural health monitoring literature: 1996–2001. Los Alamos National Laboratory Report LA-13976-MS.

Song, BK, Jayasuriya, S (1993) Active vibration control using eigenvector assignment for mode localization. Proceedings of the American Control Conference, pp 1020-1024.

Tang, J, Wang, KW (1999) Vibration control of rotationally periodic structures using passive piezoelectric shunt networks and active compensation. Journal of Vibration and Acoustics 121: 379-390.

Tang, J, Liu, Y, Wang, KW (2000) Semi-active and active-passive hybrid structural damping treatments via piezoelectric materials. Shock and Vibration Digest 32: 189-200.

Tang, J, Wang, KW (2000) High authority and nonlinearity issues in active-passive hybrid piezoelectric networks for structural damping. Journal of Intelligent Material Systems and Structures 11: 581-591.

Tang, J, Wang, KW (2001) Active-passive hybrid piezoelectric networks for vibration control: comparisons and improvement. Smart Materials and Structures 10: 794-806.

Tang, J, Wang, KW (2003) Vibration delocalization of nearly periodic structures using coupled piezoelectric networks. ASME Journal of Vibration and Acoustics 125: 95-108.

Tang, J, Wang, KW (2004) Vibration confinement via optimal eigenvector assignment and piezoelectric network. ASME Journal of Vibration and Acoustics 126: 27-36.

Taylor, DV, Damjanovic, D (1997) Evidence of domain wall contribution to the dielectric permittivity in PZT thin films at sub-switching fields. Journal of Applied Physics 82: 1973-1975.

Tsai, MS (1998) Active-passive hybrid piezoelectric network based smart structures for vibration controls. Ph.D. dissertation, The Pennsylvania State University.

Tsai, MS, Chen, JS (2003) Robust tracking control of a piezoactuator using a new approximate hysteresis model. ASME Journal of Dynamic Systems, Measurement, and Control 125: 96-102.

Tsai MS, Wang KW (1996) Control of a ring structure with multiple active-passive hybrid piezoelectrical networks. Smart Materials and Structures 5: 695-703

Tsai, MS, Wang, KW (1997) Integrating active-passive hybrid piezoelectric networks with active constrained layer treatments for structural damping. Proceedings of ASME IMECE Conference.

Tsai, MS, Wang, KW (1999) On the structural damping characteristics of active piezoelectric actuators with passive shunts. Journal of Sound and Vibration 221: 1-22.

Utkin, VI (1993) Sliding mode control design principle and applications to electrical drives. IEEE Transactions on Industrial Electronics, 40: 23-36.

Vakakis, AF (1994) Passive spatial confinement of impulsive response in coupled nonlinear beams. AIAA Journal 32: 1902-1910.

Vakakis, AF, Kounadis, AN, Raftoyiannis, IG. (1999) Use of non-linear localization for isolating structures from earthquake-induced motions. Earthquake Engineering & Structural Dynamics 28: 21-36.

von Flotow, AH, Beard, A, Bailey, D (1994) Adaptive tuned vibration absorbers: tuning laws, tracking agility, sizing, and physical implementations, Proceedings of NOISE-CON 94, pp 437-455.

Wang KW, Kahn SP (1997) Active-passive hybrid structural vibration controls via piezoelectric networks. In: H.S. Tzou et al. (eds) Structronic systems, smart structures, devices and systems. World Scientific Publishing Company.

Wong, CN, Zhu, WD, Xu, GY (2004) On an iterative general-order perturbation method for multiple structural damage detection. Journal of Sound and Vibration 273: 363-386.

Wei, ST, Pierre, C (1988) Localization phenomena in mistuned assemblies with cyclic symmetry part I: free vibrations. ASME Journal of Vibration, Acoustics, Stress and Reliability in Design 110: 429-438.

Wolf, A, Swift, JB, Swinney, HL, Vastano, JA (1985) Determining Lyapunov exponents from a time series. Physica D, 16: 285-317.

Wu, S (1996) Piezoelectric shunts with a parallel R-L circuit for structural damping and vibration control. Proceedings of SPIE, Smart Materials and Structures 2720, pp 259-269.

Wu, S. (1998) Method for multiple mode shunt damping of structural vibration using a single PZT transducer. Proceedings of the SPIE, Smart Materials and Structures 3327, pp 159-168.

Wu, TY, Wang, KW (2007) Periodic isolator design via vibration confinement through eigenvector assignment and piezoelectric circuitry. Journal of Vibration and Control 13: 989-1006.

Xie, W-C, Ariaratnam, ST (1996a) Vibration mode localization in disordered cyclic structures, I: single substructure mode. Journal of Sound and Vibration 189: 625-645.

Xie, W-C, Ariaratnam, ST (1996b) Vibration mode localization in disordered cyclic structures, II: multiple substructure mode. Journal of Sound and Vibration 189: 647-660.

Xue, X, Tang, J (2006) Robust and high precision control using piezoelectric actuator circuit and integral continuous sliding mode control design. Journal of Sound and Vibration 293: 335-359.

Yu, H, Wang, KW, Zhang, J (2006) Piezoelectric networking with enhanced electromechanical coupling for vibration delocalization of mistuned periodic structures – theory and experiment. Journal of Sound and Vibration 295: 246-265.

Yu, H, Wang, KW (2007a) Piezoelectric networks for vibration suppression of mistuned bladed disks. ASME Journal of Vibration and Acoustics 129:559-566.

Yu, H, Wang, KW (2007b) Vibration suppression of mistuned coupled-blade-disk systems using piezoelectric circuitry network. ASME International Design Engineering Technical Conferences & Computers and Information in Engineering Conference, DETC-34443.

Zhang, Q, Slater, GL, Allemang, RJ (1990) Suppression of undesired inputs of linear systems by eigenspace assignment. Journal of Guidance, Control, and Dynamics 13: 330-336.

Zhou, F, Fisher, DG (1992) Continuous sliding mode control. International Journal of Control 55: 313-327.

Copyright Permission Acknowledgment

We thank the ASME, Elsevier, Sage Publications, and Institute of Physics Publishing for allowing us to use the relevant materials in our papers published.

Figures 2.1 to 2.6, 2.8 and 2.9 are reprinted from Smart Materials and Structures, 10, Tang, J, Wang, KW, Active-passive hybrid piezoelectric networks for vibration control: comparisons and improvement, 794-806 (2001), with permission from Institute of Physics Publishing.

Figures 3.1 to 3.7, Figures 3.9 to 3.18 and Table 3.1 are reprinted from Journal of Sound and Vibration, 221, Tsai, MS, Wang, KW, On the structural damping characteristics of active piezoelectric actuators with passive shunts, 1-22 (1999), with permission from Elsevier.

Figure 3.8 is reprinted from Shock and Vibration Digest, 32, Tang, J, Liu, Y, Wang, KW, Semi-active and active-passive hybrid structural damping treatments via piezoelectric materials, 189-200 (2000), with permission from Sage.

Figures 4.1 to 4.14 and Tables 4.1 to 4.4 are reprinted from Journal of Vibration and Acoustics, 124, Morgan, R, Wang, KW, An active-passive piezoelectric vibration absorber for structural control under harmonic excitations with time-varying frequency, 77-89 (2002), with permission from ASME.

Figures 4.15 to 4.22 are reprinted from Journal of Sound and Vibration, 255, Morgan, R, Wang, KW, Active-passive piezoelectric absorbers for suppression of multiple non-stationary harmonic excitations, 685-700 (2002), with permission from Elsevier.

Figures and tables in Chapter 5 are reprinted from Journal of Sound and Vibration, 293, Xue, X, Tang, J, Robust and high precision control using piezoelectric actuator circuit and integral continuous sliding mode control design, 335-359 (2006), with permission from Elsevier.

Figures 6.1 to 6.7 and Tables 6.1 to 6.7 are reprinted from Journal of Vibration and Acoustics, 126, Tang, J, Wang, KW, Vibration confinement via optimal eigenvector assignment and piezoelectric network, 27-36 (2004), with permission from ASME.

Figures 6.8 to 6.18 and Table 6.8 are reprinted from Journal of Vibration and Control, 13, Wu, TY, Wang, KW, Periodic isolator design via vibration confinement through eigenvector assignment and piezoelectric circuitry, 989-1006 (2007), with permission from Sage.

Figures 7.1 to 7.14 and Tables 7.1 to 7.2 are reprinted from Journal of Sound and Vibration, 295, Yu, H, Wang, KW, Zhang, J, Piezoelectric networking with enhanced electromechanical coupling for vibration delocalization of mistuned periodic structures – theory and experiment, 246-265 (2006), with permission from Elsevier.

Figures 7.15 to 7.24 are reprinted from Journal of Vibration and Acoustics, 129, Yu, H, Wang, KW, Piezoelectric networks for vibration suppression of mistuned bladed disks, 559-566 (2007), with permission from ASME.

Figures and tables in Chapter 8 are reprinted from Journal of Sound and Vibration, 309, Jiang, LJ, Tang, J, Wang, KW, On the tuning of variable piezoelectric transducer circuitry network for structural damage identification, 695-717 (2008), with permission from Elsevier.

Index

Printed in the United States of America